ESSENTIALS OF MINERAL EXPLORATION AND EVALUATION

ESSENTIALS OF MINERAL EXPLORATION AND EVALUATION

S.M. Gandhi

B.C. Sarkar

ELSEVIER

AMSTERDAM • BOSTON • HEIDELBERG • LONDON • NEW YORK • OXFORD
PARIS • SAN DIEGO • SAN FRANCISCO • SINGAPORE • SYDNEY • TOKYO

Elsevier
Radarweg 29, PO Box 211, 1000 AE Amsterdam, Netherlands
The Boulevard, Langford Lane, Kidlington, Oxford OX5 1GB, UK
50 Hampshire Street, 5th Floor, Cambridge, MA 02139, USA

British Library Cataloguing-in-Publication Data
A catalogue record for this book is available from the British Library

Library of Congress Cataloging-in-Publication Data
A catalog record for this book is available from the Library of Congress

ISBN: 978-0-12-805329-4

For Information on all Elsevier publications
visit our website at http://www.elsevier.com/

Publisher: Candice G. Janco
Acquisition Editor: Amy Shapiro
Editorial Project Manager: Marisa LaFleur
Production Project Manager: Mohanapriyan Rajendran
Designer: Matthew Limbert

Typeset by MPS Limited, Chennai, India

Contents

8. Drilling

9. Sampling and Analysis

10. Geographic Information System and Common Earth Model

11. Conventional and Statistical Resource/Reserve Estimation

12. Geostatistical Resource/Reserve Estimation

Preface

The standard of living, livelihood and security of a Nation depends, to a great extent, upon the availability of continuous flow of minerals, metals and fuels. Mineral exploration is of utmost importance, to discover new mineral resources, in order to compensate the fast depletion of resources owing to year on year mining activity and also to aggrandize the mineral resource base of a Nation. A nation's economic growth is dependent to a large degree on the discovery of new mineral deposits and continued renewal of reserves. Mineral exploration is potentially a high risk and high reward activity.

We had pedagogic interest when we worked together for Industry and we used to give guest lectures/courses on various topics in Mineral Exploration and Geostatistics to various Universities, Industries and Research & Training institutes. The courses were well-received and the attendees felt that the course materials given were comprehensive for their understanding the subject better. The positive response and encouragement received from the students and other participants were seminal to put together all the updated course notes material in a book form. Although this book is intended for a reader who already has some basic geological concepts and understanding of ore deposits and their likely origins, we have tried to assemble, in summary form, the comprehensive account of concepts of mineral exploration and evaluation in mineral industry and bits of other information that the readers would seem likely to appreciate. We earnestly believe that this will help explorationists to widen the horizons to locate new mineral deposits.

The book is addressed not only to students who are pursuing Mineral Exploration course in the Universities and Colleges but also to the geologists engaged in mineral exploration, to get a glimpse of various exploration techniques and a source of general information. This book is intended to provide balanced coverage of all the relevant and related topics to give the reader an overall view of Mineral exploration concepts and their applications. The arrangement of chapters of this volume depicts the sequential manner of carrying out an exploration campaign.

Mineral exploration has ever been challenging due to its multifaceted difficulties involving the geology, economics, legal issues, administration, market value and potential for development. In general, exploration of mineral deposits is structured under a broad framework of activities with sequences and duration. The purpose of mineral exploration is to find and acquire maximum number of economic mineral deposits at a minimum cost and within minimum time. The important ingredients of a successful exploration include selection of the right geological terrain, optimum level of exploration funding and keeping pace with new technology. Modern day mineral exploration requires collaborative, integrated team approach that combines geology, geophysics, geochemistry and GIS, to meet the discovery challenges. Of late, greenfield discoveries are very less despite the high exploration spending. Many of the new discoveries are basically historical discoveries which hitherto were unworked but revived due to higher metal prices. The systematic use and development of 3D models for each mining camp will definitely lead to the discovery of new resources in many old mining areas.

The increased processing power of computers and myriad interpretative software packages available today in geological, geophysical and geochemical exploration techniques have revolutionized in processing and analyzing vast amount of data. A combined R&D effort of Universities, Industry and Government agencies in geoscientific exploration techniques will help to achieve the desired result of better success rate in future exploration efforts.

Some of the ideas and philosophies, mentioned herein, on various exploration techniques were from fellow geologists and explorationists in different parts of the world, with whom we had worked and discussed on various occasions. We have tried to give pertinent and adequate references accompanying every chapter, to facilitate readers who may like to delve further in the topics of interest. Our overall goal was to produce an appealing, informative textbook that will provide all the nuances of mineral exploration.

Many experienced fellow geologists, from the multinational exploration companies, academic institutes and government agencies, encouraged and inspired the authors by sharing their time and experience, during the preparation of this book. We would like to thankfully acknowledge the significant help and counsel of many of them; many professional friends and publishers, for permitting us to reproduce the figures and also those who contributed unpublished data and participated in discussions and engaged in critical reading and reviewing of the manuscripts.

We are grateful to the publisher M/s Elsevier, for accepting our proposal for the Book. We are especially thankful to Marisa LaFleur, Amy M. Shapiro, and Tasha Frank from Elsevier, who worked with us closely, offering guidance, support and encouragement throughout this project which cannot be expressed in just few words. We are indebted to them from bottom of our hearts.

Lastly, we would like to thank our families, friends and well-wishers for their unstinted support, cooperation and encouragement for our attempt. We are thankful to Dr Anup K. Prasad, Associate Professor, Mr Rahul Kumar Singh, Research Student, both of Department of Applied Geology, ISM Dhanbad, and Mr S.V. Ramkumar of Krishna Systems, Rajapalayam, for their untiring assistance during various stages of this project.

Dr. S.M. Gandhi
Consultant, Mineral Exploration
and Evaluation, (Formerly,
Chief, Exploration Wing,
Hindustan Zinc Limited, India),
83, P.S.K. Nagar, P.O. Rajapalayam
626108, Tamil Nadu, India

Dr. B.C. Sarkar
Professor, Department of Applied
Geology, Indian School of Mines,
Dhanbad 826004, Jharkhand, India

Some Abbreviations and Acronyms Used in the Text

AAS	Atomic Absorption Spectrometry
AD	Anno Domoni (year)
ALOS	Advanced Land Observation Satellite
ASTER	Advanced Space borne Thermal Emission and reflection Radiometer
ASTM	American Society for Testing Materials
BC	Before Christ (year)
BOH	Bottom of Hole
CAD	Computer-Aided Drafting
CCT	Computer Compatible Tapes
CIF	Carriage, Insurance, and Freight
CIPEC	Conseil Inter-governmental des Pays Exportateurs de Cuivre (Intergovernmental Council of Copper Exporting Countries)
CIS	Commonwealth of Independent States (includes many of the countries formerly in the USSR)
d.c.	Direct current
DCF ROR	Discounted cash flow rate of return
DDH	Diamond drill hole
DEM	Digital elevation
DGPS	Differential global positioning system
DTH	Down-the-hole (logs)
DTM	Digital Terrain Model
EIS	Environmental Impact Statement
EGNOS	European Geostationary Navigation Overlay Service
EM	Electromagnetic
EOH	End of Hole
EPMA	Electron Probe Micro Analyzer
ERTS	Earth Resources Technology Satellite
EU	European Union, sometimes still referred to as "EEC"
Ga	Giga (10^9) or Billion (age-years)
GIS	Geographical Information Systems
GPS	Global Positioning System
GRD	Ground Resolution Distance
g/t	grams per tonne
HRV	High Resolution Visible
INPUT	Induced Pulse Transient System
IP	Induced Polarization
IR	InfraRed
Ma	Million (10^6) age (years)
MIP	Magnetic IP
MMR	Magnetometric Resistivity
mRL	meter Reduced Level

MS	MultiSpectral
MSS	MultiSpectral Scanners
MT	Magnetotelluric survey (in Geophysics)
NAA	Neutron Activation Analysis
NPV	Net Present Value
OECD	Organization for Economic Co-operation and Development
OPEC	Organization of Petroleum Exporting Countries
PFE	Percent Frequency Effect
PGE	Platinum Group Elements
PGM	Platinum Group Metals
ppb	parts per billion
ppm	parts per million
ppt	part per trillion
RAB	Rotary Air Blast
RC	Reverse Circulation
REE	Rare Earth Elements
ROM	Run-of-Mine
RQD	Rock Quality Designation
SEDEX	SEDimentary EXhalative
SEM	Scanning Electron Microscopy
SG	Specific Gravity
SI units	Systȇme International Units
SLAR	Side-Looking Airborne Radar
SP	Self-Potential
SPOT	Satellite Probatoire l'Observation de la Terre
TEM	Transient Electromagnetics
TIR	Thermal InfraRed
TM	Thematic Mapper
TM	Transverse Mercator
TMI	Total Magnetic Intensity
tpa	tonnes per annum
tpd	tonnes per diem
USBM	United States Bureau of Mines
USGS	United States Geological Survey
UTM	Universal Transverse Mercator
VHR	Very High Resolution
VLF	Very Low Frequency
VMS	Volcanogenic Massive Sulfide
VNIR	Visible and Near InfraRed
WAAS	Wide Area Augmentation System
XRD	X-ray Diffraction
XRF	X-ray Fluorescence

WEIGHTS AND MEASURES

Length

	Centimeter	Meter
1 angstrom	10^{-8}	10^{-10}
1 millimicron	10^{-7}	10^{-9}
1 micron	10^{-4}	10^{-6}
1 millimeter	0.1	0.001
1 centimeter	1	0.01
1 meter	100	1
1 kilometer	1000	1000
1 fathom	182.88	1.8288

Surface and Area

	Square Centimeter	Square Meter	
1 square centimeter	1	0.0001	
1 square meter	10,000	1	
1 hectare	10^8	10,000	2.4710 acres
1 square kilometer	10^{10}	10^6	0.3861 square mile
1 square inch	6.4514	0.000645	
1 square feet	929.0	0.9290	
1 square yard	–	0.83613	
1 acre	–	4047	
1 square mile (640 acres)	–	2,590,000	

Weight

	Gram	Kilogram
1 milligram	0.001	10^{-6}
1 gram (1000 milligrams)	1	0.001
1 kilogram	1000	1
1 metric tonne	10^6	1000
1 ounce (troy)	31.1035	0.311035
1 ounce (avoirdupois)	28.3495	0.02835

1 short ton = 2000 pounds = 0.907 tonne.
1 pound = 16 ounces = 0.454 kilogram = 14.5833 troy ounces.

Volume

	Cubic Centimeter	Cubic Meter
1 cubic centimeter	1	10^{-6}
1 liter	1000.027	0.001
1 cubic meter	10^6	1
1 gallon (Imperial)	4.546 liters	
1 gallon (US)	3.785 liters	

CHAPTER

1

Metals and Minerals: Global Trends, Outlook, and Mineral Exploration

OUTLINE

1.1 GENERAL

From the prehistory to the present, mining has provided materials for fuel, shelter, and the acquisition of food. From the earliest man into the period of the first civilization, the

DOI: http://dx.doi.org/10.1016/B978-0-12-805329-4.00008-9

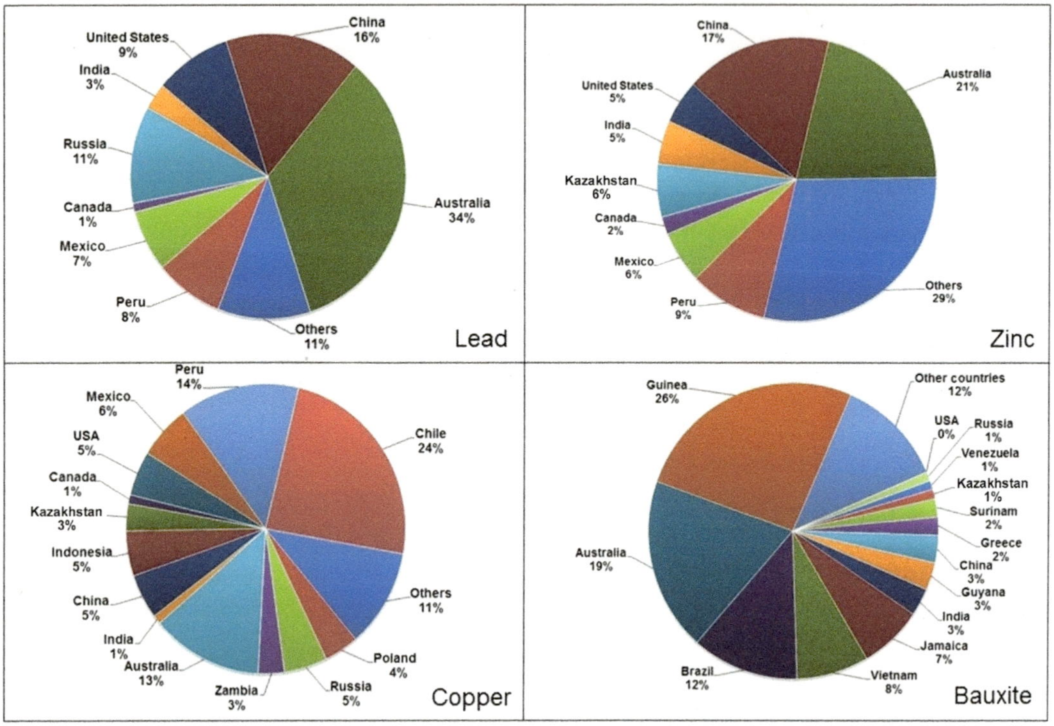

FIGURE 1.1 Global distribution of lead, zinc, copper, and bauxite in 2011.

products of the earth won by mining were essential for the requirement of the people for making implements, buildings, trade, jewelry, cosmetics, and treasure. Today, the standard of living of the people of the world is compared on the basis of per capita consumption of various metals. Transportation, communication, and construction have been revolutionized partly because of the materials that have been derived from Earth resources. For the foreseeable future, the metal and mineral products will continue to play their vital role in meeting society's needs.

Mineral deposits are not evenly distributed and a few nations in the world are self-sufficient in all commodities. Individual mineral deposits become valuable for several reasons. They may be very rich, they may be more accessible than other deposits, the location may be more favorable, or the commodity is in great demand. Obviously technological advances have made deposits of lesser mineral content into economic deposit. Mineral deposits are a wasting asset. The minerals are irreplaceable, hence, the deposit, as it is mined, decreases in its value. Dependence of man in his mineral heritage long established will continue indefinitely. Global distribution of some metals is given in Fig. 1.1.

As per Worldsteel, the share of world steel production in 2014 and 2013, by country, is given in Fig. 1.2. China is the top producer both in 2013 and 2014, followed in decreasing order by European Union countries, Japan, the United States, Russia, South Korea, etc. The annual consumption of various metals in 2012 is given in Table 1.1 (Wagner and

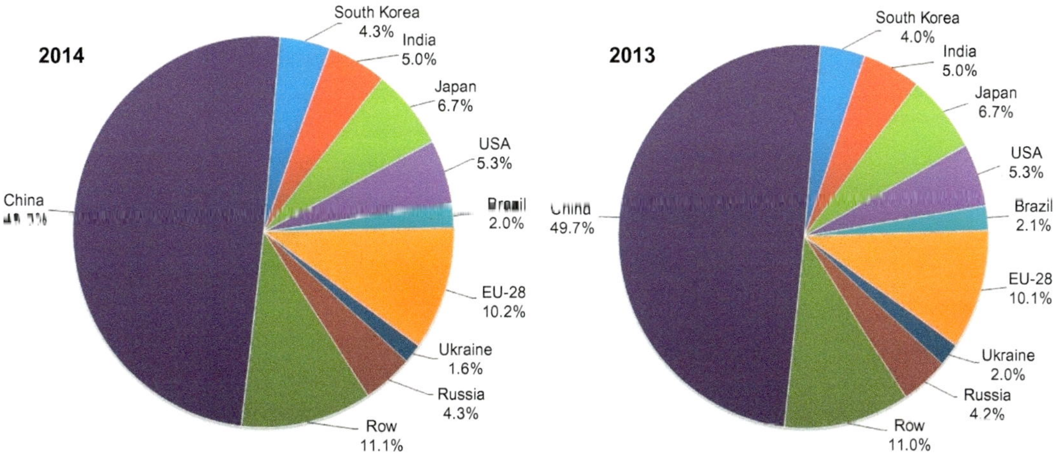

FIGURE 1.2 Share of world steel production in 2014 and 2013.

TABLE 1.1 Annual Consumption of Metals, in tonnes, in 2012

BASE METALS	
Copper	14,427,000
Zinc	9,104,000
Lead	3,648,000
Nickel	1,277,000
PRECIOUS METALS	
Silver	26,000
Gold	3100
Platinum	198
ELECTRONIC METALS	
Silicon metal	915,000
Rare earths	81,400
Selenium	1400
Indium	220
Tellurium	160
Gallium	110
Germanium	58

Geosciences & Natural Sciences Federal Institute (BGR), Germany, Database.

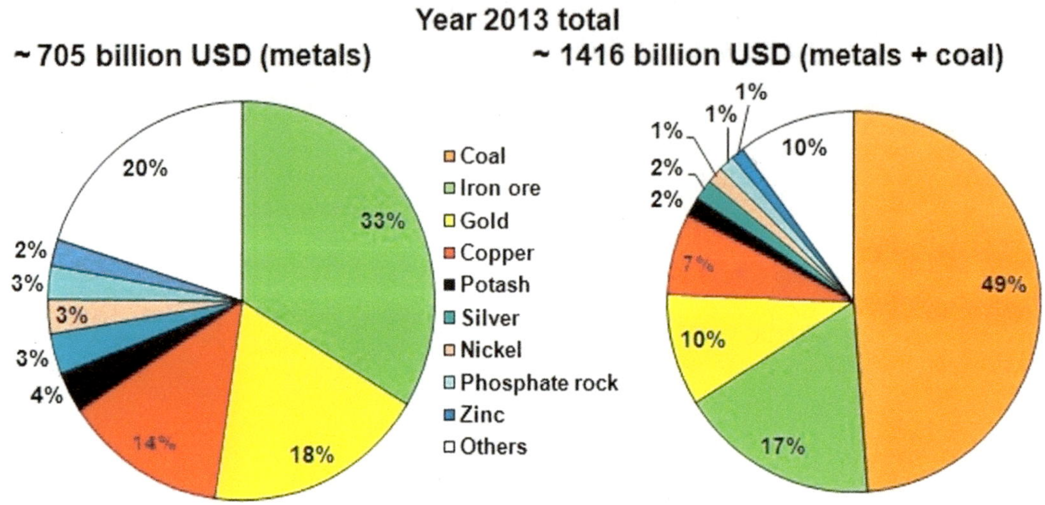

FIGURE 1.3 The value of global production, by metal, in 2013. *Source: Raw Materials Group, Sweden.*

Wellmer, 2012). Copper tops the list followed by other base metals viz. zinc, lead, and nickel. The metals used for electronic industry are being consumed increasingly owing to vibrant growth in that sector.

Human beings have been extracting the resources like minerals and metals from nature. The use of minerals and metals increased proportionately with the increase of population and economy. Such a growth in the consumption of minerals and metals has serious implications in both fronts of availability of resources and the degradation of environment, owing to the usage of metals. With the increase in population and industrialization in many nations, the demand of minerals and metals has become insatiable. Urbanization and spread of population areas and the increased pressure for environmental sustainability have compelled the exploration teams to look for economic prospects/deposits in more remote areas. Exploration companies have been forced to maximize the returns on the capital spent on exploration and concentrate on the cost consciousness, owing to the severe competition and investor's cost consciousness. The procuring and processing of lower grade ores will not only facilitate metal availability over longer periods of time but also pose new challenges in their environmental impacts. The value of global production, by metals, in 2013 is given in Fig. 1.3.

1.2 EMERGING ECONOMIES OF THE NEW WORLD

The goal of industry innovators in the global metals industry is to ride out the economic slump and position themselves for an eventual market turnaround. Emerging economies, led by Asia, have driven demand growth for raw materials in recent years. History shows that economic cycles cause fluctuations in demand for raw materials. Until the turn of this century, the mining industry experienced a prolonged period of declining commodity prices. This was a result of low demand growth and reducing production costs due to increasing scale and efficiency of mines, bulk processing of ore, and significant near surface mineral discoveries in the most accessible parts of the earth's crust. With high demand growth for products and resources now being depleted at unprecedented rates, every nation and major exploration companies are looking to secure a foundation for future growth. This growth will come from expanding existing operations, new mine developments in emerging regions, through technology, exploration, and mergers and acquisitions.

The big companies focus on large, long-life, low-cost, high-quality assets with wide diversification by commodity, by countries, and by markets which will safeguard them to remain profitable, despite the ups and downs of the economic cycles. North America, South America, Australia, and South Africa are the resource-rich regions in the world hosting many world-class mineral assets. In addition, they have unparallel expertise in exploration, development, and operation of mines in prospective regions elsewhere in the world facilitating to aggrandize the overall resource base of minerals.

The world population will double in about three decades. The industrial revolution will likely have phased into a scientific revolution. Coping with the growing dimension of mineral consumption will challenge the ability of the international mining industry to meet its most basic responsibility, the supply of metal and mineral products. Per capita consumption as well as the population is climbing at significantly higher rates. If many nations have to maintain the current ratio of production and consumption, the production index of the mining industry needs to be multiplied by a factor of 2 or more in ensuing three decades. If the demand does indeed escalate to the levels predicted by the study groups of various nations, there is no question about the technical ability of mining industry to meet it. If future supply problem develops, they will originate not from scarcity of reserves/ resources, but from a tangle of political, social, and business circumstances which may cause shortage of mineral production centers (Tilton, 1990). Because of the long lead time required to find, delineate, and develop unknown ore bodies (in a time span of 15–25 years), it is utmost important to launch expansion plans of the undeveloped mineral resources. The studies imply for an enormous growth of mining industry. New materials will be needed to build, drill platforms, nuclear plants, gas pipes, petro refineries, mining equipment, and other components. The basic ingredients are the refined products of natural ores.

If the nations turn to increased imports, it will be faced with a steadily rising negative balance of payments in international mineral trade. There are many constraints, which if uncorrected could lead to vexing mineral supply problems in the days to come. The rising of third world is pushing for new economic order and redistribution and wealth. It is to be expected that nations will seek to exercise control over the production of resources from their lands. In the last decade, mines and oil producing facilities have been expropriated or nationalized with compensation based on a negotiated definition of back value—usually a fraction of true market value or replacement costs.

The needs of the hour are (1) to assemble knowledgeable exploration crews and risk the seed capital for searching out new deposits and (2) to thrash out the methods of mining and extraction, and finally, (3) to mobilize the heavy financing for development and a competent management team. The undirected force of minerals nationalization has led to a situation in which (1) the exploration level abroad is greatly reduced, (2) the world geography for prudent exploration effort has shrunk in the alarming way, and (3) many nations may have locked themselves into an age-old technology with the mines and plants that have conveniently acquired. Many of the third world nations lack human resources for progressive management and thin in business talent.

1.3 PROGRAM FOR PROGRESS

A new level of statesmanship is necessary among nations of the world to avoid future stress in the comfort levels of mankind. Mineral deposits are unevenly distributed, so is technology, business acumen, and financing ability. Man may as well forget about solving global mineral and energy problems if they are viewed only in nationalistic terms and if nations fail to understand their mutual interdependence. The per capita consumption of aluminum, copper and steel (as per FIMI, 2013), and zinc (as per IMMMF, 2012) in different countries is given in Figs. 1.4 and 1.5, respectively.

World's resources are used more by the citizens of rich industrialized nations and they also produce more waste, thereby depleting their own environment besides other people. As per Lennon (2012), "USA is the world's largest consumer for many resources (copper, lead, zinc, tin, aluminum, rubber, oil seeds, oil and natural gas) and U.S.A. is also the world's largest per capita consumer for many more." There have been two trends causing

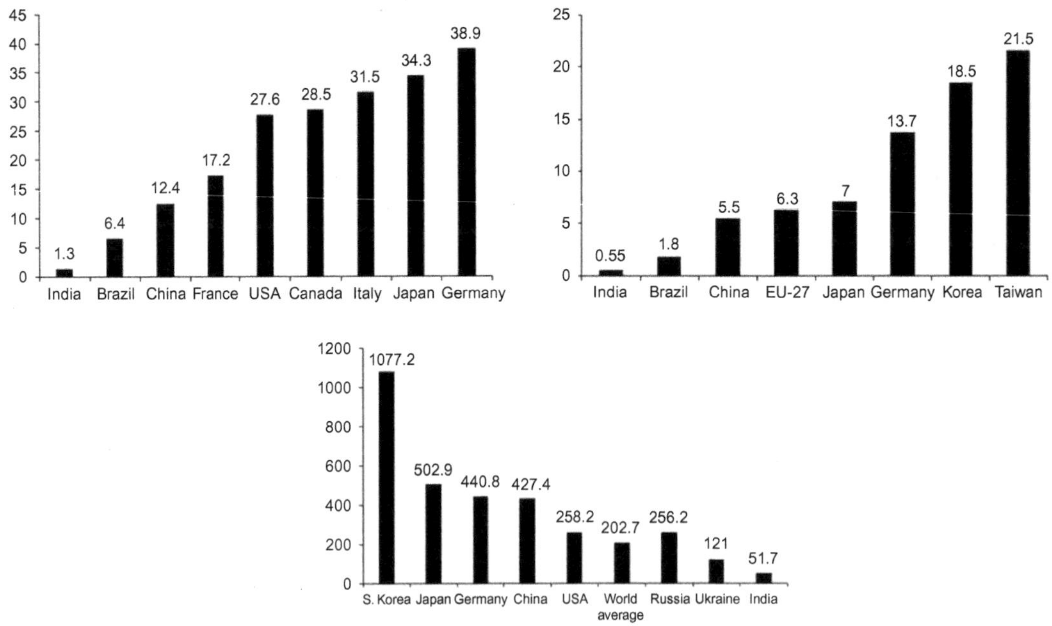

FIGURE 1.4 Per capita consumption (in kg) of aluminum (top left), copper (top right), and steel (bottom) in 2010.

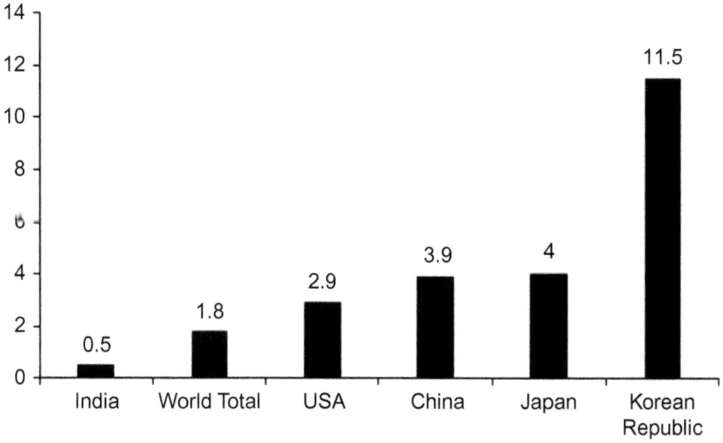

FIGURE 1.5 Per capita consumption (in kg) of zinc in 2013.

the individuals, companies, and nations alike to assess their use of scarce natural resources. In the 1970s, there has been an increasing realization that many resources, notably metals and fossil fuels, are unavailable for ever ("finite"). Since the 1980s, there has been growing concern about the environmental downslide owing to greedy/unmethodical exploitation of natural resources thereby causing pollution, environmental degradation/conservation of land.

It is said that out of the energy produced by burning coal in a power station, only 2–3% is eventually used, owing to the chain of inefficiencies at every stage of production of electricity, its transmission, and the final use by the consumer. Every company and Government are adopting strategies to reduce environmental degradation in the world. These are done by reducing the amount of material and energy used in providing their services and by reusing and recycling materials wherever possible. In the United States, the average gasoline consumption has been halved since the 1970s and by better insulation; European homes have reduced heat loss by 50% or more, during the same period.

The world consumption of zinc, lead, and copper metals in 2010 (as per FIMI, 2013) is given in Fig. 1.6.

1.4 RECYCLING AND CONSERVATION

If the mineral deposits are exploited in the same pace as it is done currently, they are bound to get exhausted since these are all nonrenewable resources. If we consume more and more, it is obvious, the coming generations will get less. Despite many innovative technologies are paving ways to the use of various substitute materials in the industry, "recycling" is likely to play a pivotal role in the time to come. The amount of recovered and renewed materials in the way of recycling is usually much less compared to the vast amount of consumption of primary resources. The depletion of a commodity to a certain extent could be reduced by conserving a commodity by using less of that. Increasing cost

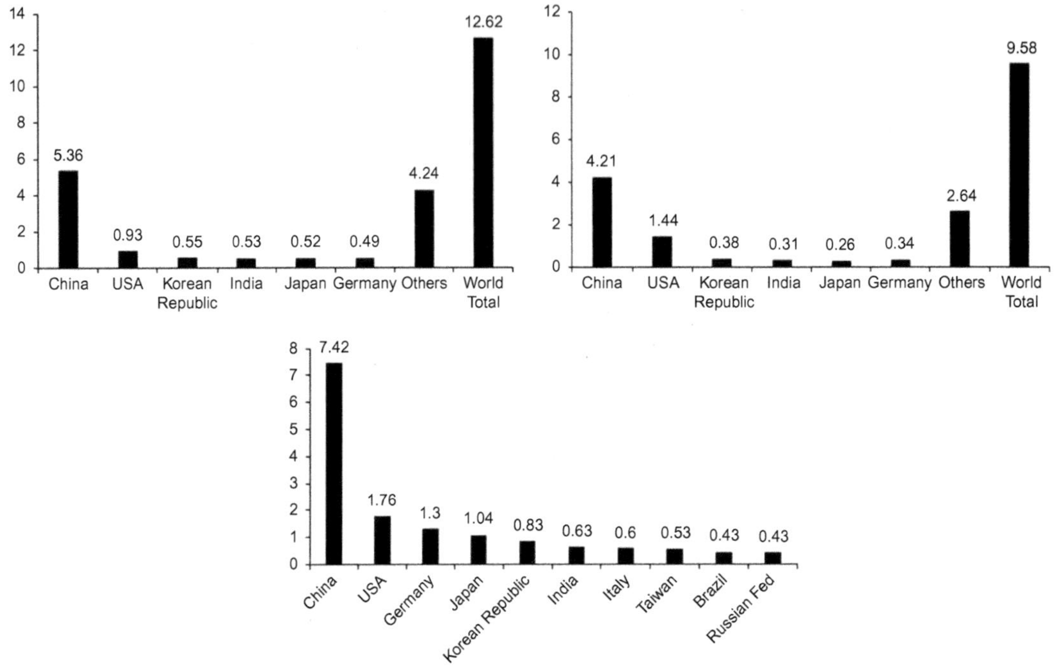

FIGURE 1.6 World zinc (top left), lead (top right), and copper (bottom) metal consumption (in million tonnes) by country in 2010.

and demand of the commodities is normally dictated by how quickly recycling and conservation efforts accelerate.

The disposal of metal waste has a profound impact on the environment owing to slow dispersal of metallic particles in the ecological system. Even inefficient or poor recovery of metal values from industrial/consumer waste indirectly causes concern on the resources of the Earth, from where they have been extracted. Metals are unique, in that, they can be recycled without loosing their important properties and recycled indefinitely resulting in a friendlier environment. Metal recycling is different from plastics, because the latter can be restored fully, irrespective of the physical/chemical form although it might probably be difficult and/or involve higher cost. The cost of retrieving and processing normally reflects the success of the recovered metal markets.

Large quantity of energy is needed for production of aluminum from its ore and the process to free "aluminum" from "alumina" is a complex one. Compared to the energy consumption of producing virgin aluminum metal, the recycling aluminum uses only about 5% of energy and very less (~5%) CO_2 emission, besides, reduction of amount of waste to the landfill. It is stated that "Recycling 1 kg of aluminum saves up to 6 kg of bauxite, 4 kg of chemical products, and 14 kWh of electricity. In Britain, the film processing industry reuses 5 million film cassettes a year, retailers reuse 40 million clothes hangers, and the aluminum industry recycles some 2 billion cans a year. The latter saves

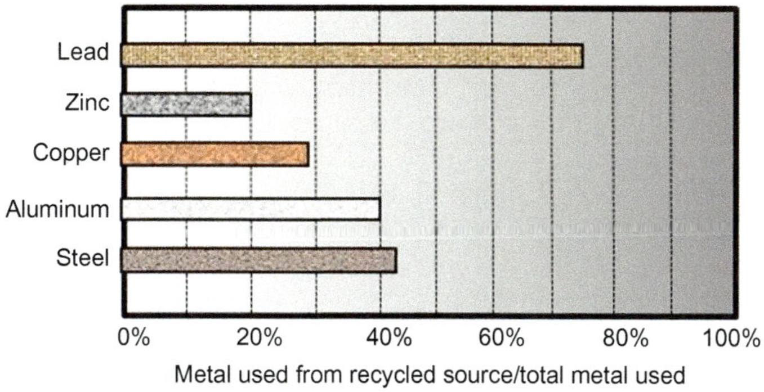

FIGURE 1.7 The percentage of recycled source of metals.

sufficient electricity, which would otherwise go to smelting new aluminum" (Hartman, 2009). The percentage of recycled source of some of the metals as per British Recycling Association is given in Fig. 1.7.

1.5 SOME ENVIRONMENTAL BENEFITS OF METALS RECYCLING

The industry of "Metals Recycling" not only aids to conserve the world's natural resources to a certain extent but also provides employment to many people who are engaged in various stages of recycling process. Global recycling industry directly employs more than 1.5 million people, annually processes over 500 million tons of commodities, and has a turnover exceeding US$160 billion. Secondary raw materials have been recovered and fed back into manufacturing process owing to "the recyclers," who collect products which have been dumped out by the users. But for the metal recycling industry, many valuable materials might have been lost forever to the production cycle and probably consigned to landfills. Metals like copper, nickel, brass, lead, gold, silver, etc. can all be recycled because these metals are all quite valuable and less in circulation.

Statistics show that "over 11 million tons of iron and steel scrap are produced each year, out of which 70% is recovered. Of the remaining quantity 2/3 is land filled. Each ton of steel recycled can save 1.5 tons of iron ore, 0.5 tons of coal, 40% of the water used in production, 75% of the energy needed to make steel from the raw material, 1.28 tons of solid waste, reduction in air emission 86%, and reduction of water pollution by 76%. Energy saved using recycled material versus virgin ore: 74% for iron and steel scrap, 95% for aluminum scrap, 85% for copper scrap" (Hartman, 2009).

1.6 SUBSTITUTION

There is a difference between "a substitute and a replacement." A substitute normally implies to a change with another or equivalent or better value whereas a replacement implies the change with another equivalent or having better properties, form, and/or function. The new materials have to be cheaper, lighter, better, and of high quality. Plastics, woods, bamboos have been found to be replacing in various application areas. Material substitution has its own challenges. For some metals like steel and aluminum, substitutes are available whereas for many others such as copper, chromium, manganese, and lead, no good substitutes are available. Many a times, it is noticed that steel and aluminum can substitute each other. Many other rare metals like "rhenium" (used for turbo engines, radiation screens, etc.), "thallium" (used for medical imaging, etc.) have very low potential substitutes. If the speciality metals supply (like rare earth metals) is to be maintained, the best way is to recycle them, which involves efforts in different levels viz., product design, waste collection, separation to recycling (Schulman, 2013).

Material scarcity and the price rise are normally seminal to look for substitutes. With the advancement of technology, scientists have been designing newer and better materials that can be used as substitutes. Material design of higher efficiency, ie, reduction in quantity per function, is expected as an immediate measure. The advancement of nanotechnology has opened many new horizons for the design and use of innovative materials in many speciality applications. Many precision components are being made now by "injection molded plastics" and act as substitutes to items which were hitherto made of expensive metals like brass, copper, stainless steel, aluminum, etc. Many of these substitutes are cheaper, lighter, better, and high quality.

1.7 GLOBAL FLOW OF METALS AND MINERALS

World Bank Group (WBG) (2006, 2008) has carried out excellent synthesis of the global flows of metal and minerals and much of the account given below is extracted from these studies.

1.7.1 Metal Prices in Nominal Terms

WBG (2008) has indicated that "International metals prices have risen substantially in the recent years and are at all-time highs in nominal terms, and in some cases match or exceed the highest real levels seen in the last few decades. Prices have been driven up by many factors. Rising costs have raised equilibrium prices for all commodities." In China, the demand growth soared up sharply and caught the mineral industry by surprise, resulting in low levels of inventories. The prolonged period of low metal prices with uncertain prospects and poor returns to the investors discouraged many companies to invest more funds and expand the mining capacity. In the recent times, the technical and labor problems, the delay in bringing the projects to the desired capacity, drove the metal prices even higher.

1.7.2 Chinese Demand Growth

China is one of the few countries that contains deposits of almost all known minerals and metals. China's mineral resources vary widely in both quality and quantity in an abundance capable of meeting most domestic demand for various metals. China has

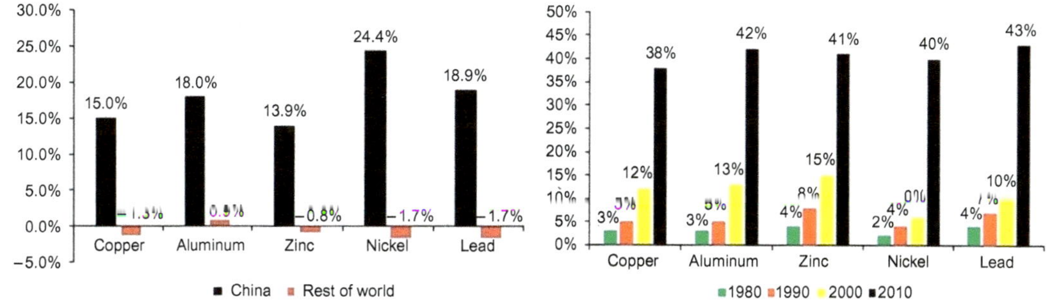

FIGURE 1.8 (Left) China's average growth in consumption (2000–2010); (right) China's growth of metal consumption (by decades).

played a dominant role in both demand and supply growth. It is the world's largest producer of a wide range of minerals and metals (antimony, bismuth, tin, tungsten, zinc, gold, etc.); surplus metals are sold in international market often at prices that Western producers find impossible to compete with. China is now the world's largest steel producer, steel consumer, steel exporter, and iron ore importer, the world's third largest iron ore producer, and the third largest steel importer. China has excelled in smelting/refining due to (1) extremely low capital costs of building new smelters (about one-third or one-fourth of the rest of the world), (2) short lead times to build (takes 12–18 months compared to 3–4 years elsewhere), (3) low environmental standard (slowly changing now), and (4) access to cheap finance and other incentives. China has moved from decades of economic isolation to bring one of the world's fastest growing markets for minerals and mining machinery. The rapid growth in GDP, hence demand, shows no sign of slowing. China's domestic GDP has gone up 10-fold making it the second largest economy in the world, overtaking Japan. China is poised to regain the historical role as the leading economy in the world (Kahn and Yardley, 2007).

Lennon (2012) in his study highlighted that "Chinese metals demand growth has averaged 10% per year, and in the last decade it has accelerated to 17% per year. Chinese metal demand has been driven by fast growth in industrial production, and investment in infrastructure, construction, and manufacturing. For a number of metals, China accounted for 70% or more of global demand growth in last decade, and the country is now the world's single largest user of almost all metals." China's average growth in consumption during 2000–2010 and China's growth of metal consumption by decades (as per Lennon, 2012) are given in Fig. 1.8.

Substantial investments were made by China in both metals mining and processing which led to strong metals production base. China has become one of the world's largest miner and refiner in many metals. It has also grown as a major source of demand for raw and refined materials in the global markets.

1.7.3 Higher Prices Bring Benefits to Producer Countries and Investors

Metals producing countries and the shareholders have received substantial benefits, owing to higher metal prices. The profits of many companies have soared up, so also the tax revenues. Since the mining companies are cash rich and have high share values, their "exploration spend" has risen significantly and also encouraged the companies looking for further acquisition of properties in various parts of the world. Many new entrants were attracted to international financial markets from diverse countries (Russia, India, Kazakhstan, China, etc.). Due to various reasons, the future path of developing countries to developed country level seems rather uncertain. But when the economic growth is fast, as in the case of China, the metal demand grew phenomenally fast yielding high revenues to the metal producer countries and also to the investors.

1.7.4 Market-Driven Industry Gives Grounds for Optimism

The nature of the mining industry is competitive and international, which caters to global demand for metals. In a span of two decades, despite a low key role of the private sector compared to State enterprises, both the sectors have slowly become interdependent. Private sector dominates now in the world mining sector and international metal trades continue from the countries like Russia, China, Brazil, South Africa, and elsewhere and it is anticipated that their market shares are likely to increase. The process of "continuing globalization" of mining industry is not expected to have a smooth run, but will have its own share of reversals. By adding value to their resources than exporting raw materials, some Governments are concerned about sharing the benefits between themselves and private companies. In some countries (in Africa, PNG, etc.), resurgence of economic nationalism is aimed at restricting investments from foreign companies in mining sector. In developing countries, where the Governments are weak, there are also concerns about fair competition for access to mineral resources and the application of "common good practice standards" for social and environmental issues. It is likely that the Industry will move in the coming days with a broad trend of more "private sector market-based approach" that will be able to meet the global demand for metals at a reasonable cost.

1.7.5 Long-Term Price Trends

It is uncertain whether the long-term price trend of metals will return back to its normalcy once the cyclical high is over. The complex forces of demand and supply of the market will dictate the overall outcome. The accelerated world economic growth in developing countries will pose challenges. With the adequate resource base of metals and minerals available, it can be reasonably guessed that a competitive international industry will continue to meet its growth in metal demand at a reasonable cost as it used to be done in the past. However, it is possible that the trend decline in real prices that has been a feature of the last four or five decades could be stopped and reversed to some extent. Since the cost pressures will remain upwards, prices may not return to previous levels. Major rise in operating and capital costs is unlikely to unwind due to high energy and steel prices, keeping capex and operating costs near current levels.

1.7.6 Policy Implications

When compared to the size of the world economy and international trade, the size of metals industry is small despite it gets affected by higher metal prices and thereby inflation. Higher metal prices can definitely bring in foreign exchange, various taxes, and also the inflow of new investments in some resource-rich economies. The metal prices, in general, follow a cyclical path and have their impacts in the flow of investors into a country. Whether metal prices show rise or fall, they need to be managed efficiently. In short-term metal supply management, Governments have very little say. But in order to facilitate a well-regulated long-term policy for mining industry and also to access the resources for development, a fair Government policy needs to be formulated.

Metals industry can meet the demands placed upon them, "making the markets to work," seems to be a fair policy. Once the mineral industry becomes international and the nations are more interdependent, Government's focus should be to help to address issues like inappropriate barriers to local versus foreign private investors, different subsidies, and other protections for the domestic producers. In order to ensure that the producing countries and the local communities receive appreciable, sustainable benefits from the development of their mineral resources, the Governments, the investors, and the international community should work in unison, owing to their common interest. This practice will not only contribute to the welfare of these nations and communities but also help to ensure more effective development and operation of the fair metals industry.

1.8 GLOBAL DEMAND OF MINERAL RESOURCES UP TO 2050

The demand of metal resources is increasing very rapidly in the world. In the 1960s, economic growth was limited to a relatively few countries that had about 20% of world population but the economic growth is presently occurring in heavily populated countries on a global scale (Rogich and Matos, 2008). The estimated per capita GDP for each country and up to year 2050 is shown in Fig. 1.9, while the population estimates for the countries are given in Fig. 1.10.

Halada et al. (2008a,b) have forecasted the consumption of many important metals up to 2050 in the BRICs (Brazil, Russia, India, China) and the G6 countries (Japan, United States, United Kingdom, France, Germany, Italy). The predicted consumption of Cu, Pb, Zn, Fe, and Al, by different countries, from 2000 through 2050, is given in Fig. 1.11. As per authors, "the forecasts were based on the linear decoupling model of the relation between per capita metal consumption and per capita GDP." According to their forecasts, "the overall consumption of metals in 2050 will be five times greater than the current levels, and demand for metals, such as Au, Ag, Cu, Ni, Sn, Zn, Pb and Sb, is expected to be several times greater than the amount of their respective reserves. Demand for Fe and Pt, which is considered to be optimistic about the resource exhaustion, will also exceed the current reserves." They further add that "Urgent measures are needed to find alternatives from common resources and to shift into sound materials circulation society. This will make it impossible to meet the demand solely with existing resource reserves. Furthermore, it is predicted that the consumption of some metals will even exceed their reserve base, so demand will rise to reconsider the use of these resources, and develop revolutionary new means of acquiring and recycling resources."

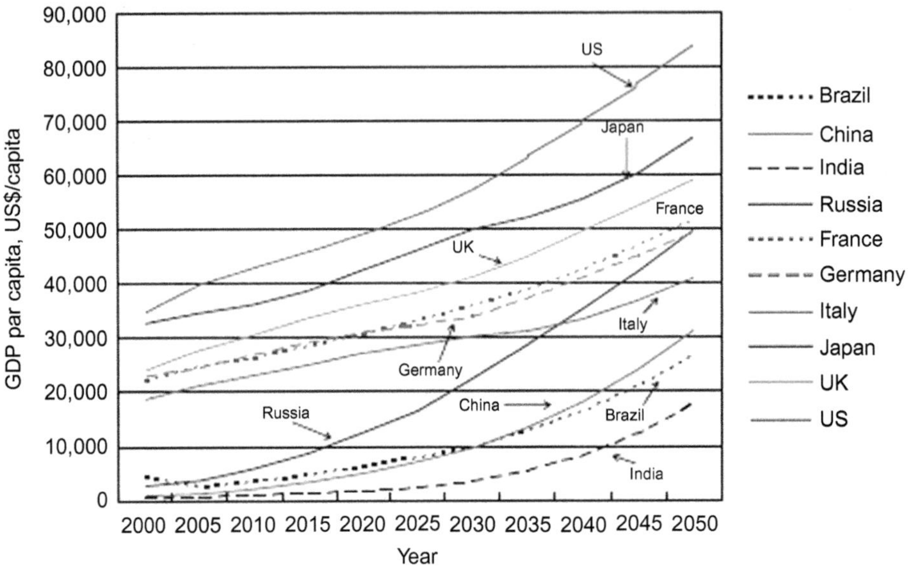

FIGURE 1.9 Per capita GDP of the countries predicted by Goldman Sachs. *Source: From Halada et al. (2008a).*

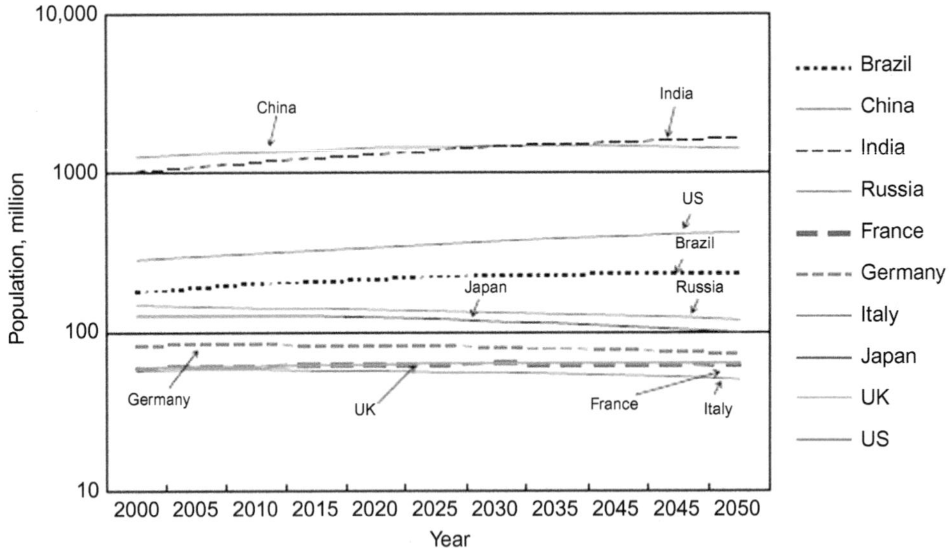

FIGURE 1.10 Population estimates of the countries predicted by Goldman Sachs. *Source: From Halada et al. (2008a).*

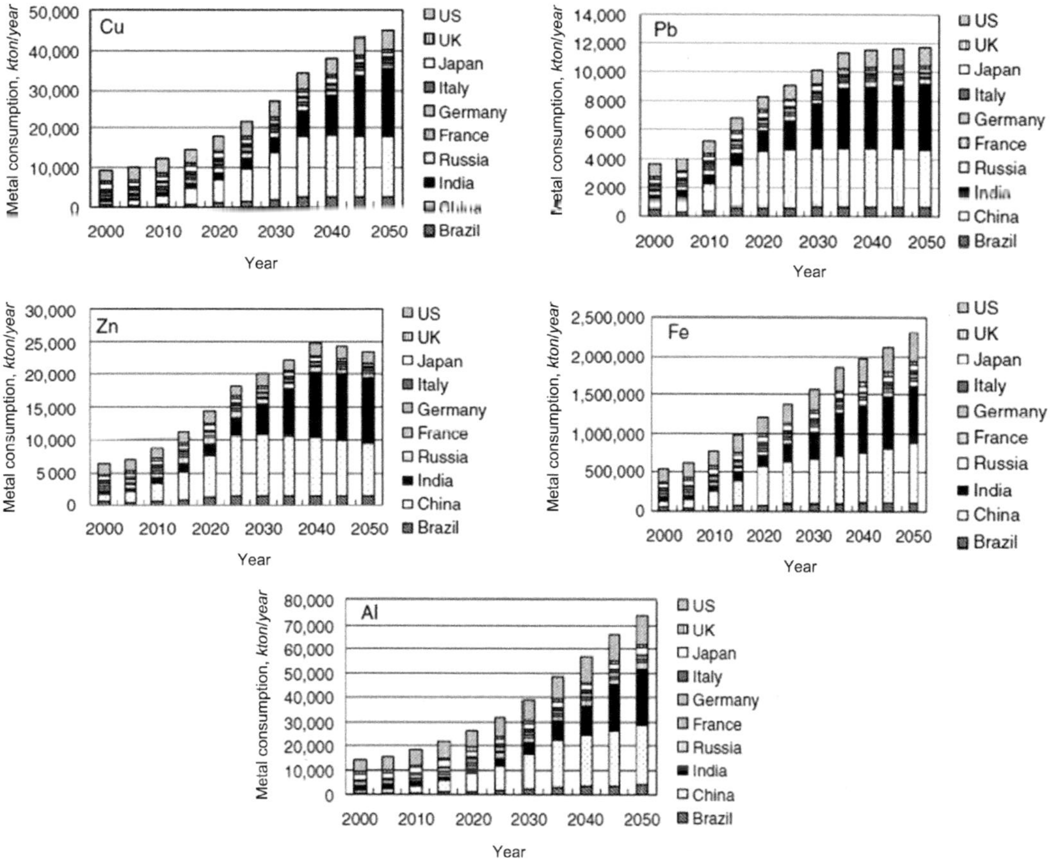

FIGURE 1.11 Predicted consumption of various metals (Cu, Pb, Zn, Fe, and Al) by different countries, from 2000 through 2050. *Source: From Halada et al. (2008a).*

1.9 SHIFTING MARKETS CREATE NEW OPPORTUNITIES

The slowdown in economic activity in the mature markets—particularly the Eurozone—coupled with new trends in customer demands is creating new challenges and opportunities for today's metals organizations. China and Japan are to account for the majority of their sales growth over the coming years. While the United States continues to be the top market for growth expectations, it is worth noting the rise of the emerging markets; India already ties the United Kingdom as the fourth most likely growth market and outpaces some of the more mature markets like Germany, where it is expected to see the majority of their sales growth as per Economic Intelligent Survey Report of November 2012 (Fig. 1.12). Emerging markets are growing in importance as a source of raw materials both from China and India. While this reflects growth in local demand more than international demand, it does indicate a continuing shift in the market towards Asia and other

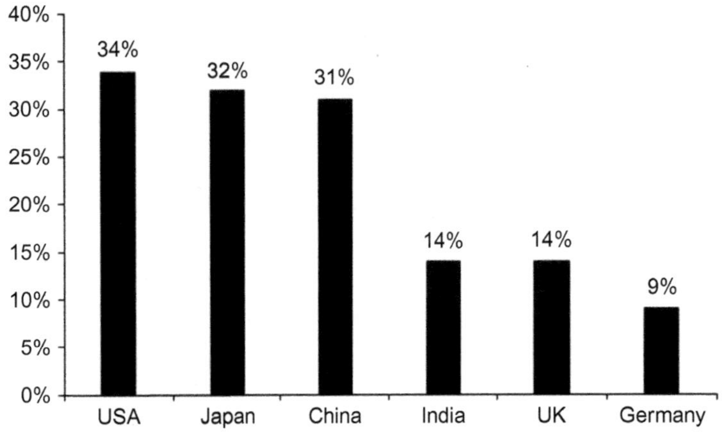

FIGURE 1.12 Growth expectations shift to emerging markets.

high growth economies. Metals use per capita is still very low in countries such as China and India, but they are still at a stage where metals consumption relative to GDP is rising and large populations make them more than significant forces on the market.

Given that China makes up almost 45% of all global metals demand, it is not surprising that many metals organizations expect to conduct much of their manufacturing processes in China. China ranked above the United Kingdom and Germany—and only slightly below India—as a top destination for the production of those goods that require significant investment potential. China may now be struggling with a slowing economy and overcapacity in the metals sector, there is an expectation that China's share of the global manufacturing process may slowly level out and—eventually—decline as the country's relative economic balance shifts from manufacturing to services (IBEF, 2006). At the same time, organizations around the world are starting to demand evermore sophisticated products such as lightweight materials, environmentally friendly manufacturing processes, and lower cost products. "Shifting market demands and evolving customer expectations are placing renewed pressure onto metals organizations to rethink their core markets and R&D strategies in order to meet future demand in growth markets and product lines." In many cases, metals organizations will find that the most cost-effective and lowest risk approach to new market entry lies in creating partnerships or joint ventures with existing market players."

Metals use per capita is still very low in countries such as China and India, but they are still at a stage where metals consumption relative to GDP is rising and large populations make them more than significant forces on the market. Demand will continue to be strongly linked to Asian growth and high rates of increase are expected. Once the recovery from the recession is completed, capacity is expected to just keep up with growth in demand in the long term. A large share of output growth will take place in developing countries (Africa and Latin America), where there is now strong investor and exploration interest.

As per IMF report of October 2013, the pace of growth of demand and the capacity of the industry to develop supplies are likely to dictate the future average price of metals, despite an adequate availability of geological base. The key determinants of the pace of

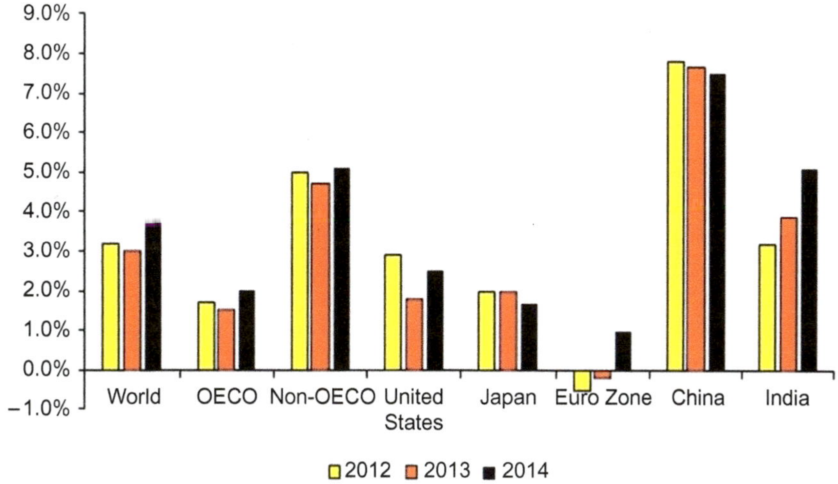

FIGURE 1.13 Economic growth in different countries.

growth of metal demand will largely depend upon the rate of growth of the developing countries and also on the rate of growth of global economy. It is presumed that China is likely to grow strongly for some time, in the case of fast growth scenario, followed by India, due to the continuance of the recent trend of fast growth. Besides these two giants, many developing countries will contribute a total population of about 3 billion people. Metal demand will continue to grow at a faster rate than the past two decades. The metal demand will be slow, if there is a slump in global growth, slowdown in China, and failure of India to take off as anticipated. Owing to India's liberalization policy and economic reforms, its economic growth has shown a good acceleration. Presuming that the trend continues, the future rate of growth of India could be higher than in the past and probably could match that of China. It is anticipated, that by 2030, India's population growth could take over China (Fig. 1.13).

1.10 THE GREAT FALL OF CHINA AND GLOBAL VOLATILITY

Many newspapers and weeklies of the world like Times, Economist, Forbes, Fortune, Al Jajeera, etc. had highlighted the great fall of China and its global aftermath. The summary of the observations of these periodicals is "China was world's second bright economy and was regarded as powerful engine of growth for the rest of the world with its unsatiating appetite for raw materials, energy, and consumer goods. The recent devaluation of major currencies followed by current depreciations in a large number of Asian emerging markets raises the risk of complete devaluation." The structural down turn in Chinese economy was forecast by many international experts, as China ran out of cheap labor and export market. The unexpected sudden burst of Chinese stock market in mid-June 2015, and within weeks, resulted in a loss of a whopping sum of about $3.4 billion. It was rather unfortunate that about 80% of Chinese households had invested directly or indirectly in the stock market before its crash. The crash necessitated Chinese Government to take drastic measures to ward off further fall of the market. China could probably overcome the recent shocks of financial markets but the confidence of their political leadership had a major hit. Efforts are on by the Chinese Government to regain confidence and rejuvenate the historic glory of China. Chinese economy relied heavily on public finances, infrastructure spending, and state-owned enterprises.

The magazines continue to report that the declining oil prices, recession and deflation in Europe, and Chinese great debt, worry investors around the globe. China was on borrowing spree and the total debt (Government, Corporates, and households) went up by 100% of its GDP since 2008 and reached now to a level of 250% of its GDP. The market crash could mean an economic downturn. Most of the financial crashes elsewhere in the world are preceded by a frenetic rise in borrowings (Japan, South Korea in the 1990s; America and Britain in 2008). The sum of real output and inflation, in China, has doubled from an average of 15% in a year in the 2000s to 8.5% and now is likely to fall further. China's devaluation of its currency has created ripples in global market and has triggered a global financial turmoil, hurting stocks, and currency markets worldwide. Competitive currency devaluation, at a time when global demand is sluggish, is a major threat to stability. And obviously being part of the globally integrated economy, this will leave some transient impact on the Indian economy as well. There are powerful deflationary forces at work in the global economy. Global volatility, with a lot of uncertainty, is not going to go away in a hurry. India, in the short term, will get as much affected, by global volatility, as everywhere else. Global economy will slow down and Chinese economy will also slow down.

Owing to its economic growth and robust underlying demand trends, India now occupies a noteworthy spot in the global economy. The one real challenge that looms large for India's economy appears to be price inflation. The global volatility will benefit India in the medium to long term because India will stand out, as growth is still very solid and all the macroeconomic parameters in India are very strong. India is likely to touch a growth rate of 8% for the financial year of 2015 and the future seems to be brighter.

1.11 CHALLENGES OF DEVELOPING ADEQUATE SUPPLY OF MINERALS

In spite of diminishing ore grades over the years and stricter environmental/social constraints in the mining operations, the amount of supply and flow of metals to world markets is continuing despite the falling real costs of the products. The reduction in the production cost could be achieved owing to diverse range of technological improvements in mining and processing operations. Besides, in the recent past, the efficacy of leaching technologies has made a great headway to bring down the cost of processing operations. The future demand growth will be very large which have to be met within diverse constraints. Generally, the resource base is not likely to pose a problem but the main other challenges like high exploration costs, diminishing ore grades, ever-increasing environmental and social concerns, water issues, etc. are likely to add to the problems. The chances of reaping real benefits are minimal, unless there is a development of a large open pit mine or opening up of new brownfield mines nearby. Mines located far away locations are likely to add heavy infrastructure costs which will not be economical in the days to come.

Among multitude of factors, the ability of mining sector to access the best possible resources for development will be the key factor for sustained supply of mineral commodity. It is likely that the current trend of more open and globally competitive metal market will continue for some years to come. Many regions and countries (like Central Asia, Russia, India) will continue to encourage the liberalization policy in metals sector and invite investments from local and foreign companies. Elsewhere in the world, Africa will be bogged down with the problems of political insecurity, lack of infrastructure, resource nationalism, etc. Latin-American countries are deciding to have greater "state control" in their mineral industry. The leading exploration companies world over expand their operations and try to spread over internationally by acquisition, investment, expertise, and competitive international mining. Governments from developed countries will try to maintain a balance with industry needs, with constraints and safeguards. In such a positive scenario, optimization of long-term investment will ensure comfortable supply to meet the demand of mineral commodities. Alternatively, in a protected and over-restrictive world, competition is slackened and distorted, thereby the best resources will not be accessible hence the supply of mineral commodity will be less than the anticipated demand (ICMM, 2012).

1.12 THE NEED FOR INTENSE MINERAL EXPLORATION

In order to sustain industrial growth, concerted and integrated approach encompassing various earth sciences was required to enable exploration of known occurrences and at the same time locate new deposits. Experience has shown that ore deposits are becoming increasingly more difficult to find since those that outcropped have largely been found, and almost all new ones are expected to be deep-seated and concealed. Most of the commonly applied geological, geochemical, geophysical, and remote sensing techniques are almost totally ineffective in areas of significant cover. Unlike in the past, it will require highly dedicated professional team using all the modern exploration tools available today to have the best chance of making an economic discovery.

While discussing the trends and economic issues of world mineral exploration, Tilton et al. (1988) observed that "with global economic growth and demand for mineral resources showing no signs of slackening, exploration technology is now more important than ever. There has never been a better time to be an earth explorationist." The face of mineral exploration industry is changing owing to the advanced exploration technology that takes less time in carrying out undercover deeper exploration. The demand for mineral resources is ever growing in spite of escalated commodity prices (Tilton, 2006). The industrializing countries like China and India escalated the demand of minerals and metals in tune with the accelerated growth, irrespective of higher cost of mineral commodities. The place to locate the resources to satiate this ever-growing demand has always been a conundrum. Tilton et al. (1988) commented that "finding new resources is not easy—but there are solutions to help." The problem of finding new resources is challenging governments and businesses around the world.

In order to maintain smooth functioning of a nation's economy and its security, it is essential to have an uninterrupted supply of minerals and metals in the world which is frequently interrupted by military and political conflicts. In such cases the things worth considering are: mineral conservation, commodity recycling, reuse, and adequate substitution of metals. These have limited roles to play especially in the prevailing government policies, better practices and attitudes, to meet current and projected demand of metals in future. Increasing difficulties and challenges prevailing elsewhere in the global scenario, it is for national interest one must consider the likely opportunities for discovery, development, and production of mineral commodities domestically.

It is anticipated that the upcoming "greenfields" projects are likely to be in remote areas, far away from market centers necessitating huge expenditure on infrastructural development of the area. Besides, ore grades are also likely to decline along with diminishing discovery rate coupled with huge exploration expenditure. In order to meet the demand of metals and also to balance the total ore reserves, it is necessary to discover more prospects/deposits in the days to come.

The optimum target in the metal sector must be high-quality deposit of good optimal location. If the location is not optimal, then the acceptable exploration target will have to be of an exceptional quality. For example, a poly metal or gold exploration target, in a remote, inland area, a large mineral deposit must be sought with a 100 or more million tonnes of high-grade resource or even higher tonnage of near surface low-grade resource. High unit value mineral resource alone can be explored to offset the ever-increasing cost of transportation, infrastructure, and production costs. On the contrary, selection of smaller targets, close to the operating mines, will be ideal. Such brownfield target finds will be very important to the operating mine which already has all established downstream facilities for milling, smelting, refining, etc.

The scarcity of mineral resources will slowly lead to diminishing returns and ultimately lead to the scarcity of metal supplies at the prevailing metal prices. Exploration activity is an ongoing process. Successful exploration campaigns and their economic returns will definitely encourage more and intense exploration in the time to come, especially when the metal scarcity will escalate the prices of metals. Since most of the near surface resources have been worked out, efforts are on for developing innovative better exploration techniques to search for resources from deeper areas of the crust. Presently, many companies are engaged in a lot of research in this front to focus to develop better understanding of search for deeper resources. Besides a lot of efforts in exploration technology, etc., a viable and a useful strategy seems to be with the reduced consumption of metals, their recycling and better substitutions. Hence a combination of exploration innovation, coupled with reduced consumption of metals, will facilitate to sustain the global population in terms of mineral commodity supply, for the years to come.

References

FIMI, 2013. India—the next landmark for Mining and Metals. Presented in PDAC 2013, Toronto, Canada, March 2013.

Halada, K., Shimada, M., Kiyoshi Ijima, K., 2008a. Forecasting of the consumption of metals up to 2050. Mater. Trans. 49 (3), 402–410.

Halada, K., Shimada, M., Kiyoshi Ijima, K., 2008b. Decoupling status of metal consumption from economic growth. Mater. Trans. 49 (3), 411–418.

Hartman, R.A.. 2009 Recycling: www/wikipedia.org/wiki/ferrous recycling etc.

IBEF, 2006. Metals: Market & Opportunities. India Brand Equity Foundation Publication, Gurgaon, p. 24 <www.ibef.org/download/Metals_210708.pdf>.

ICMM, 2012. Trends in Mining and Metal Industry. ICMM Publication, London, 16 p. <www.icmm.com/document/4441>.

IMMMF, 2012 India Metals and Minerals Forum (Ferrous and Non Ferrous) Indian Chamber of Commerce Publication, New Delhi, 28 p.

Kahn, J., and Yardley, J., 2007, As China roars, pollution reaches deadly extremes: New York Times, August 26, 2007, p. 9.

Lennon, J., 2012. Base Metals Outlook—Drivers on the Supply & Demand Side. Macquarie Commodities Research, London, p. 38., Ppt.

Rogich, D.G., and Matos, G.R., 2008, The global flows of metals and minerals: U.S. Geological Survey Open-File Report 2008-1355, 11 p.

Schulman, A. 2013: Metal substitution: finding practical solutions with technical compounds. Compounding World Forum 2013, Philadelphia, USA, December 2013.

Tilton, J.E. Editor, 1990. World Metal Demand Trends and Prospects. In: Colorado School of Mines Report (1990).

Tilton, J.E., 2006, Depletion and the Longrun Availability of Mineral Commodities. Society of Economic Geologists, Special Publication 12, pp. 61–70.

Tilton, J.E., Eggert, R.G., Landsberg, H.H. (Eds.), 1988. World Mineral Exploration: Trends and Economic Issues. Resources for the Future, Washington, DC.

Wagner, M., Wellmer, F.W., 2012. Global mineral resources, occurrences and distribution. Env. Eng. Geo. v.III, 26.

World Bank, 2008: The Commodity Boom: longer term prospects. In: World Bank Economic Prospects 2009. Commodity markets at the cross roads. Washington, DC, pp. 51–93.

World Bank Group, 2006: The Outlook for Metals Markets. Background Paper Prepared for G20 Deputies Meeting, Sydney, September 2006.

CHAPTER

2

Mineral Deposits: Types and Associations

OUTLINE

Essentials of Mineral Exploration and Evaluation
DOI: http://dx.doi.org/10.1016/B978-0-12-805329-4.00009-0

2.1 INTRODUCTION

In order to widen the horizons to locate new ore resources/deposits, an exploration geologist will need to have a good appreciation of basic geological concepts and understanding the geological mechanisms by which metals are transported from large regions of the Earth's crust, where they are present in trace concentrations, to much smaller regions of the crust, where they are present in quantities that can be mined at a profit. Voluminous literature is available in detailing various concepts (Bateman, 1962; McKinstry, 1962; Park and MacDiarmid, 1970; Rose et al., 1979; Roberts and Sheahan, 1988 Kesler 1994; Evans, 1998; Moon et al., 2006). Since it is beyond the scope of this volume, only salient features are discussed hereunder.

2.2 DEFINITIONS

Mineral: A mineral is a "naturally occurring, inorganic, homogeneous element or compound with a definite chemical composition and an ordered atomic arrangement." Most minerals have distinctive crystalline habit and may occur in well-formed crystals or crystalline aggregates, but a few species are characteristically noncrystalline/amorphous. Minerals can be metallic (gold, silver) or nonmetallic.

Rock: A rock is a naturally found material composed of an aggregate of grains or crystal of one or more minerals. Some rocks may be monomineralic; limestone consists of largely of the mineral calcite and many sandstones are dominantly quartz. Most rock varieties contain a number of mineral species and their classification depends upon the minerals present, their relative abundance, etc.

Ore: An ore is a naturally occurring accumulation or deposit of one or more minerals of economic value (more or less mixed with gangue) that can be exploited at a profit. The term is usually restricted to the description of mineral deposits that are of value for their metal content. Nonmetallic deposits although similar to metallic ores generally are valued for the minerals or components present (as with rock salt; Halite) rather than for individual contained elements, though such minerals as "elemental sulfur" are of much value.

Mineral deposit: A mineral deposit is any valuable mass of ore. It is a concentration of useful minerals or metals of sufficient size and grade, in the Earth's crust, that can be exploitable economically, under the prevailing technology and economic conditions.

Ore deposit: A mineral deposit that has been tested is known to be of sufficient size, grade, and accessibility to be mined at a profit. Testing commonly consists of surface mapping and sampling, as well as drilling through the deposit. An ore deposit is an economic term.

Metal: A metal is any of a class of substances (such as gold, iron, and aluminum) that typically are fusible, opaque, and are good conductors of electricity and show a metallic luster. Most metals are malleable, ductile, comparatively heavy, and all are solid (except mercury) at ordinary temperature.

Alloy: An alloy is a metal made by the fusion of two or more metals (an inferior metal mixed with a more valuable one) or a metal with a nonmetal.

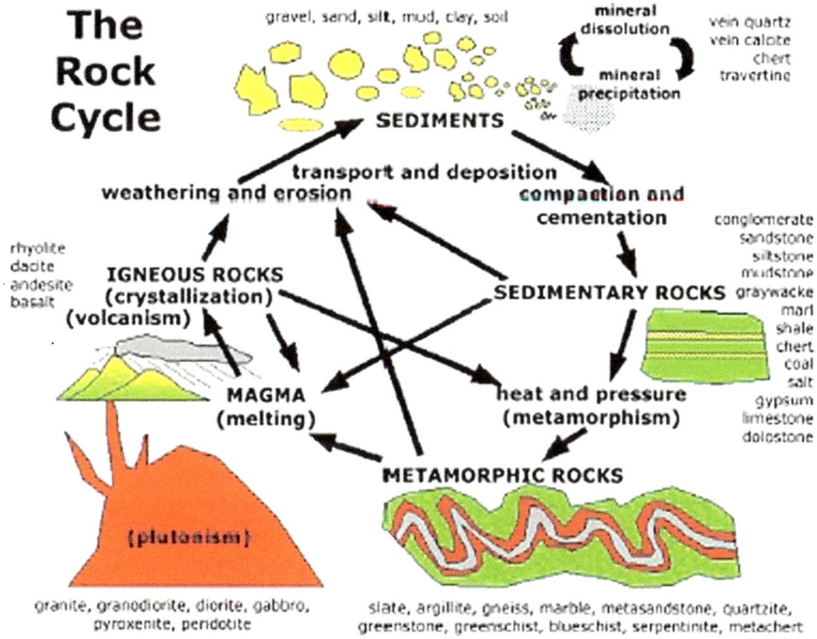

FIGURE 2.1 A sketch map of the rock cycle. *From USGS (2005).*

2.3 THE ROCK CYCLE

Rocks are grouped as (1) igneous rocks, (2) sedimentary rocks, and (3) metamorphic rocks. Igneous rocks are formed by the crystallization of magma (molten rock) and are described as "plutonic" if formed at great depths, "hypabyssal" if formed at intermediate depths, and "extrusive" if found at or on the surface of the earth. Igneous rocks are broken down into tiny particles/sediments by surface process of weathering/erosion and are carried away by wind/water/gravity to be deposited once again at a lower elevation to form sedimentary deposits. The action of softening, disintegration, partial solution of the mineral matter, etc. takes place depending upon the mineral composition of igneous rocks. When the sediments are buried by more sediment and subjected to pressure and temperature or both, finally lithify as sedimentary rocks. Continued burial of sedimentary rocks (and igneous rocks) added with temperature and pressure or both recrystallize and become metamorphic rock. These rocks that are exposed at the surface will also weather to form sedimentary rocks. When the metamorphic rocks become very hot, they melt and form magma again, completing the rock cycle. A sketch map of a typical rock cycle is shown in Fig. 2.1.

2.4 ROCK-FORMING MINERALS

The Earth's crust consists of eight common elements, viz., oxygen, silicon, aluminum, iron, calcium, magnesium, sodium, potassium in the order. Among myriad known minerals, a few hundred minerals of common occurrence account the bulk of the Earth's crust. The rock-forming minerals in all three petrographic categories—igneous, sedimentary, and metamorphic—are largely silicates and oxides, many of which contain one or more of the elements cited above. Quartz, feldspars, amphiboles, pyroxenes, micas, clays, and carbonates are the main rock-forming mineral groups. In sediments, additional mineral groups assume importance (especially carbonates, represented by calcite and dolomite) and sulfates such as gypsum and anhydrite. The carbonate minerals are also rock formers in metamorphic group, where they occur as recrystallized sediments (marbles). Mineral components normally determine the color of the rock: quartz, feldspars, carbonates, some micas are light colored, tan and pinkish; pyroxenes, amphiboles, some micas are dark green to blackish owing to their high iron and magnesium contents.

2.5 ORE BODY

Generally, an ore body is an accumulation of a solid and fairly continuous mass of ore with gangue, distinctly distinguishable by form and character from the enclosing host rocks. An ore deposit might include single or several ore bodies. The ore bodies are distinguished by their shape: (1) *isometric ore bodies* are accumulations of mineral substances that are approximately equal in all measurements, (2) *flat ore bodies*—sheets, veins, and lenses—have two long dimensions and one short dimension. The *sheet*, the most common shape in which sedimentary deposits occur, is a tabular body separated from other rocks by bedding planes. *Veins* are ore bodies formed when a mineral substance fills fracture cavities or when there is metasomatic substitution of mineral substances for rocks along cracks. The plane of contact between the vein and the enclosing rocks is called the selvage. A *lens* is a lenticular geological body that tapers out markedly in all directions; its thickness is slight compared to its length. In terms of morphology, lenses and lenticular beds are transitional formations between isometric and flat ore bodies. (3) Ore bodies elongated in one direction are called *ore pipes or pipes* which are oval in cross-section. They form when an ore substance from magmatic melts or hydrothermal solutions is concentrated; the melts or solutions penetrate from the abyssal parts of the Earth's crust along the line where tectonic fractures intersect or along fractures that intersect easily penetrated rock strata.

2.6 FORMATION OF MINERAL DEPOSIT

In order to form a mineral deposit, there are four basic geological requirements for any mineral deposit to form as suggested by McQueen (2009). They are (1) a source for the ore components (metals and ligands); (2) a mechanism that either transports these components to the ore deposit site and allows the appropriate concentration or removes nonore components to allow residual concentration; (3) a depositional mechanism (trap) to fix the components in the ore body as ore minerals and associated gangue; (4) a process or geological setting that allows the ore deposit to be preserved. The four basic geological requirements for any mineral deposit to form are shown in Fig. 2.2

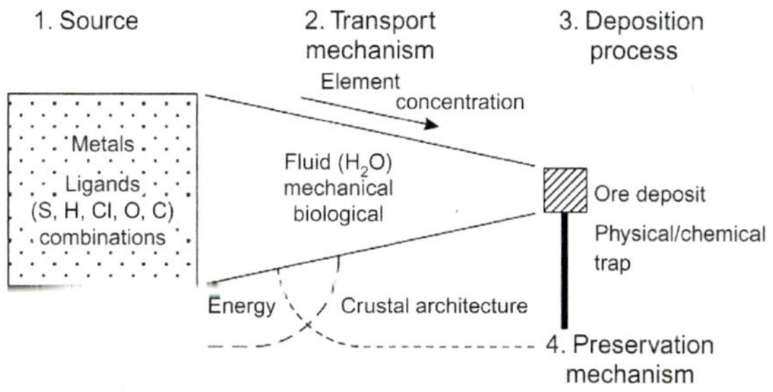

FIGURE 2.2 The four basic geological requirements for any mineral deposit to form. *From McQueen (2009).*

2.7 CHEMICAL AND PHYSICAL CONTROLS OF ORE DEPOSITION

Ores are deposited in response to changes in physical and chemical conditions. Consequently, environments localizing such changes are attractive exploration targets. Specific targets are geological conditions where one possible physical–chemical process of deposition could be effective or several concomitant processes may cause deposition simultaneously. Depositional processes often cause two- or three-dimensional mineral zoning; some metals are concentrated nearer to source, others far away (Park and MacDiarmid, 1970). Zoning data serve in predicting metal content variations (in milling operations) and in exploration for centers of mineralization and specific metals.

In order to power the transport mechanism and a suitable crustal structure to focus ore forming components and accommodate their deposition, energy (which can be thermal, gravitational, and/or deformational) is the essential requirement. Partial melting and fluid-related leaching of the mantle or crust yield metallic elements. Ligands can be generated from the same sources or from the atmosphere, hydrosphere, and biosphere. Mechanical or mass transfer mechanism and fluids are the main sources of metal transport. Biological processes can facilitate either to concentrate and transport ore components or remove nonore components, on or near the Earth's surface. Major medium of transport for many ore systems is hydrothermal fluids which are essentially water with less or variable amounts of CO_2, H_2S, SO_2, CH_4, N_2, NaCl, and other salts, as well as dissolved metal complexes. McQueen (2009) opines that "they are derived from a variety of sources, including (1) water-rich silicate melts, (2) circulated sea, connate, and meteoric waters, and (3) formational, diagenetic, and metamorphic fluids. At upper crustal levels, fluids are typically hotter than the rocks they traverse and in which they deposit their ores; they have variable pH and Eh, and they may be charged with a range of metal complexing agents including Cl- and HS-."

The changes in physicochemical parameters (temperature, pressure, pH, redox state, and total concentration of "ligands") normally result in ore mineral deposition. Addition of components by contamination, phase separation, cooling over a temperature gradient, decrease of pressure, fluid mixing, and reaction with host rocks are the normal changes associated with such processes. The essential requirements for ore deposit formation and concentration of ore elements are a variety of geological processes (or geochemical cycle, etc.). The complex interplay of favorable combinations of source, transport, and depositional variables usually determines vast array of ore deposit types and their elemental compositions. Ore formation is an intrinsic part of crustal evolution; coincidence of favorable combinations of processes and source parameters results in large and super large ore deposits.

Ore localization is due to the interaction of several geological factors and recognition of the specific combination of such "controls" which made ore in a given district provides the primary basis for exploration and development. Detailed accounts of ore controls are given by McKinstry (1962), Park and MacDiarmid (1970), Guilbert and Park (1986), and McQueen (2009).

Ore commonly occurs exclusively in some particular litho unit. This favorable horizon may be the result of original sediment accumulations of ore components. The ore shoots are generally related to sedimentary structures. Mineral localization also may result from crystallization of mafic, ultramafic rocks by concentration of gravitative settling of crystals of magnetite, chromite, and other heavy minerals or immiscible melts. Rocks of all classes may localize ore where cut by fractures which served as channel ways for aqueous ore-bearing fluids. Such localization may result from chemical reaction between rock and ore fluid, owing to permeability of host rock. Permeability differences might be the original rock fractures or may have been induced by selective fracturing of the favorable horizon because of greater brittleness.

All gradations exist between localization of ore entirely within fractures (vein) and entirely within the litho-unit cut by fractures. Ore shoots tend to be related to the maximum dilation produced by offset of irregular fault surfaces, intersection of fractures, or brecciation along the fault planes. The tension components of a conjugate fault system produce wider openings than do the shear components. The crests of folds may localize ore because of increased permeability, either through dilation or the presence of minor fractures resulting from folding.

Contact between contrasting rock types (including fault contact) often provides favorable environments through (1) localization or deflection of fractures, (2) steep chemical gradients, or (3) imposition of a relatively impermeable barrier upon the "plumbing system." The water table imposes a strong chemical rather than strictly physical interface which may localize both residual and secondary enriched deposits and in some cases, deposits formed by ascending solutions. Such water table is related to the surface topography at the time of deposition.

2.8 ORE DEPOSIT TYPES

Economic mineral deposit types are known to occur in all geological environments. Mineral deposits have been historically classified in many different ways according to a number of geological parameters, viz., geological setting, host rocks, conceptual model, formation, etc. They are: (1) composition of the deposit (contained minerals/metals); (2) form of the deposit (size, shape, orientation, and ore mineral distribution) and tectonic setting in which the deposit occurs; (3) host rocks or geological setting (the rocks enclosing the deposit and ore associations); and (4) interpreted genesis of the deposit (geological processes, controls to form the deposit). No two mineral deposits are alike and many are known to have formed by more than one process hence classifying into a single type is difficult. Hence there is considerable change of opinion among geologists as to the exact mode of genesis of most mineral deposits. Despite varying opinion, the classification which has a genetic connotation is accepted by many.

2.9 COMPOSITION OF THE DEPOSIT

Classification based on chemical or mineral content emphasizes its use. They distinguish mineral deposits worked for their metal contents (copper, zinc, iron ore deposits, etc.) from those worked for other components (coal, petroleum, etc.). Sometimes the term "ore" is also referred to nonmetal (eg, elemental sulfur). The four main types are: (1) metalliferous deposits, (2) industrial or nonmetallic deposits, (3) coal, and (4) petroleum and natural gas (mineral fuels), and (5) renewable resources.

2.9.1 Metallic Deposits

Naturally occurring local concentrations of metal-bearing minerals are called metallic deposits. In nature, metals occur in all kinds of rock types in small quantities which are difficult to extract. Generally, the metallic minerals can be melted to obtain new products. Metal-bearing mineral deposits become economic when they are mined at a profit which is dependent upon myriad economic factors. Metallic deposits are divided into (1) ferrous (where iron is the dominant metal) and (2) nonferrous (where deposits contain concentration of other metals). The ferrous metals include iron, steel, cobalt, and nickel. The nonferrous metals include gold, silver, copper, lead, zinc, aluminum, tin, titanium, vanadium, manganese, platinum tungsten, etc. Metallic minerals are usually hard and have luster, sheen of their own, ductile, and malleable. Depending upon a particular metal-bearing mineral, the metal may be chemically combined with a variety of compounds (oxides, sulfides, carbonates, and silicates). This is true for both ferrous and nonferrous deposits.

2.9.2 Nonmetallic Resources

Nonmetallic or industrial minerals are mineral materials which though valuable are not used primarily as sources of metal or as fuels. Several rocks and minerals have a dual or multiple uses (eg, rutile is an ore of "titanium" metal and is also an industrial mineral when used as source of TiO_2 for pigment). Nonmetallic resources are important to a nation's economy. The occurrence of industrial minerals and rocks embraces all geological occurrences. Many minerals/rocks may be used in their natural state after mining with or without processing, but also may be used after some complex treatment. Nonmetallic minerals are not so hard and have no shine, luster of their own. Since these resources occur in abundant quantities, they are less expensive.

Igneous, sedimentary, and metamorphic rocks are used for buildings (dimension stone) or as crushed rock and for certain special purposes. Allogenic minerals (clastics) found in sedimentary accumulations include ilmenite, rutile, zircon, quartz, monazite claya, and some gems. Minerals resulting from chemical or biogenic action include limestone and phosphorites. These may involve chemical precipitation due to environmental changes, such as evaporation, change in pH, temperature changes, or other factors disturbing equilibrium, as well as diagenesis. The evaporate minerals or chemical sediments include salts (NaCl & KCl), gypsum, anhydrite, borides, and nitrates. The minerals of metamorphic rocks include the calcite, dolomite of marble; refractory minerals such as kyanite, sillimanite, and asbestos; abrasives and/or gems such as garnets; and other minerals such as talc and soapstone.

Gemstones, besides their value as jewelry, have many important industrial applications. Gemstones include a range of minerals, viz., diamonds, rubies, emeralds, beryl, garnet, topaz, and zircon and some of them are frequently used as abrasives owing to their natural hardness. Diamonds, being the hardest substance, are used in drill bits and saws designed to cut through rock and steel.

The nonmetallic minerals and their use in various industries can be grouped as follows:

Construction materials: stone, gravel, sand, granite, limestone, marble (for construction activity, cement, concrete, etc.), volcanic material (gravel source), gypsum (plaster, wall boards, etc.)

Agriculture: nitrate, rock phosphate, potassium compounds (important source of fertilizers)

Industrial use: graphite (lubricant), rock salt (food preservative), sulfur (fertilizer, acid), barite (drilling mud), fluorite (toothpaste, teflon, steel, plastics), bentonite (cement, oil well drilling), pumice (abrasive)

Household items: silica sand (glass ware), talc (lubricating, cosmetic powder), borate (cleaning).

2.9.3 Energy Resources

There is a close link between availability of energy and future economic growth of a nation. An energy resource is defined as "something that can produce heat, power life or with which one can produce electricity." Coal, petroleum and petroleum products, uranium (nuclear materials), and geothermal resources are the key energy resources. Currently, fossil fuels (coal, oil, natural gas, and nuclear materials) cater various proportions of energy needs of different countries in the world. The byproducts of petroleum are also important in the production of plastics, asphalt, and other related products.

2.9.3.1 Coal

Coal is a combustible layered rock by the accumulation and preservation of vegetable matter in a densely resistant environment. Plant materials were interlayered as peat in deposits of clay mud and sand which on "lithification" formed a sequence of sedimentary rocks consisting of coal, underclay, shale, and sandstone. Coalification normally involves two steps: (1) biochemical conversion of peat to low rank coal and (2) progressive alteration to higher rank coal by interaction of time, confining pressure and temperature from weight of overlying strata and/or stress from mountain building forces. In the second step, devolatilization takes place by decrease in volatiles and moisture, increasing carbon content and calorific values. Different kinds of coal (peat, lignite, subbituminous,

bituminous coal) result from different degrees of compaction and depth of burial. These varieties of coal get converted to black, shiny form of coal "anthracite" due to metamorphism. Most coal occur in relatively flat lying beds in basins, but locally it may be completely folded and faulted. From coal, gas and oil can also be produced.

2.9.3.2 Petroleum

Petroleum is a highly inflammable liquid, composed primarily of hydrocarbons (90–98%) and the rest, organic compounds like O, N, S, and traces of organometallic compounds. Petroleum is used mainly as motive power, lubricating agent, and a source of raw material for manufacturing various industrial chemicals. Petroleum formation involves: (1) organic matter from organisms must be produced in greater abundance; (2) the organic matter must be buried before oxidation takes place, and (3) slow chemical reactions transform the organic material into hydrocarbons found in petroleum. The organic matter that eventually becomes petroleum is derived from photosynthetic microscopic organisms like plankton and bacteria, originally deposited along with clays in the oceans. The resulting black shales form the "petroleum source rock." With the depth of burial of these strata, the organic material breaks down into waxy "kerogen." Continued heating breaks down "kerogen" into different compounds in different temperatures like oil and gas, gas, etc. Nature separates the oil and gas. As a result of compaction of sediments containing petroleum, the oil and natural gas are forced to migrate out into "reservoir rock." In order to form petroleum, four features are necessary: (1) formation of source rock, (2) formation of migration pathway so that petroleum can move upwards, (3) filling a suitable reservoir rock with petroleum, and (4) development of oil trap to prevent the oil from migrating out of reservoir. Structural traps include faults between permeable and impermeable rocks, thrust faults, folds, etc. Because of these factors, the development of an oil reserve is geologically rare hence they are geographically limited. The presence of oil pools in a number of traps in an area is called "oil field." The largest known reserves are currently in Persian Gulf.

Through a series of drilled wells, crude oil and gas are recovered. Either the high confining pressure or pumping will help the petroleum to rise to surface. Sometimes steam/water is also pumped into oil pool to increase the pressure on oil. The recovered crude is shipped to a refinery for distillation and separation of various products, viz., natural gas, gasoline (petrol), diesel, kerosene, other oils and chemicals, asphalt. An array of petrochemicals are also produced which are used in myriad manufacturing industries. When the terrestrially discovered oil fields become less productive or dry, the petroleum companies moved to offshore to explore along continental shelves and discovered many productive oil fields across the world.

2.9.3.3 Oil Sands (Tar Sands) and Oil Shales

The most important deposits of solid hydrocarbons are oil sands and oil shales from which oil can be extracted. Oil sands are basically sandstone deposits cemented with asphalt/tar, which are widespread in the United States (La Brea Tar Pits) and Canada (Athabasca oil sands). These deposits are strip-mined and processed for oil extraction. Oil shales are organic-rich shale formations. "Kerogen" or pyrobitumen is a finely divided solid bituminous substance in oil shales that yields oil when the shales undergo destructive distillation to yield a mixture of compounds of high molecular weight compound of C, H, N, O, and S. In American terminology, "an oil shale is a shale or a marlstone that when distilled yields approximately 60 or more liters of oil per tonne." Oil shales are of three types based on origin and composition: (1) shales of humic origin—lignite, bituminous, and carbonaceous; (2) sapropelic shales—cannel, boghead, and kerogen; and (3) secondary shales—redeposited shales saturated with petroleum, asphalt, and carbonized impregnates. Oil shale deposits are generally low grade and are difficult and costly to mine. Research is on to find ways out to extract oil from the "rocks-in-place" by heating with microwaves to separate oil.

2.9.3.4 *Uranium and Geothermal Sources*

Uranium is the only fissionable material occurring in nature. It is useful in explosive devices and in the generation of power. Thorium isotope (Th^{232}) must be converted into fissionable isotope U^{233} to be utilized as a source of power. Uranium and thorium are found in many mineral species; some of which contain appreciable amounts of these elements. Based on their chemical affinities such as oxides, carbonate, sulfate, phosphate, arsenate, silicates, multiple oxides, the uranium and thorium minerals are classified. The hypogene mineral "uraninite" (pitchblende) and the supergene mineral "carnotite, torbernite" are important uranium minerals occur in sandstone, organic black shales, and phosphate deposits in many parts of the world. Good resources of uranium are in Canada, Australia, Russia, the United States, etc. Electricity can also be generated using geothermal source, but it accounts for negligible portion of the energy consumption of the world. Geothermal energy is limited to known areas of recent volcanic areas. It can be a good local resource, but it is unlikely to play a major role as an energy resource. If deeper heat sources across the world can be located and exploited, more geothermal power may be used in future.

2.9.3.5 *Renewable Resources*

The renewable sources of energy are that of natural resources that are inexhaustible and can be used to produce energy again and again. Solar energy, wind energy, tidal energy, geothermal energy, water energy, bioenergy, and burning hydrogen from the breakdown of water are some of the examples. As the supply of nonrenewable resources dwindles, these resources assume greater importance, since they are renewable. In fast breeder reactor technology, the atomic minerals are also inexhaustible source of energy.

2.9.3.6 *Metallic and Nonmetallic Ore Minerals*

The list of minerals in Table 2.1 includes those minerals most commonly of economic value in the metallic type of ore deposit. Many ore deposits, however, contain uncommon minerals in sufficient quantities to justify their recovery. The grouping of minerals in the table is based on chemical composition of minerals and not on their geological association. Besides, important uses are also given.

TABLE 2.1 Metallic, Nonmetallic Ore Minerals and Some Other Common Minerals and Their Uses

Ore Mineral	Formula	Composition (%)	Main Uses
FERROUS METALS			
Iron			
Magnetite	Fe_3O_4	Fe 70; O 30.0 (minor Ti)	Iron and steel industry. Construction, manufacture of machinery, alloys, automobiles, etc.
Hematite	Fe_2O_3	Fe 72.4; O 27.6	
Goethite	$HFeO_2$	Fe 62.9; O 27.0; H_2O 16.1	
Siderite	$FeCO_3$	(Fe 48.2) FeO 62.1; CO_2 37.9	
Pyrite	FeS_2	Fe 46.6; S 53.4	
Pyrhotite	$Fe_{1-x}S$	($x = 0-0.2$)	
Manganese			
Pyrolusite	MnO_2	Mn 63.2; O 36.8	Batteries, steel making, alloys, alloying agent, drier in paints, coloring in bricks, decoloring in glass, pottery, car wheels, axels, etc.
Manganite	Mn O (OH)	Mn 62.4; O 27.3; H_2O 10.3	
Psilomelane	Mn_5O_{10} (Ba, H_2O)	Besides Mn, small amounts of Ni, Co, Cu	
Rhodochrosite	$MnCO_3$	MnO 61.7; CO_2 38.3	
Braunite	Mn_2O_3	Mn 60–69	
Haulandidte	MnO Mn_2O_3	Mn 65–72	
Chromium			
Chromite	FeO Cr_2O_3	Cr_2O_3 68.0; FeO 32.0	Electroplating, alloys, stainless steel, refractory bricks, pigments, and dyes
Magnochromite	Comp. of chromite variable; Cr_2O_3 50–60		
Titanium			
Rutile	TiO_2	Ti 60.0; O 40.0	Alloying agent, turbine blades, alloys for aerospace industry, paints and pigments, pyrotechnics
Ilmenite	FeO TiO_2	Ti 31.6; Fe 36.8; O 31.6	
Ilmenite leucoxene		TiO_2 96	
Sphene leucoxene		TiO_2 67	
Vanadium			
Vanadinite	Pb_3 $Cl(VO_4)_3$	PbO 78.7; Cl 2.5; V_2O_5 19.4	Alloys, catalysts, steel additive, jet engine parts
Patronite	V S_4	V 29	
Roscoelite	$KV_2AlSiO_3O_{10}$ $(OH)_2$; V 19–29		
Descloizite	(Zn,Cu) Pb $(VO_4)OH$; V 20–23		

(Continued)

TABLE 2.1 (Continued)

Ore Mineral	Formula	Composition (%)	Main Uses
NONFERROUS METALS			
Aluminum			
Bauxite	$Al_2O_3.2H_2O$	Al_2O_3 73.9; H_2O 26.1	Many construction items, electrical cables, wires; transport, consumer durables, packaging, castings, aircraft structures
Gibbsite	$Al_2O_3.3H_2O$		
Bohemite	$Al_2O_3.H_2O$	Al_2O_3 ~85	
Diaspore	$Al_2O_3.H_2O$		
Copper			
Native copper	Cu	Cu ~92	Copper, as a pure metal in electrical wires (60%), roofing and plumbing (20%), and industrial machinery (15%) combined with other elements to make an alloy (5%) brass and bronze; production of compounds for nutritional supplements and fungicides in agriculture
Chalcopyrite	$CuFeS_2$	Cu 34.6; S 35; Fe 30.4	
Covellite	CuS	Cu 64.4; S33.6	
Chalcocite	Cu_2S	Cu 79.8; S 20.2	
Cuprite	Cu_2O	Cu 88.8; O 11.2	
Bornite	Cu_5FeS_4	Cu 63.3; S 25.5; Fe 11.2	
Bournonite	PbCuSb S_3	Pb 42.4; Cu 13.0; Sb 24.9; S 19.7	
Tenorite	CuO	Cu 79.9	
Enargite	Cu_3AsS_4	Cu 88.6	
Atacamite	Cu SO_4 $CuCl_2$	Cu 59.5	
Malachite	$Cu_2CO_3(OH)_2$	Cu 57.4; CuO 71.9; CO_2 19.9; H_2O 8.2	
Zinc			
Sphalerite	ZnS	Zn 67; S 33	Galvanizing, die casting, alloys, dry batteries, cosmetics, pharmaceutical, micronutrient for human, animals, and plants
Zincite	ZnO	Zn 80.3; O 19.7	
Willemite	$Zn_2(SiO_4)$	ZnO 73.0; SiO_2 27.0	
Smithsonite	Zn CO_3	ZnO 64.8; CO_2 35.2	
Hemimorphite	$Zn_4(Si_2O_7)(OH)_2$	ZnO 67.5; SiO_2 25.0; H_2O 7.5	
Franklinite	(Zn, Fe, Mn) (Fe, $Mn)_2O_4$	Zn, Fe, Mn, substitution takes place	
Lead			
Galena	PbS	Pb 86.6; S 13.4	Storage battery, electrodes, ceramic glazes, stained glass, radiation shields, paints, power and telephone cables, chemicals, etc.
Anglesite	$PbSO_4$	Pb 76.3; SO_4 23.6	
Cerussite	$PbCO_3$	PbO 83.5; CO_2 16.5	
Jamsonite	$Pb_4FeSb_6S_{14}$	Pb ranges from 20 to 40	

(Continued)

TABLE 2.1 (Continued)

Ore Mineral	Formula	Composition (%)	Main Uses
Cobalt			
Cobaltite	(Co,Fe)AsS	Co 35.5; As 45.2; S 19.3	High temperature alloy, steel tools, viz., bits, cutter, pigments, paints, ceramic glazes
Skutterudite	$(Co,Fe)As_3$		
Smaltite	Co As_2		
Erythrite	$(Co_3AsO_4)\ 8H_2O$	CoO 37.5; As_2O_5 38.4; H_2O 24.1	
Linneite	Co_3S_4	Co ranges from 57 to 96	
Glaucodot	(CoFe) As S	Co 23–85	
Nickel			
Niccolite	NiAs	Ni 43.9; As 56.1	Ferrous and nonferrous alloys, stainless steel, electrical resistant heating element, food processing equipment, chemicals, etc.
Millerite	NiS	Ni 64.7; S 35.3	
Pentlandite	$(Fe\ Ni)_9S_8$	Fe:Ni is close to 1:1	
Garnierite	NiO SiO_2 H_2O	NiO 46	
Polydymite	Ni_3S_4	Ni 40–54	
Molybdenum			
Molybdenite	MoS_2	Mo 59.9; S 40.1	Alloying agent, heating elements, thermocouples, welding electrodes, cathode ray tubes, ferroalloys, etc.
Wulfenite	$PbMoO_4$	PbO 60.8; MoO 39.2	
Powellite	$CaMoO_4$	Mo 48	
Wulfenite	$PbMoO_4$	Mo 46	
Ferrimolybdite	$Fe_2O_3.2MoO_3\ 7H_2O$	Mo ~60	
Tin			
Cassiterite	SnO_2	Sn 78.6; O 21.4	For food containers, soldering alloy, tin platings, wrapping food products, etc.
Stannite	Cu_2S FeS SnS	Sn 27.5; Cu 29.5	
Tillite	PbSn S_2	Sn 30.4	
Tungsten			
Wolframite	(Fe,Mn) WO_4	WO_3 76.3 in ferberite; 76.6 in huebnerite	Incandescent lamp filaments, alloying agent, cutting material, defense projectiles, etc.
Scheelite	$CaWO_4$	WO_3 63.9	
Tungstite	H_2WO_4	WO_3 74	

(Continued)

TABLE 2.1 (Continued)

Ore Mineral	Formula	Composition (%)	Main Uses
Antimony			
Antimonite	Sb_2S_3	Sb 71.4	Alloys with lead, textile, fiber, etc.
Gudmundite	FeSbS	Sb 57.8	
Berthierite	$FeSb_2S_4$	Sb 57	
Cervantite	Sb_2O_4	Sb 79.2	
Bismuth			
Native bismuth	Bi	Bi 99.9	Chemicals, pigments, additives, alloys, cosmetics, etc.
Bismuthinite	BiO_2	Bi 80	
Tetradymite	Bi_2Fe_2S	Bi 59.37	
Bismite	Bi_2O_3	Bi 89.6	
Mercury			
Naïve mercury	Hg	Hg 100	Chemicals, medicine, electronic applications, thermometer, lamps, etc.
Cinnabar	HgS	Hg 86.2	
Cordeorite	$Hg_3S_2Cl_2$	Hg 82	
RARE ELEMENTS			
Lithium			
Spodumene	$Li\ Al(Si_2O_6)$	Li 6–7.5	Battery, ceramics, heat resistant glass, nuclear fusion, rocket propeller
Lepidolite	$KLi_n(FeMg)_m\ Al_p\ (Si_4Al_2O_{10})(F\ OH)$		
Petalite	$LiAl(Si_4O_{16})$	Li 3.5–4.5	
Beryllium			
Beryl	$Be_6Al_2Si_6O_{18}$	BeO 10–12	Alloys, electronics, electrical devices
Phenacite	$Be_2\ (SiO_4)$	BeO 40–44	
Bertrandite	$Be_4\ (Si_2O_7)(OH)_2$: BeO 40–42		
Niobium and tantalum			
Wodginite	$(MnFe)(Ta\ Sn)_2O_6$ Ta_2O_5 70		Electroceramics, super alloys, superconductor
Fergusonite-Y NbO_4—Brochanite $CaNbO_4$ Nb_2O_5 38–58			
Zirconium			
Zircon	$ZrSiO_4$	ZrO_2 65	Refractories, insulation, electroceramics
Beddeleyite	ZrO_2		

(Continued)

TABLE 2.1 (Continued)

Ore Mineral	Formula	Composition (%)	Main Uses
RARE EARTH METALS AND YTTRIUM			
Monazite	(Ca, La, Y, Th)PO_4; Th is (0.0–20%)		Catalysts, metal alloys, electronics, magnets, x-rays, nuclear control rods, lasers, pigments, polishing, phosphors, ceramics, etc.
Xenotime	Y (PO_4)	RE/Y 51–60	
Bastnaesite	(La,Ce)(CO_3)F	RE/Y 65–75	
Churchite	(Ce,Y) PO_4 $2H_2O$ RE/Y 42–59		
Scandium			
Thortveitite	$Sc_2(Si_2O_7)$	Sc_2O_3 53	
Sterrettite	$Sc(PO_4)$ $2H_2O$	Sc_2O_3 39	
Thallium			
Loranelit	$TlAsS_2$	Tl 59	Rodent killer, glasses with high index of refraction
Hutchinsonite	$PbTlAs_5S_9$	Tl 18–25	
Verbaite	$TlAs_2SbS_5$	Tl 29–32	
Cadmium			
Greenockite	CdS	Cd 77.8	Batteries, electroplating, nuclear fission, paints, and pigments
Otavite	$CdCO_3$	Cd 65.8	
Cadmoselite	CdSe	Cd 47	
Indium			
Native indium	In	In 100	Semiconductor, low temperature alloys, batteries
Indite	Fe In_2S_4	In 59.3	
Roquesite	Cu InS_2	In 47.4	
PRECIOUS METALS			
Gold			
Native gold	Au	Au ~98–100	Jewelry, electronics, bar holding, investment, dentistry
Electrum	Au Ag		
Auricupride	Au_2Cu_3		
Sylvanite	$(Au,Ag)Te_2$	Au:Ag = 1:1; Te 62.1; Au 21.5	
Rodite	Au(Pt,Rh,Ir,Pd)		
Caleverite	$AuFe_2$	Au 44.03; Fe 55.97	

(Continued)

TABLE 2.1 (Continued)

Ore Mineral	Formula	Composition (%)	Main Uses
Silver			
Native silver	Ag	Ag 80–100	Jewelry, dentistry, coinage electronics, investment, photography, chemical compounds
Argentite	Ag_2S	Ag 87.1; S 12.9	
Cerargyrite	AgCl	Ag 75.3; Cl 24.7	
Pyrargyrite	Ag_3SbS_2	Ag 59.7; Sb 22.5; S 17.8	
Proustite	Ag_3AsS_3	Ag 65.4; As 15.2; S 19.4	
Tetrahedrite	$Cu_{12\ 64}Sb_4S_{13}$	Solid soln. series	
Tennantite	$Cu_{12}As_4S_{13}$	Cu, Ag, Fe present	
Polybasite	$(AgCu)_{16}Sb_2S_{11}$	Ag 62.1–84.9	
Stephanite	$Ag_3Sb\ S_4$	Ag 68.3	
Platinum group			
Braggite	(Pt, Pd, Ni)S–Pt 32–58, Pd 17–38		Autocatalyst, electrical and electronic uses, jewelry, glass, investment
Cooperate	PtS	Pt 79–86	
Sperrylite	$PtAs_2$	Pt 56.5	
Vysotskite	(Pd,Ni) S	Pd 59.5	
Laurite	(Ru, Ir, Os) S_2	Ru 61–65	
Hollingworthite	$(Rh,Pt)(AsS)_2$	Pt 20; Rh 25	
RADIOACTIVE ELEMENTS			
Uranium			
Carnotite	$K_2((UO_2)_2\ V_2O_8)\ 3H_2O$; U 64		Nuclear fuel for electricity, medical treatment, research, germicide, etc.
Autunite	Ca $(UO_2PO_4)_2\ 12H_2O$; U 60		
Torbernite	Ca $(UO_2PO_4)_2$ 8–12 H_2O; U 61		
Uranophane	$CaH_2\ (UO_2\ SiO_4)_2\ 5H_2O$; U 67		
Brannerite	$(U,Ca,Th,Y)(TiFe)_2O_6$ U 28–44		
Thorium			
Thorianite	ThO_2	Th 88	Nuclear fuel, catalysts, heat resistant ceramics, glass, etc.
Thorite	Th SiO_4	Th 81.4	
Broggerite	(U,Th) O_2 Th 6–15		
Uranothorite	(Th,U) SiO_4	Th 50–70	
Th-bearing monazite	$(Ce,Th)(P,Si)O_4$; Th 3.5–10, up to 40		

(Continued)

TABLE 2.1 (Continued)

Ore Mineral	Formula	Composition (%)	Main Uses
SOME OTHER MINERALS			
Quartz	SiO_2	Si 46.7; O 53.3	Glass, building material, gemstone, porcelain, paint, mortar industry, and acid flux in smelting furnaces
Limestone	$CaCO_3$	Cement industry, as flux in steel industry	
Marble	$CaCO_3$	CaO 56	Building and decorative stone
Apatite	$Ca_5(PO_4)_3(FClOH)$	P_2O_5 40–45	Fertilizer, gem variety also
Sylvite	KCl	K 52.4; Cl 47.6	Source of potash, fertilizer
Rock phosphate	$3Ca_3(PO_4)_2$	P_2O_5 15–35	Fertilizer
Barite	$BaSO_4$	Ba $(OH)_2$ 65.7; SO_3 34.3	Drilling mud, filler, paper glazing, medical, rubber
Diamond	Pure C	C 100; Gem, abrasive, cutting tool, drill bit	
Dolomite	$CaMg(CO_3)_2$	Ca 21.7; Mg 13.2	Source of magnesium, ornamental stone, refractory bricks, glass making
Graphite	C	C 70–85 steel making, crucibles, refractory, pencil lead, electrodes, etc.	
Gypsum	$CaSO_4$ $2H_2O$	CaO 32.5; SO_3	Cement, plaster of Paris, insulation board
Anhydrite			
Halite	NaCl	Na 39.4; Cl 60.6	Salt, preservative, soap, bleaching
Kyanite	$3Al_2O_3$ $2SiO_2$	Al_2O_3 63.2; SiO_2 36.8	Heating element, ceramic industry, refractories, gemstone
Andalusite	Al_2SiO_5	Al_2O_3 63.2; SiO_2 36.8	
Kaolin	$3Al_4Si_4O_{10}(OH)_8$	SiO_2 46.5; Al_2O_3 39.7	Paper, coating clay, in rubber industries, bleaching
Bentonite	Al_2O_3 $4SiO_2$ H_2O	SiO_2 66.7; Al_2O_3 28.3	Drilling mud, metal casting pellets, metal casting
Talc	$3MgO$ $4SiO_2$ H_2O	MGO 31.7; SiO_2 63.5	Cosmetics, paint, ceramics, paper, plastic, detergents
Corundum			
Fluorite	CaF	Ca 51.1; F 48.9	Flux in steel manufacturing, HF acid, enamels for utensils
Asbestos	$CaMg_3Si_4O_{12}$ $(OH)_2$	Fire proof building materials, sheets, pipes, etc.	
Magnesite	$MgCO_3$	MgO 47.6; CO_2 52.4	Refractory bricks, steel industry, magnesium metal
Sulfur	S	S 100	Sulfuric acid, fertilizer
Realgar	As S	As 70.1; S 29.9	Fireworks, pigments
Orpiment	As_2S_3	As 61; S 39	Pigment

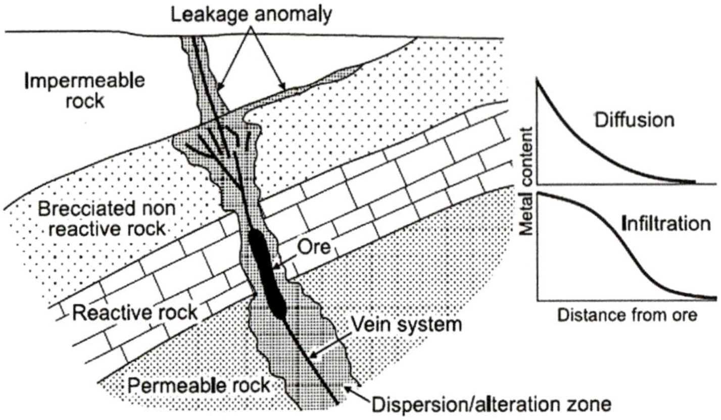

FIGURE 2.3 Primary dispersion and hydrothermal vein system. *From McQueen (2009).*

2.10 CLASSIFICATION BASED ON FORM

It is of three types, viz., *veins, replacement, and disseminations*. Veins are either filling of fissures, fractures, or faults, which are transverse to structural features such as bedding in the containing rocks. They may be parallel to such features indicating that they are open-space fillings (Fig. 2.3). Mineral-bearing body may have the form suggesting that the economic mineral and its associated waste mineral (gangue) or both were deposited from solutions that simultaneously dissolved away the country rock and laid down the economic mineral and its gangue, giving a replacement. Gradations between veins that are fillings and those that are replacements are found. Replacements may be highly irregular in form giving "chimneys" or "pipes" and filling in around or replacing irregular fragments of country rock. Disseminations consist of minute or large units of economic mineral with or without an associated gangue, scattered irregularly through the country rock. The latter may be rather coarsely granular igneous rocks such as granite; fine-grained igneous rock (eg, lava or porphyry); an unconsolidated gravel or sand or consolidated sediment such as limestone, sandstone; a highly compressed and recrystallized quartzite, slate, schist or marble, or a greatly crystallized igneous rock such as gneiss.

2.11 CLASSIFICATION BASED ON THE THEORY OF ORIGIN

Despite debatable origin in many cases, the classification based on the theory of origin of deposit is most commonly used. A simplified sketch of genetic classification covering all ore deposit types is shown in Fig. 2.4. The broad categories of ore forming processes and the subsequent overprints that may affect the deposits are highlighted in the above cited scheme. Mineral deposits are also broadly subdivided into "syngenetic" (formed at the same time with the enclosed rocks) and "epigenetic" (formed later, by introduction into pre-existing rocks). Syngenetic deposits are usually "stratabound" (confined to a particular stratigraphic unit) or "stratiform" (confined to the stratigraphy and internally stratified or layered). Although the deposit form and geometry vary greatly, these features

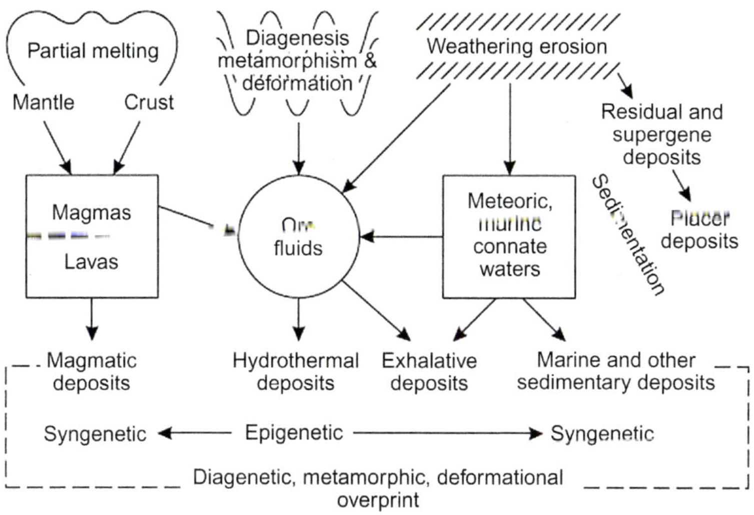

FIGURE 2.4 A simplified genetic classification of all ore deposit types. *From McQueen (2009).*

reflect the nature of ore forming processes. Deposits formed by hydrothermal solutions exhibit forms related to the geometry of fluid channel ways (like veins, fracture fillings, and stock works).

2.11.1 Magmatic

These originate by the aggregation of the desired mineral particles or crystals at one or several stages during the cooling and solidifying of molten rock. The heavier metal-rich liquids sink and concentrate at the base of the intrusive body, while lighter silicate liquid tends to rise. They are commonly disseminations locally richer, however, than the main body of such rock; they may also form continuous masses of varying form and size. Many chromite, nickel, platinum, ilmenite, and magnetite bodies have had this origin. *Pegmatites*: Late stage crystallization forms pegmatites and many residual elements (Li, Ce, Be, Sn, V) are concentrated. Pegmatites are vein-like or lenticular masses suggesting very coarsely crystalline intrusions of igneous rock in texture and composition, cooling in or near an igneous rock, into which they may locally grade at their margins. They are important sources of industrial minerals such as micas, beryl, tourmaline, and spodumene.

2.11.2 Hydrothermal

Hot water or hydrothermal solutions, probably originating from molten rocks, have been observed forming mineral deposits. The ore constituents (Cu, Pb, Au, etc.) are dissolved in a hot aqueous solution along with other deposit constituents such as Si, S, and Fe. The metallic elements are deposited to form the ore and gangue, in response to a change in the solution and very sharp decrease in temperature (Tin veins of Cornwall, Au deposits of Homestake. US). *Mesothermal, epithermal, and telethermal veins and replacements*: These are attributed to the action of progressively cooler and more liquid, less gaseous (hydrothermal) solutions similar to but cooler than hydrothermal deposits, mentioned above. They are commonly found near igneous intrusions but may be so far away that the relation of the igneous rocks to the mineralization is in doubt. The three subdivisions are based on inferences as to the temperatures, pressures, and compositions of the depositing solutions (Mesothermal—Mother Lode Au deposit of California, USA; the porphyry copper deposits of Arizona; Leadville, lead–zinc–silver, Colorado; Epithermal—gold–silver–quartz veins of Nevada, many mercury deposits; Telethermal—lead–zinc–barite–fluorite deposits of Mississippi Valley region).

2.11.3 Syngenetic

Syngenetic deposits are formed contemporaneously as the rock in which they occur. They have formed by the same process and at the same time of the geological time frame as the enclosing rocks. They are sometimes part of the succession on the Earth's surface in the form of a sedimentary layer (iron-rich sedimentary horizon). These are two types:

1. Igneous: *Magmatic layered mafic intrusion*: During the crystallization of magma, usually mafic, ultramafic heavy, metal-rich liquids settle and accumulate at specific sites, often at the base, within the intrusion (platinum group of metals; chromite; Ni–Cu (PGM)).
2. Sedimentary (deposits occur as beds/bed-like matter are conformable with underlying and overlying rocks): They include many sedimentary mineral deposits such as limestone, oil shale, bauxite, coal, phosphorite, iron, and manganese ores. Certain deposits of chromium, uranium, and vanadium belong to this type. The width and thickness of these deposits rang from a few meters to a few hundred meters and strike length running to km. Sedimentary massive sulfide (Sedex): deposits formed by hydrothermal emanations on or near the sea floor in association with the deposition of sedimentary rocks (Pb–Zn–Ag; Ba). Volcanic massive sulfides: deposits formed as massive lens-like accumulations on or near the sea floor in association with the volcanic activity (felsic volcanic hosted—Cu–Pb–Zn–Ag–Au; mafic volcanic hosted—Cu (Zn, Au); mafic volcanic/sedimentary—Cu–Zn (Au)).

2.11.4 Epigenetic

Epigenetic deposits are believed to have come much later than the host rocks in which they occur. They have been introduced into the pre-existing country rock after its formation via migration of metal-bearing fluids. A mineral vein is a good example. Fracturing and breaking of the rock along weak plane (or fault) at a depth ranging from surface to several km below surface is a prerequisite. Hydrothermal solutions pass through open spaces/fractures and deposit the minerals. These deposits have various forms like fissure veins and sheet-like form.

Some examples are as follows: (1) porphyry type deposits that are large, low-grade deposits associated with porphyritic intrusive body (Cu–Mo; Cu (–Au); Mo (–W)); (2) Skarn-type mineral deposits formed by replacement of limestone by ore and calc-silicate minerals adjacent to a felsic or granitic intrusive body (W–Cu (–Zn, Mo); Zn–Pb–Ag (–Cu, W); Cu (–Fe, Au, Ag, Mo)); and (3) fracture-filling vein type deposits which have both lateral and depth extension but are narrow (hypothermal—Cu (Au); mesothermal—Cu–Pb–Zn–Ag–Au; Epithermal—Au–Ag (Hg)).

2.11.5 Sedimentary Deposits

These originate from surface or near-surface processes such as evaporation (*Evaporites*—most deposits of gypsum, halite, etc.); from biochemical extraction and precipitation usually in enclosed water bodies (many *banded iron formations* (BIF) of Lake Superior region). More than 90% of world's iron deposits are tied up in BIF; or from physical concentration of solid particles from weathering of primary minerals and transported by stream (*placers*, yielding gold nuggets, diamond, etc.); leaching of rock leaves residual minerals behind (Laterites, Al, Ni, Fe). Coal is a sedimentary mineral deposit. Many building stones such as limestone and sandstone also belong to this group.

2.11.6 Secondary Deposits

Reworking of primary ore deposits and metals remobilized. Previously unworkable deposits, which may have had any of the origins described earlier, have been converted to economic deposits by the later process of enrichment. Secondary changes include processes such as oxidation—the effect of oxygen, water, or other near-surface agents on, eg, copper sulfides or silver compounds, converting them to oxides and carbonates. Any component elements, if they are soluble, may be transported to greater depths by descending near-surface waters where the copper or silver is again precipitated as a sulfide through "secondary sulfide enrichment." Or the primary deposit may contain such volumes of gangue minerals as to be economically not mineable, but leaching (solution and removal) of these may result in a "residually enriched" deposit at or near the surface. Thus manganese compounds, such as the carbonate, may be converted near the surface into a relatively soluble group of manganese oxides and hydrous oxides, referred to as "laterites" which accumulate while the surrounding waste material is dissolved away. This process is responsible for many workable deposits of iron, nickel, and aluminum, the latter as "bauxite." The generalized occurrence and possible relationship between primary and secondary geological dispersion patterns around ore deposits are shown in Fig. 2.5.

2.12 CLASSIFICATION BASED ON ORE FORMATION PROCESSES AND ELEMENT ASSOCIATIONS

One of the classifications of economic mineral deposits is by ore formation process and the geological setting. This classification has its own lacunae because many deposits may be formed by one or more of the basic genetic processes. Major groups of ore forming processes and typical element associations are given in Table 2.2. The key features

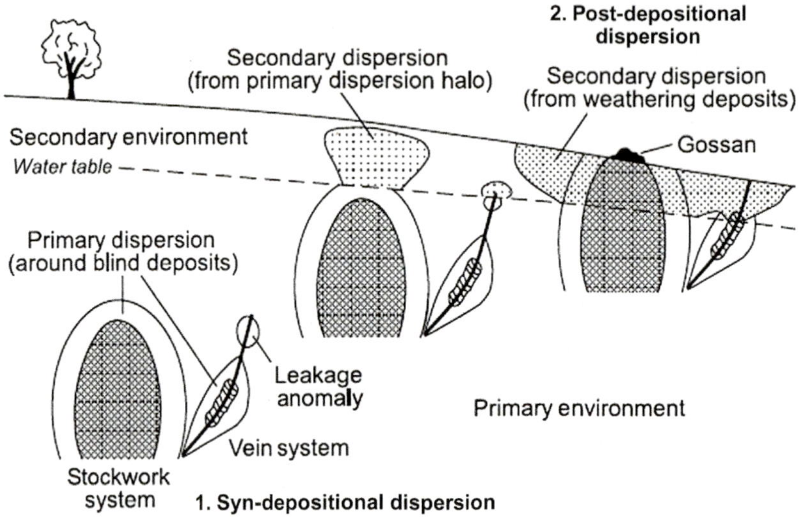

FIGURE 2.5 The generalized occurrence and possible relationship between primary and secondary geological dispersion patterns around ore deposits. *From McQueen (2009).*

for detection of geochemical signatures are composition, size, geometry of ore deposit and dispersion pattern of associated elements and wall rock alteration. The typical element associations do not necessarily lead to direct target for exploration and discovery but they can provide very useful pathfinders to the deposit.

2.13 PLACERS

The term "placer deposits" is applied to accumulations of unattached particles of metals or minerals, in erosional debris, remaining after destruction of rock in which they originally occurred. Sufficient size and metal content are necessary for the deposit to be called a placer. Placer deposits are generally made of alluvial, colluvial, and eluvial materials which contain economic quantities of valuable minerals. The loci of concentration of placers in different places are shown in Fig. 2.6. Variation in prices, new exploration methods, changes in access, transport, available power, etc. alters economic potential of placers. Minerals/metal products that have high specific gravity than rock debris, which are immune to natural agents of solution and also hard enough to resist pulverization, accumulate as placers. They usually have high values for their weight. Some diamond and gold placers are of second generation, ie, derived from erosional destruction of conglomerate, which was an earlier placer. Huge volumes of rocks containing mineralized veins (casiterite-bearing pegmatites) have been destroyed by erosion and the higher, lighter portion was carried away leaving only the richer, heavier minerals.

TABLE 2.2 Some Deposit Types and Typical Element Associations

MAGAMATIC	
Komatiite-associated nickel deposits	Ni–Cu–Co–PGE±Te (S) minor As
Layered mafic–ultramafic intrusion deposits	Cr–PGE–Ni–Cu–Co–Au (S–O)Fe–Ti–V (O)
METASOMATIC SKARN DEPOSITS	
Copper gold skarns	Cu–Au–Bi–Te (S)
Distal gold skarns	Au–As–Bi–Te (S)
Magnetite skarns	Fe–Cu–Pb–Zn–Ag (S)
Zinc–lead–copper skarns	Fe–Cu–Pb–Zn–Ag (S)
Tungsten skarns	W–Mo±Cu–Pb–Zn (O,S)
HYDROTHERMAL, EPIGENETIC	
Fractionated granitoid-associated deposits	
Tin–tungsten deposits	Sn–W±As–Cu–Zn (O–S–F–B)
Tungsten–molybdenum deposits	W–Mo±Cu–Pb–Zn–Bi–As
Pegmatite and complex veins	Sn–Ta–Nb–Li, Be–Li–Cs–Rb±U–Th REE (Si)
Porphyry-associated deposits	
Copper–gold stock works and veins	Cu–Au±Ag–Bi–Mo–Te–Re (S–K)
Copper–molybdenum deposits	Mo–Cu (S)
Breccia pipe deposits	Au±Cu–Ag±Mo–Pb–Zn (S)
Epithermal gold–silver deposits	
High sulfidation type	Ag–Au–As–Sb–Te (S–Si)
Low sulfidation type	Ag–Au–As–Sb–Mn–Ba±Hg–Te–Se (S–Si–CO_2)
Iron oxide copper gold deposits	Cu–U–Au–Ag–REE (S–F), Cu–Au–Bi
Syndeformational hydrothermal and replacement deposits	
Metamorphic copper deposits: *Mount Isa*	Cu–Co–As±Pb–Zn (S–Si)
Metasediment-hosted polymetallic deposits	Pb–Zn–Ag±Cu–Sb (S)
Replacement deposits, magmatically related	Cu–Au–Ag–Bi–Co–Hg–Mo–Se–Sn–Te–W (S–F)±Zn–Pb
Orogenic gold deposits	
Archean greenstone-hosted deposits	Au–Ag–As–W–Sb–Te±Cu–Pb–Mo (CO_2–S±K–B)
Sediment-hosted reef deposits	Au–Cu–Pb–Zn–Bi–Mo–W–Co–Ni (S–CO_2)
Slate-hosted quartz-vein deposits	Au–As–Sb±Ag (CO_2–S)
Ranitoid-associated deposits	Au–Ag–Zn–Pb–Cu–Bi (CO_2–S)

(Continued)

TABLE 2.2 (Continued)

Carbonate-hosted stratabound lead–zinc deposits	
Mississippi Valley-type	Pb–Zn–Fe±Ag–Cu (*S*-hydrocarbons)
Unconformity-related uranium deposits	U–Au–Cu–Co–Ni–Ag±Zn–Sn–Pb–Bi, Pt–Pd (Mg)
EXHALATIVE DIAGENETIC	
Volcanic-associated massive sulfide deposits	
Abitibi type	Cu–Zn–Pb±As–Sb–Bi–Sn–Mo–Se–Ag–Au–Ba (S)
Kuroko type	Pb–Zn–Ag±As–Sb–Hg–Ba, Cu±Bi–Sn–Te–Mo–Co–Au(S), Cu–Au–Ag–Zn–Te
Sediment-hosted stratiform base metal deposits	
HYC, Mount Isa (lead–zinc), Century, Broken Hill	Zn–Pb–Ag–Cd–Mn–Ba–Tl±Cu–As–Sb–Hg–In(S–K–B)
Sediment-hosted copper deposits	Cu–Zn–Pb–Bi–Co–Ni–As±Ag–Au (S)
MARINE SEDIMENTARY	
Banded iron formations (generally precursor ores)	Fe–Mg–Al–Ca–Mn–P±Ti (Si–O–S)
Sedimentary manganese deposits	Mn–Fe±Ca–Al–Zn–Li–Ba (O–Si–CO_2)
RESIDUAL AND SUPERGENE	
Bauxite deposits	Al–Fe±Ti–Nb–Ga–Mn–Zn–Zr (O)
Lateritic nickel–cobalt deposits	Fe–Ni±Co–Mn–Cr–Zn (Si–O)
Lateritic and supergene gold deposits	Au±Ag–As–W–Sb–Bi (lateritic types)
Supergene deposits associated with paleochannels	Au
Lateritic phosphate–REE–Nb deposits	P–REE–Nb±Sr–Ca–Al–K–U–Ti–Ta
Supergene iron ores and transported channel deposits	Fe–Mn±P
Residual and supergene calcrete uranium deposits	U–V (K–CO_2)
Reduced paleochannel/"roll front" U deposits	U±S–As–Se–Mo–Co–Ni–V–Cu–Zn–Pb–REE–Y (Si–O–C–P)
PLACERS	
Placer gold deposits	Au–Ag±W–Ti–Zr–Ba
Heavy mineral sand deposits	Ti–Fe–Zr–Th–REE±Cr–Sn
Placer tin deposits	Sn–Fe–Ti±W–Nb–Ta

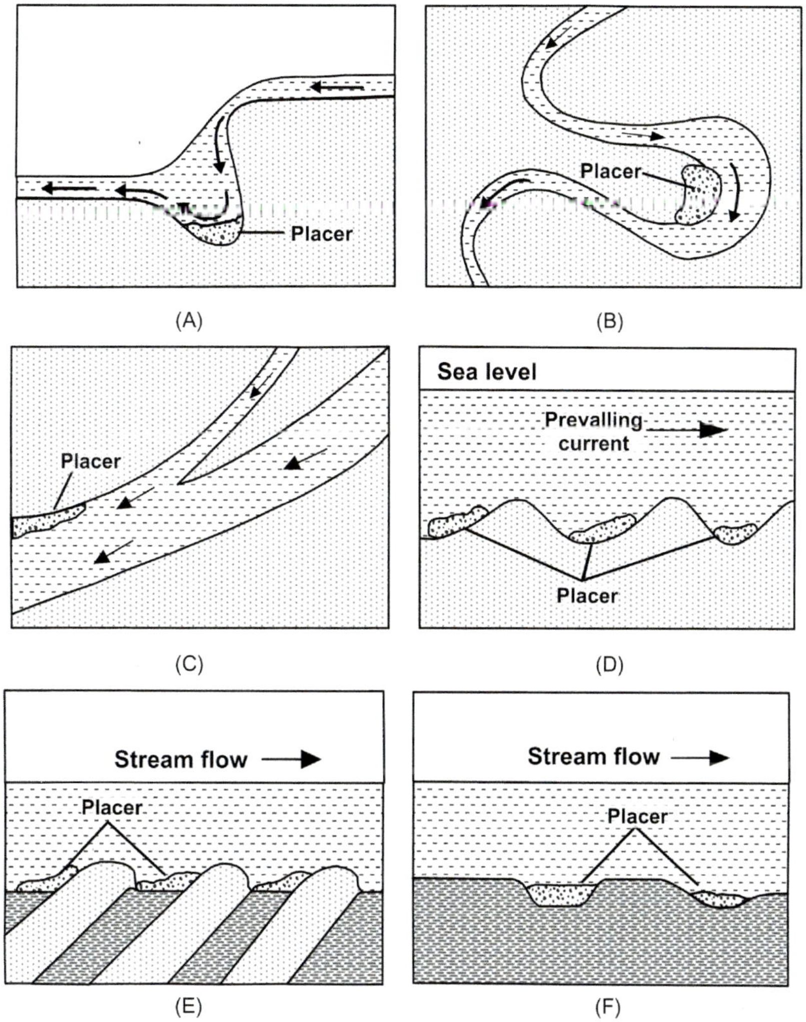

FIGURE 2.6 The loci of placers concentration (A) below waterfalls, (B) inside meander loops, (C) downstream from a tributary, (D) behind undulations on ocean floor, (E) behind rock bars, and (F) in rock holes.

Minerals and elements recovered from placers are given in the table.

Mineral Compounds	Elements/Metal
Cassiterite, chromite, columbite, tantalite, ilmenite, rutile, monazite, garnet, etc.	Diamond, gold, platinum group

Except for some kimberlite pipes and dykes in South Africa, Canada, Australia, Russia, world's production of diamonds come from placers, sometimes found at great distances from the presumed source. Sometimes, two or more valuable products, derived from the same original rock, are found in the same placer. Multiple yields include diamonds (gold); cassiterite, gold; columbite, tantalite, cassiterite, tungsten minerals.

Many placers are concentrations by wave action in beach sands (alluvial rutile deposit, Australia; beach placers (monazite, ilmenite, rutile, garnet) of India; cassiterite deposits in the ocean sands of Malaysia, Thailand). Large and rich deposits of diamonds are known beneath the sea off the west coast of South Africa and Southwest Africa, continuing up the coast of Angola. Offshore placers along and very near the shores are exploited on large scale. In most places, the values are concentrated in that portion of the gravel/detritus near or immediately on the bedrock or occasionally on intermediate beds of hard pan or cemented gravel.

2.14 METALLOGENIC PROVINCES AND EPOCHS

High concentrations of valuable mineral/metal deposits are not distributed randomly over Earth's surface. In order to explain the uneven distribution of metals and ore deposit types in geological time and space, various theories have been proposed. Metallogenic provinces are broadly defined as large regions (with one of its dimensions running to 1000 km or more) containing deposits of one or group of minerals or metals. These unusual accumulations are due to various combinations of one or several factors, viz., primordial differences in the mantle, preferential concentration of elements in the crust, and the prevalence of ore forming processes at certain times and/or places. All these deposits need not necessarily form at the same time. Most provinces have complex, multistage origins despite the controversy about the relative importance of inheritance versus processes. Guild (2001) opines that "in recent years the geographic relationship of many geologically young provinces to present-day plate-tectonic positions (accreting or consuming margins, intraplate structures, etc.) has been widely recognized, and the presumption is strong that older provinces had similar relationships to former plates." He further adds that "as most ore deposits resulted from a favorable conjunction of geological processes that are no longer operative, elucidation of their genesis requires reconstruction of the geologic history of the province, with particular emphasis on events coeval with mineralization."

An important aspect of reconstruction is the tectonic analysis. Vast amount of data from orbiting satellites have contributed immensely to this analysis. Present-day improved technology facilitates both the synoptic viewing large areas and the ability to trace faint contrasts (that have revealed linear, curvilinear, and circular features) which, hitherto, were not recognized from field studies. The development or the extent of metallogenic provinces might reflect the basement structures. The delineation and recognition of such provinces will not only be valuable to the assessment of mineral resources available but also as guides to exploration for the future ore search.

2.14.1 Metallogenic Provinces

Metallogenic provinces are areas of highly varying size containing deposits of more or less similar type and composition believed to have had the same or similar origin. The "metallo" means a deposit or mine, not solely of a metal in the literal sense but of both metallic and nonmetallic elements and minerals. Some of the deposits were formed over a relatively short period of geological time: for example, the Iberian pyrite belt of southern Spain and Portugal contains over 80 known sulfide deposits that were formed in the Devonian to early Carboniferous. Formation over a short period of geological time is not, however, a strict requirement for the delineation of a metallogenic province. In the copper provinces of the south-western United States and South America, numerous porphyry deposits of a comparable type and metal association are found, yet these have formed over a wide interval of geological time.

Metallogenic provinces containing concentrations of ore deposits of similar characteristics can also be found at different locations on the Earth. These may have at one time composed a larger single province that has been broken up by the processes of plate tectonics. They may alternatively have formed in different regions but by similar geological processes. For example, an unusually high density of vein-hosted gold deposits is found in certain areas of Precambrian shield rocks in Canada, Australia, Brazil, and southern Africa. The remarkable similarities between these deposits indicate that they were formed by identical processes. Metallogenic provinces can also overlap in areas where concentrations of more than one type of ore deposit or metal association occur. In the Precambrian shield rocks mentioned above, concentrations of volcanic-hosted massive sulfide deposits commonly overlap areas where the vein gold deposits are found, even though these two types of deposit formed at different times and by different processes.

The metal contents in common crustal rocks are too low to be mined economically. The generation of an ore deposit typically requires that metals be derived from a large volume of source rock and then concentrated many times into a relatively small volume of host rock. This is accomplished by a combination of geological conditions and processes that are active during the mineralizing event. These processes are in turn controlled by regional geological characteristics such as rock type, structural discontinuities, subsurface fluid flow, temperature, and pressure. The regional tectonic setting determines the make-up of these features, and it is readily apparent why such concentrations of particular ore deposit types exist. Metallogenic provinces could thus be said to be a manifestation of the regional control of geological processes by tectonic setting (Hancock and Skinner, 2000).

Metallogenic provinces are delineated by the occurrence of numerous concentrations of mineralization with similar characteristics and metal association. These concentrations may include economic ore bodies, subeconomic deposits, and small showings. Metallogenic provinces have no fixed size and may always be extended or confirmed by the discovery of new mineralization. They provide a valuable first step to the exploration geologist in the search for further mineral resources.

2.14.2 Metallogenic Epochs

Metallogenic epochs are the "geological times" when favorable conditions of origin of mineral deposits existed. The prominent metallogenic epochs are as follows:

- Archean: Chromite, PGM, Cu–Fe–Ni, Au, some VMS
- Proterozoic: Placer Au & U deposits, PGM, Chromite, BIF, Fe–Ti oxides with anorthosite massifs, diamonds in kimberlites, sediment-hosted base metal deposits (Cu–Pb–Zn)
- Phanerozoic: Phosphorites: proterozoic—Cambrian boundary, podiform (Alpine type) chromite, Coal: (Carboniferous), PCDs (Mesozoic), Residual (Cretaceous–Recent).

2.15 METALLOGENIC PROVINCES IN RELATION TO PLATE TECTONIC SETTING

Plate tectonics explains many features and movement of Earth's surface. It also provides the basis for a better understanding of the origin of mineral deposits and their distribution. In the early 1960s, several seminal discoveries were made that became evidence of sea floor spreading; the 1970s saw the experiments to target ore deposits with tectonic data; from the 1980s, application of the knowledge of tectonic phenomenon gained importance. The basic challenge in carrying out exploration for concealed deposits lies in the identification of favorable metallogenic provinces and judicious selection of potential target areas within those. Selection of potential target areas is to be based on identification of specific geological parameters, which are useful in developing a prototype conceptual genetic model.

Mineral deposits and their relationship with plate tectonics are significant because (1) the processes of mineral deposition are controlled by geological processes operating due to energy released at plate boundaries, (2) plate tectonics govern the particular tectonic settings wherein mineral deposits form, and (3) fragmented continents and their reconstruction can provide a basis for search for new deposits in new geographic locations.

Our understanding of basic "endogenic and exogenic" processes, responsible for mineralization and possible locations of such systems in the Earth's crust, has improved dramatically with the advent of "plate tectonic theory." This also accounts for abnormal accumulations of metallic elements in different geographic locations. Application of experimental and theoretical geochemistry, fluid inclusion geothermometry, geobarometry, hydrothermal alteration studies, computer-based simulations, etc. facilitate better understanding and modeling of the processes. Several hypotheses were arrived at, as a result of many academic exercises, which one could probably be put to ground test, to unearth concealed/buried mineral deposits.

Predictive models could be constructed from the genetic classification which incorporates elements of composition, form, and association, and used to search for geological environments wherein ore forming processes could probably have operated. Global plate tectonic system, added with the increased knowledge of planetary evolution, provides a better understanding of ore forming geological environments and processes. "A number of major metallogenic epochs are recognized and these are thought to relate to global geodynamic processes, including major periods of crustal break-up and convergence" (Jaques et al., 2002). The different plate tectonic and geodynamic settings (as per scribd.com/doc/platetectonics and mineralization and www.home.hiroshima-u.ac.jp) that control "which mineral deposit will form where," are shown in Figs. 2.7 and 2.8, respectively.

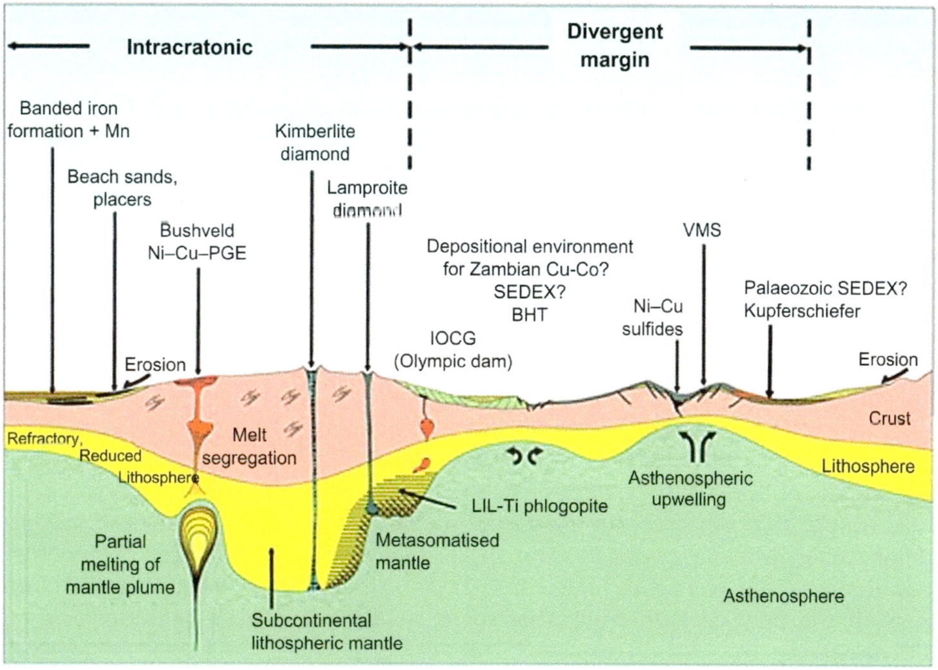

FIGURE 2.7 Geodynamic setting of mineral deposits.

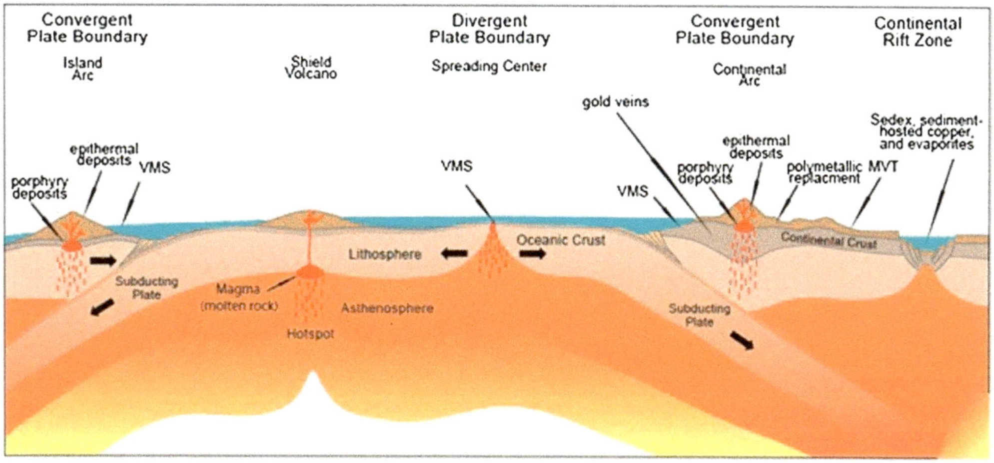

FIGURE 2.8 Likely occurrence of mineral deposits in different plate tectonic settings.

Advancements in geological concepts, genetic modeling, and their application have helped in planning new approaches for mineral exploration in terrains which, hitherto, were not considered conducive to economic mineralization. Better understanding of the physicochemical processes involved, rock types, mineral associations and their characteristics, will facilitate to recognize geological/ore environment of different types of mineral deposits elsewhere in the world so that systematic and concerted efforts could be attempted to look for new areas for metals of interest, in the right direction.

It is well known now that most of the metallic ore deposits are the result of plate tectonic activity. At divergent plate margin boundaries (such as mid-oceanic ridges) high heat flows and convection currents are active and create submarine hot springs called "black smokers" which deposit solid masses of metallic minerals. These environments were ideal for the deposition of copper, iron, zinc, lead, gold, nickel, manganese, etc. These metal-bearing solutions, being denser than water, collect in the basins of ocean floor and form rich deposits. Along converging plate boundaries, island arc systems develop and create massive sulfide deposits rich in base metals and hot spring gold–silver deposits on the flanks of andesitic volcanoes. Metallic ores occurring in intrusive rocks (such as porphyry copper deposits) are due to partial melting and rising of crustal material along subduction zones. Metallic deposits from these systems might have also been contributed from sea floor, which were subducted and became part of the new magma. Ultramafic intrusions in the new oceanic crust, that form, at divergent plate boundaries, host rich chromite deposits.

References

Bateman, A.M., 1962. Economic Mineral Deposits, second ed. Wiley, New York.

Evans, A.M., 1998. Ore Geology and Industrial Minerals—an Introduction, third ed. Blackwell Scientific Publ. Inc., Oxford, p. 389.

Guild, P.W., 2001. Tectonics and metallogenic provinces. Adv. Space Res. 3 (2), 9–16.

Guilbert, J.M., Park, C.F., 1986. The Geology of Ore Deposits. Freeman and Co., New York, p. 985.

Hancock, P. and Skinner, B.J. 2000. Metallogenic provinces. The Oxford Companion to Earth, 700 + words.

Jaques, A.L., Jaireth, S., Walshe, J.L., 2002. Mineral systems of Australia: an overview of resources, settings and processes. Aust. J. Earth Sci. 49, 623–660.

Kesler, S.E., 1994. Mineral Resources, Economics and Environment. MacMillan, New York, p. 391.

McKinstry, H.E., 1962. Mining Geology. Asia Publishing House, New Delhi, p. 680.

McQueen, K.G. 2009: Ore deposit types and their primary expressions, www.crcleme.org.au/RegExpOre/1-oredeposits.pdf, 14 p.

Moon, C.J., Whateley, M.G., Evans, A.M., 2006. Introduction to Mineral Exploration. Blackwell Publishing, p. 34.

Park Jr., C.F., MacDiarmid, R.A., 1970. Ore Deposits, second ed. W.H. Freeman, San Francisco, p. 456.

Roberts, R.G. and Sheahan, P.A. 1988: Ore deposit models. Geoscience Canada, Reprint Series 3, Geol. Assn. Canada, 194 p.

Rose, A.W., Hawkes, H.E., Webb, J.S., 1979. Geochemistry in Mineral Exploration, second ed. Academic Press, London, p. 657.

USGS, 2005. Fundamental Principles of Geology (E-publication: 3dparks.wr.usgs/gn/nyc/common/geologicbasics.html).

CHAPTER

3

Reconnaissance and Prospecting

OUTLINE

Essentials of Mineral Exploration and Evaluation
DOI: http://dx.doi.org/10.1016/B978-0-12-805329-4.00010-7

FIGURE 3.1 A field geologist examining an outcrop.

3.1 RECONNAISSANCE

Geological reconnaissance is the examination or survey of the general geological features and other characteristics of a region. Normally, the activity of mineral exploration begins with "reconnaissance" and advances to "detailed reconnaissance" followed by selection of target/targets. If this stage is successful then it leads to "advanced exploration" and finally to "evaluation." The work in the initial reconnaissance stage involves a visit by a geologist or a small group of geologists with limited support, to look for rock outcrops in the area of interest. While traversing, small samples from rock outcrops, soils, or streams will be collected for chemical analysis (Fig. 3.1). Since time is the essence, it is important to carry out reconnaissance of the whole area using methods that will give the most information at the earliest date. Depending upon the terrain, one has to decide where to look first. In a hilly tract and along water courses, one has to make traverses up and down. Low-level flying helps one to scan larger areas and then select areas where one can concentrate more. The preliminary examination will necessarily be hasty and superficial and the selection will be based on incomplete evidence. There is always a risk of rejecting a ground that contains ore. The selection and rejection cannot be infallible, but must be based on the most intelligent guess that can be made with the available knowledge.

Preliminary reconnaissance of a large tract of an unfamiliar terrain is carried out by flying over it and selecting from the air the main geological and geographic features. Areas of gossan exposures, physiographic features like subsidences, old workings if any, could be seen and their locations are recorded. The activities like geological mapping, sampling, widespread preliminary geochemical sampling, and geophysical surveys are carried out. Once an area of interest is selected, it is acquired through permit/concession/claim and will warrant more detailed exploration involving detailed geological mapping, sampling, geophysical and geochemical surveys with a good gridded survey control. Normally the works cited above will yield targets that may necessitate testing by pitting or trenching or drilling (rotary, percussion, or diamond) involving the use of heavy equipment.

If the results of the project exploration at this stage are sufficiently encouraging, then advanced exploration will be carried out, which might involve intensive drilling, trenching and if need be, an exploratory decline or drift or an adit, to gain underground access to the deposit. In order to evaluate the economic potential of the mineral deposit it might be necessary to collect a large or bulk sample from underground excavation.

3.2 RECONNAISSANCE MAP

For systematic investigation of an area, a map is indispensible for showing geological features or to plot observed ore and mineralization. If Federal and Provincial governments made topographic and geological maps, they form excellent base map for recording the results of broad-scale reconnaissance and prospecting. Aerial photographs form the most satisfactory and economical base map. A reconnaissance map will normally show, broader geology, indications of any old mine, prospect, outcrops of mineralized rock. If sufficient information is available, such a map will at once bring out the trends of ore-bearing belts and show their relation to rock structures, intrusive bodies, etc.

In exploration work, one should keep in mind that the final photo interpretation is not a finished complete map and that should be subject to confirmation by the field checks. If one or two geologic relationships of exploration significance can be interpreted and portrayed in photogeologic maps, the photo and interpretative work are worth their relatively low cost. A good photo interpretation requires considerable geologic skill. The work is based upon visual recognition of photographic tone contrasts, textures, characteristic shapes in patterns, shadowing, etc. The skill and photogeologic interpretation is limited by ingenuity and field experience of the geologist. Sufficient ground checking shall be done to establish validity. In recent years, color aerial photos have improved in quality and are useful in many phases of regional and district-scale exploration. Definition is good and interpretation of some features is made easier because of the reasonably true color rendition. A good pocket stereoscope or a small desk model can be used to interpret geological features of interest directly transferred to available base map or clear overlays.

After reconnaissance, portions of tract can be chosen in the following way: (1) ground in which there are known ore bodies or promising indications, (2) ground in which favorable structural conditions are known to exist, (3) ground in which no favorable conditions are known to exist, and (4) ground in which favorable conditions are believed to be absent. A distinction should be made between "mere absence of positive indications and actual presence of negative indications." Presence or absence of rock alteration is often useful guide. After rejecting least promising areas, emphasis should be shifted to the active search for ore within the selected claim. Preliminary geological, geophysical, and geochemical surveys can be of assistance in pointing to the most promising places. Actual discovery must be made either by finding ore, naturally exposed or expose it artificially by pick and shovel or drill.

Some mineral finds are accidental. (Ag in Cobalt, Cu–Ni in Sudbury, both in Canada, were discovered while excavating for railroad cut.) Despite many anecdotes of accidental discoveries, that many mines that are known to day have been discovered by purposeful prospecting.

3.3 RECONNAISSANCE SURVEY

Initial broad reconnaissance exploration survey is made using fixed wing aircraft. Later short aerial inspections may be made touring over the project, locating water for drills, preliminary plant site inspections. The "early morning hour air photos" are often preferred, since low sun angle enhances many features of interest. Careful planning should precede aerial reconnaissance and there should be a clear idea of the specific features sought. Standardization of methods (particularly, where more than one field crew is involved) is much to be desired. At all stages of reconnaissance work, there should be clear understanding of the objectives of work and the constant review and appraisal of results, to insure that the desired data are being obtained.

The major pitfalls to be avoided in preliminary reconnaissance are:

- Insufficient planning with respect to both geologic criteria to be investigated in the layout of the work
- An inadequate record of observation was made and what was observed
- Too few samples taken
- Premature evaluation of data, resulting in termination of the project or excessive modification of original plan.

3.4 GEOLOGICAL SURVEY

Geological survey usually involves investigating in an area of interest, for the purpose of building some base maps showing rock unit distribution, stratigraphic and structural details, landforms, models, etc. The type of geological survey employed depends upon the scale, purpose and conditions, traverse area to be covered, etc. The base maps can be used as the basis for further studies. Depending upon the purpose of geological survey, the method of study might involve various activities, viz., visual inspection, land surveying, outcrop studies, structural elements, soil distribution, rock, mineral and soil sampling, some amount of drilling, geochemical analysis, remote sensing, some amount of geophysical survey, etc. If needed, activities like air-borne geophysics, subterranean profiling (sonar) are also undertaken. Geological surveys encompassing various techniques are normally carried out by federal (Central), provincial geological survey organizations, to maintain geological inventory and also to provide useful geological data for the benefit of a country.

The primary aim of a geological survey is to develop an understanding of regional and local geology in order to identify environments likely to host economic quantities of minerals. This, in turn, influences exploration/exploitation of minerals for the benefit of society. Fundamental and basic geological information spanning the entire country is normally collected by Government agencies through systematic mapping on topo sheet scale (1:50,000) over many years. However, the useful life of a geological map is generally considered as 10 years before it needs revision due to new data generation and evolving ideas. Thus geological mapping should be a continuous process to develop geological modeling and draw inferences to favorable areas for ore search.

A "geological map" is a visual record of geological facts (rock units and associated structures, etc.), usually on a plane surface. Once an area of interest is selected, geological field mapping is done which will include identifying all geological aspects of the area with a purpose of preparing a detailed geological report accompanied by an accurate geological map. Hence, a geological map will normally show geological formations and their age relationship, various rock types, structures, leached outcrops, subsidences, any mineral showings and their distribution, fossils, etc. can be superimposed over a topographic map to get an overview of the area. The quality of geological map usually depends upon the accuracy and precision demanded. It also depends upon completeness with which geologic and geographic data are presented on the map and also on the care with which scale, conventions, colors are chosen. Geological maps of the present days are more precise than ever before, owing to technological advancement and application of a combination of accurate satellite imageries, aerial photos, high-tech geological equipment, and Geographic Information System (GIS).

3.4.1 Equipments Used in Geological Field Mapping

Maps: For geological field mapping of the area chosen, a latest available geological map will be handy as a reference. Besides, topographic, physical, relief, road maps of the area of interest will be useful.

Aerial Photographs and Stereoscope: Aerial photos are very useful for the geological field work especially where large geological features like folds, faults, unconformity, volcanoes, craters, etc. are involved. Aerial photos are studied with the aid of "stereoscope" (which gives 3D view), both before the field work (for planning) and during field studies (for confirmation purposes). Wherever aerial photos of desired quality are not available, satellite imageries will be useful despite they may not possess finer details as seen in the aerial photos. Many recently launched satellites help to provide imageries of very high resolution.

Geographic Positioning System (GPS): GPS is a global radio navigator satellite system that allows land, sea, and air-borne users to determine exact location and time, round the clock, in all weather conditions, anywhere in the world. It is most applicable in surveying and mapping and is an indispensible tool for a field geologist. It is used to locate one's position, locate lithology, structural elements, sample points, elevations, tracing the lithological contacts, etc. It helps to log geological traverses, store locations "en route." GPS facilitates speedy geological mapping operation covering larger area within small time.

Field Camera: In addition to field notebook, camera will help to record interesting features (geological, structural, etc.), ensuring a scale. Besides, videos may also be taken. The GPS recordings of the exact location of each photo are to be recorded. These will come in handy for descriptive purposes in report writing and also for making any presentations. Many modern day cell phones also have very high precision cameras and can be used in the field.

Compass/clinometers, geological hammer, hand-held lens, sample bags, field notebook, marker pens, first aid kit are the usual accompaniments, in the field work kit of a field geologist.

The most expensive equipment of all is the "geologist's imagination." It took the imagination and ingenuity of earlier geoscientists to provide several models which we have today. Later day geologists have tested many of these models over a period of time and have been proved to be true. Hence, it is necessary that an exploration geologist uses the imagination when viewing and mapping geological and structural details.

3.5 GEOLOGICAL MAPPING

For any mineral exploration, good basic geological mapping is an absolute necessity. Accurate portrayal of geology is vital to the interpretation and communication of initial results of exploration geology, geochemistry, drilling and other investigations, irrespective of how simple the concept of ore target might be. Since many decades, geological mapping has been used extensively and it is anticipated this will continue to be essential, despite the vast and rapid improvement in the technology to record, compile, and synthesize data, both in office and in the field. Identification skills of an exploration geologist to record salient geological data are more important. The old conventional method of mapping, using paper-, pencil-based mapping, allows diverse data, viz., lithology, mineralogical features, rock alteration, structural features, mineral showings, etc., plotted on a base map. With the advancement in technology, the paper-based recorded geological mapping data are commonly converted to digital format in the office. In the mine life cycle, mapping is an ongoing process, right from reconnaissance to mine closure. For exploration of new mineral deposits and for subsequent mining, many types of essential information are provided by geological mapping, viz., outcrop mapping of litho units, the morphology and the interrelationship of host rocks, etc. Structural details (attitude of veins, postore faults, etc.) gathered during mapping can be used to predict the geology in the subsurface or laterally under postore rocks and help one to improve the utility of geophysical data of subsurface targets.

Collecting and recording the data on the host rocks, ore minerals, leached rocks on the surface will be very useful. The map containing the above data of an area, in conjunction with geochemical data map, many a times, helps to generate zonation pattern to focus on the potential ore zone. Barnes and Lisle (2004) opine that "in general, geological mapping underpins the construction of three-dimensional geological models or hypotheses that guide exploration and discovery and, when geological time is considered, produces the four-dimensional space—time models necessary for understanding of primary ore formation processes and postdepositional modification by secondary surficial and tectonic processes."

3.5.1 Digital Mapping

Traditional paper-based detailed techniques for recording data relevant to mineral deposits as developed by mining companies (such as the Anaconda system) can now be adapted to direct digital mapping systems. Digital mapping is fully feasible and allows geological field mapping to incorporate a wealth of information in digital base maps in real time at any scale, and thus may increase efficiency and interpretive power. Furthermore, use of online GPS can support mapping in areas where previous work was limited by remote access or unavailability of base maps for location purposes, thus opening new opportunities for discovery. However, there are two limitations to GPS usage related to the attenuation of GPS signals by objects between the satellite transmitters overhead and the ground GPS receiver. The most common obstacles encountered while mapping are excessively dense tree cover and steep walls in valleys (Goodchild, 1992, Kramer, 2000; Barnett and Williams, 2006).

Field-based digital geological mapping is now practicable with the advent of Global Positioning System (GPS), portability of high speed and high data storage capacity computers to the field, laser-ranging devices, etc. Digital electronic instrumentation facilitates the field geologist to access rapidly, the digital databases which include geological maps, aerial photographs, remote sensing imageries with automatic registration, and scale independence. The facilities cited above will give a lot of opportunities to improve the efficiency of mapping and support a field geologist in all respects. While mapping an area, it is possible now to view geophysical and geochemical data together with geology and synoptic aerial imagery and integrate database which can aid to identify crucial geological relationships.

In future, by using the "Information Technology" effectively, the field geologist can access endless datasets elsewhere in the world for reference and comparison that will help to synthesize data and generate geological models for the area being mapped and investigated. "However, the central challenge remains the training and nurturing of highly skilled field geologists. It is imperative to the continued success of mineral exploration that an investment is made to attract, train, and retain new practitioners of field geology regardless of which mapping system they choose to use. The digital mapping technology may help attract an increasingly computer-literate cadre of new practitioners of mapping into mineral resource exploration" (Kramer, 2000; Brimhall et al., 2006).

The technology has advanced so much now that digital computing of mapping data can be done in the field and produce maps more quickly than ever before. Normally, within a week of finishing field work, a geological map in digital form can be made. The digital mapping data can then be integrated into 3D environment (LeapFrog!) to enable generation of cross-sections and then drill planning. As and when the data on drilling are available, they can be incorporated into surface mapping data to generate 3D geological models which will help in better understanding the area of interest and form base for reliable resource estimation.

3.5.2 Structural Mapping

Besides geological mapping, structural mapping needs to be carried out and the data obtained from both give a better understanding and facilitate to define target areas for geophysical and geochemical works, cost-effective drill hole planning and also to establish a testable geological and/or structural model. The cost involved in geological and structural mapping is insignificant when compared to the costs involved in drilling and regional geophysical and geochemical surveys. Geological and structural mapping by geologists in the field continues to generate useful new data and understanding of geological relationships and scientific insights difficult to gain otherwise. Mapping plays an important role in different stages of mine life cycle.

3.6 DIGITAL ELEVATION MODELS

Digital Elevation Model (DEM) is a digital cartographic dataset in three (*XYZ*) coordinates and has been derived from contour lines or photogrammetric methods. The terrain elevations from ground positions are sampled at regularly spaced horizontal intervals. Representation of DEM can be (1) as a raster or (2) as a vector (Triangular Irregular Network, TIN). The latter is also called as "primary (measured) DEM" whereas the "raster DEM," as "secondary (computed) DEM." Li et al. (2005) opine that "the DEMs can be acquired through techniques such as photogrammetry, LiDAR, IfSAR, land surveying. The data collected using remote sensing techniques, in addition to land surveying help to build DEMs." For landscape and city modeling, and visualization applications, DSMs are used widely.

DEMs are essential for doing radiometric and geometric corrections for terrains on the remotely sensed imageries. DEMs can be generated from remotely sensed data in an efficient and cost-effective way. A variety of sensors and methodologies are available to generate such models for mapping applications. The two primary methods of generating elevation data are (1) stereogrammetry using air photos or radar data and (2) radar inferometry. Elevation models help to generate contour lines for topographic maps, slope, and aspect models.

3.7 PROSPECTING

In examining a region, the first and foremost task is to know the geology well. Until that is complete, prospecting is of no value. Prospecting involves physical labor, viz., taking traverses across an area of interest, inspecting the rock outcrops, panning, sifting, and observing for indications for any ore mineralization. Prospecting discloses information of ore occurrence/mineralization which helps to round about the geological picture. Hence prospecting and geological work should go hand in hand in an area. A good prospector works methodically inspecting every outcropping rock cracking open pieces of float. Gravel and soil are also panned to reveal heavy minerals. Prospecting requires not only specialized experience but also an unusual temperament, like patience and optimism—patience to go over the ground methodically, leaving no stone unturned; optimism to believe that even though today's work was disappointing, tomorrow will surely strike it rich.

Prospecting is the first link in a chain of events which hopefully leads to a mineable deposit. A prospect is a potential ore occurrence which has been confirmed by geological, geophysical, and geochemical studies to the degree that it can be tested. The value of a prospect hinges on the probability that hopes will materialize. The question of whether or not ore exists and how it can be found are to be solved by better insight into structure that earlier people who did not dare to try. The speculative element will continue to play an essential part in the development of prospects. In addition to observations that can be made in the field, the historical data of the past to be collected. The kind of observation and history varies in its proportion according to stage of development that the prospect has attained. Some companies concentrate their operations in a particular district or region, then covering a narrowed field but studying it exhaustively in places where ore or signs of ore already found. The others investigate the tract systematically undertaking the actual prospecting in the places that seem most promising. Large mining companies with enough manpower are in a position to follow up prospects regardless of location.

In a previously unexplored terrain, the search for mineral occurrence starts with gathering basic geological data and information on any known showings or data of any earlier study, etc. Normally, basic geological/scientific information is available from the Provincial Government Geological Department/Survey. In many countries, the provincial Governments have also developed "mineral potential maps" which may assist exploration geologists in selecting favorable area for specific type of mineral deposit. The deposit type under the consideration of an exploration geologist will determine the likely favorable geology in order to focus on such terrains. In order to further localize an area of interest, regional scale reconnaissance programmes are undertaken which might include helicopter-supported or ground geochemical sampling (stream sediments, soil, etc.), air-borne magnetometer, gravity or electromagnetic surveys with ground follow up. Based on the results of the regional programme, areas that demand further work are normally staked to take up more detailed exploration programme, involving rock and soil sampling, detailed mapping, drilling, trenching, etc. Detailed account of many of these activities, to be taken up, are given by McKinstry (1962), Kreiter (1968), Chaussier and Morer (1987), and Sillitoe (1999).

Whenever a prospector finds a small occurrence of interest (or "show"), it is necessary to work in and around that area intensively. Many a times, such shows are short-lived and abandoned forcing the prospector to move on further. If a prospector strikes, by chance, a "rich show" it will soon be swarmed by many other prospectors. If the find proves positive and economical, it will lead to mining. Despite these methods are thought of "archaic and old," these are still in practice in many parts of the world.

3.7.1 Float Sample Tracing

In areas of poor exposures or in mountainous tracts, tracing mineralized boulders (floats) is very valuable. Since the mineral-bearing boulders have moved down the slope under gravity, the probable source of them must be in a nearby cliff/mountain. Float mapping, float tracing combined with panning (stream sediment sampling) have yielded successful results in identifying potential prospects in many areas (Lindley, 1987). Mineral-bearing floats in glaciated terrain are normally moved very far from the source. Such boulder trains when traced back have led to mineralized areas in Canada and Scandinavian countries. When float and heavy minerals are traced to their source, the next step is to uncover mineralized rock by digging a series of trenches.

3.7.2 Panning

Panning is a rough and ready method of testing in preliminary exploration. In exploring for precious metal deposits, some of the significant flakes of metals/minerals migrate

FIGURE 3.2 Panning of samples while prospecting.

downhill or downstream, sometimes several kilometers away and forms the guide that may be followed back to the source, parent outcrop. Precious metals (gold, silver, etc.) and many characteristic heavy minerals derived from ores/parent rocks might appear in panning (Fig. 3.2). The prospector can figure out that they have come from uphill and he proceeds up, the indications become more abundant. Panning and float tracking succeed best in unglaciated terrain. In regions of glaciated territory, they are of little help, since the material of the till might have been transported from very long distances and become scattered and mixed, hence tracing their source becomes difficult.

3.7.3 Pitting and Trenching

In areas of poor exposures and thick covers, a pit or trench is necessary to confirm the bedrock source of an anomaly picked up by an earlier examination. Trenching is a cheap method and is useful not only in ore search but also in elimination of unpromising ground. Once the indicative minerals have been traced to their source or when other local indications of ore have been found, the obvious next step is to uncover the mineralized zone by digging a series of trenches at right angles to the rock structure or to the possible strike of the vein (Fig. 3.3). If the vein is found, trenching is continued so as to expose its continuity, etc. if needed, cross-trenching is to be done to expose the ore zone. If the overburden is deep, the bedrock is to be tested by a series of close-spaced pits. Despite shallow depth, pitting and trenching have some advantages, over drilling, in that geological logging can be done in greater detail and collection of large and undisturbed sample. If the indications pointing to ore at depth are too great for trenching, it may be tested by drilling using light weight portable drill machines. The indications of ore below reach of trenches/test pits may consist of geophysical anomalies, structural guides, leached gossans, capping, etc. The alternative underground work, tunnel, or shaft is to be undertaken where drill is impracticable. Once the presence of ore is confirmed, whether by trenching, pitting, drilling, or sinking, further systematic investigations can be carried out.

FIGURE 3.3 Trenching in gold prospects in West Africa. *Courtesy: BURKINA 95.*

3.8 PROSPECTING TYPES

Prospecting is of two types, which may vary in the degree of success obtained. "Professional geologist" is the first type, who gets paid irrespective of his work bears economic fruit or not. The second type of prospector is by the "individuals" who have the knack to find interesting mineral occurrences, stake the ground, and later sell the "claim" to a mining company. If the second type of prospector finds something interesting and sells it, he gets good money. On the other hand, if he does not find anything worth selling, he does not earn. The unpaid prospector, many a times, has the opportunity to make a fortune for his discoveries.

3.9 PRELIMINARY FIELD TRIP

Quick preliminary trip over surface and underground (if any) is done to gain general impression of whole property to decide whether any further examination at all is worthwhile. One can observe general geological setting, amount of development done, shape of the ore bodies if any—vein-like, massive, etc., the geographical setting with respect to other mines. Mapping of the area, detailing geological features as seem essential, especially veins, faults, apparent ore limits and rock contacts, underground exploration if any. Enough number of samples is to be collected. If previous sampling results are available, a few spot checks will confirm the reliability of work done earlier so that the new work can be initiated. Extension of existing ore or chance of finding new ore depends upon correct interpretation of the structure, grade, etc. Besides, evidence from dumps, mineral, and rock collection in a mine dump will tell a great deal about the history and geology.

Sources of information can be from published and unpublished reports, testimony of old residents, etc. Generally the detailed examination of the prospect leads to an appraisal of its value or recommendation whether it merits further work or development. Depending upon the type of problems the prospect presents, the nature and extent of work will vary. Hence all the prospects are likely to fall in one of the following types, viz., (1) "ore hunting prospect" (the conditions are encouraging but more and better ore has to be found than has been encountered thus far. Where is the ore, if any?); (2) "ore extension prospect" (ore and grade are good enough to provide sufficient tonnage, hence can be developed. Is there any more ore?); and (3) "ore exploitation prospect" (whether the ore can be treated at a profit. What can be done with it?).

3.10 PROSPECTING METHODS

3.10.1 Preliminary Proving

The purpose of preliminary proving is to find out the quality of mineral (reserves) and its quality in relation to current requirements of industry in a particular area. The information from preliminary proving should be able to provide the following details of the ore deposit: viz., (1) size and shape, (2) attitude of ore body, (3) thickness variation along dip and strike, (4) nature of enclosing and overburden rocks, (5) degree of uniformity of economic mineral within the deposit, (6) quality and distribution of country rock, (6) mineralogical and chemical composition of the deposit, (7) change in quality of mineral both in depth and area, etc.

3.10.2 Detailed Proving

This will involve the activities of full-fledged exploration (described in chapter: Geological Exploration).

3.11 GUIDES FOR PROSPECTING

If one knows where to look for an ore deposit, then it will be the simplest and cheapest method. But in reality, it is not that simple. A prospector has to take the help of many "sign posts/clues" (guides), like geological, structural, and other features, to locate the ore bodies. The commonest and simplest guide is a halo of altered rock around ore body (ringed targets). The limits of a particular mineral or group of minerals in a vein, ratio of metals within the vein matter, etc. sampling and examining them microscopically will facilitate to recognize variation in rock or mineral. Besides, carrying out chemical analyses will bring out the nature of distribution of metals or other elements. If an ore body is localized within the stratum by a line, one need to explore only one dimension and if ore body is localized within the stratum by intersection lines one will know exactly where to look for it.

3.11.1 Evidence from Outlying Areas

A. Regional guides: Regional guides are broad and general in nature. These do not help for direct use in development in a district but are helpful in selecting regions for exploration for ore deposits of some type. These regional guides can be:
   i. Batholiths and major igneous rock bodies associated with certain ores. Coast range batholiths of British Columbia, Canada.
   ii. Volcanic rocks of specific types and ages: Greenstone belts of Canadian shield, Kalgoorlie, Australia, West African Birmian greenstone belt, Barberton greenstone belt, S. Africa.
   iii. Igneous rocks, characteristically associated with certain ores. Acid intrusive—tin, tungsten, molybdenum ores; Basic intrusive—nickel; Ultrabasic intrusive—chromium, nickel, platinum.

B. Age relations with metallogenic epoch (Precambrians of India; Cobalt–silver at Cobalt, Canada)

C. Ore zones associated with major fault zones: (Mother Lode of California, USA, Great boundary fault (Rajasthan), Singhbhum thrust belt (Bihar) in India)

D. Sedimentary rocks of specific age; occurrence of iron formations at definite horizons

E. Climatic and topographic conditions

- Tropical: lateritic weathering—Manganese ores of India, Laterite, bauxite of Guyana
- Arid and semi-arid with deep water level favoring supergene enrichment: Porphyry coppers of Chile, Peru, and southwest United States
- Long period of deep weathering followed by erosion, conducive to accumulation of placer gold, placer rate earths, etc.

3.12 CLASSIFICATION OF GUIDES

McKinstry (1962) has grouped the guides into three categories based on the genesis. They are (1) features that were in existence before the ore was deposited and served to localize it (eg, fractures, beds subject to replacement breccia pipes); (2) features which came into existence with ore (eg, alteration holes, barren parts of veins, area of subsidence due to mineralization stoping); and (3) features resulting from the presence of ore or from presence of mineralization (eg, gossans, iron or manganese stain below an oxidized ore body, oxidation subsidence, ancient workings)." He further classified the guides based on "the nature of the guiding features" as (1) physiographic guides; (2) mineralogical guides (alteration, mineralization, oxidation products); (3) stratigraphic and lithologic guides; and (4) structural guides (fracture patterns, contacts, folds).

A number of more important mineral belts are seen to elongate topographic lows. It appears that structural breaks, intrusive activities, alteration, weathering effects upon ore, etc. may combine to differentially erode the belts/mineralized portions of the world. Any topographic depression, even where filled with postore sediments, volcanic, alluvium, or partially covered by desert pediment veneer or glacial deposits should be carefully analyzed as a possible outface expression of ore.

3.12.1 Physiographic Guides

Eminences and Depressions: Most of the important deposits do not crop out boldly at the surface and a number of mining districts as in pockets and low basins within mountainous terrain. Direct and indirect evidence of presence of ore can be from the physiographic features. Surface expression of an ore body is the direct indication although there may be valuable indirect evidences. Geological features like fault scarps, hogbacks are seen as clue. In many regions, the ore outcrops are seen as a conspicuous feature. The presence of massive quartz veins stand up as ridges in many areas while the unsilicified rocks between them weathered away. In many places, more soluble beds (like calcite veins) form well-defined depressions in contrast to resistant quartz veins ridges. The weathering of soluble minerals/veins might give deceptive broad outcrop due to close network of stringers eroded, shattered, and spread around.

Oxidational Subsidence: Erosion of soft material may result in the development of depressions. Removal of support through shrinkage of ore bodies during oxidation may also cause subsidences similar to mine subsidences.

Topography as a Guide to Iron Ore: Topography may be considered as a guide in ore search although many a times, it does not by itself indicate the presence of ore. Some ore deposits are topographically related to structures/rock types or surficial deposits that are exposed. Major topographic expressions like ridges, mountain ranges, etc. will be useful in locating large ore bodies. It is well known that iron ore, bauxite deposits occur in large masses in many parts of the world. Since these ore deposits have to be large to be commercially viable, topographic expressions will be useful as guides in locating them. Examples: Hills, ridges, and ranges—Iron ores of Lake Superior region, USA and Western Australia; Iron ores of Madhya Pradesh, Bihar, Karnataka states of India.

Physiographic Relations of Placer Deposition: In the accumulation of placer deposits of gold, platinum, diamond, and other heavy minerals, physiographic events play a critical role. Individual heavy mineral/metal particles are released from their host rocks due to prolonged weathering on mature topography, which are later concentrated by sorting agents. Major ore locus is the stream channel. The favorable areas are determined by the stream velocity and the stream gradient. An analysis of physiography provides significant information related to placers. The points of enrichment of placers are in the mouths of side streams, rock bars, rock holes, ox-bow lakes, terrace levels, etc. Many placers are concentrations by wave action in beach sands (alluvial rutile deposit, Australia; beach placers (monazite, ilmenite, rutile, garnet) of India; cassiterite deposits in the ocean sands of Malaysia, Thailand). Large and rich deposits of diamonds are known beneath the sea off the west coast of South Africa and Southwest Africa).

3.12.2 Physiography, in relation to Oxidation and Enrichment: Residual Ores

Topography to an extent controls the rock weathering. Those types of ores which owe their value due to the removal of undesirable element through the process of weathering occupy definite erosion surfaces. Weathering under tropical conditions facilitates to form Bauxite, nickel silicate, some manganese ore and lateritic iron ores. Tropical conditions are not essential in some gossans which are mineable for the content of gold, silver, lead, iron,

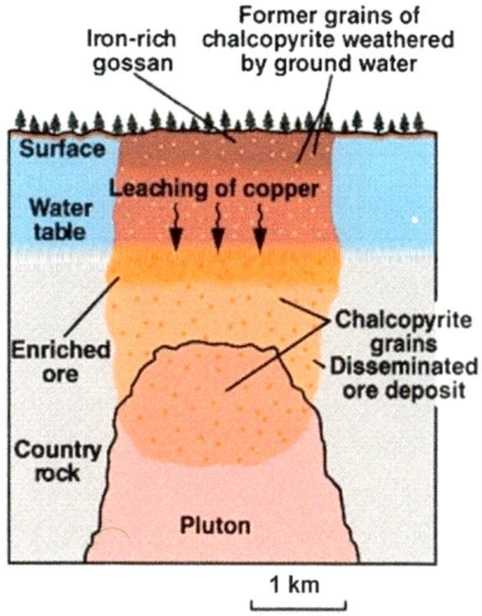

FIGURE 3.4 (Left) Gossan relict mount, Rajpura-Dariba mine, India; (right) sketch map of vertical section of supergene sulfide enrichment.

etc. The production of ore of these types by the process weathering proceeds rather slowly and reach completion to reasonable depths in different parts of the world. Fig. 3.4 shows the massive oxidized gossan relict mount at Rajpura-Dariba zinc–lead mine, India and also a sketch map of vertical section of supergene sulfide enrichment.

Supergene Sulfide Zones: The formation of supergene sulfide enrichment of copper and silver ores is similar to those physiographic conditions which produce the residual deposits. Extraction of metal from a thick overlying zone of leaching results in the concentration of sufficient metal to form a thick and high-grade zone of enrichment. Sillitoe (2005) opined that "If enrichment were carried to completion under static conditions, this would mean an extremely deep water table, but more probably the zone of leaching and the zone of enrichment below it descend progressively as erosion lowers the surface, thus keeping pace with enrichment without overtaking it." In the state of postmaturity/old age of a physiographic surface, the balance between erosion and enrichment is typical. Groundwater normally circulate more actively in greater physiographic relief (youthful/mature topography) resulting in rapid descending of oxidized zone, owing to speedier process of leaching of ore. The panoramic view of Cerro Colorado porphyry copper deposit of BHP Billiton in Chilean Copper belt exhibiting various zones is shown in Fig. 3.5.

3.13 MINERALOGICAL GUIDES

Minerals that are present and their relative abundance serve as very practical guides in ore search. Target rings might be present by the variation of properties of minerals whether in the wall rock/vein matter and whether apparent in plan/vertical sections. Such target rings are characteristic of deposits. The presence of oxidized minerals on the surface gives clue as to what lies beneath.

FIGURE 3.5 Panoramic view of Cerro Colorado porphyry copper deposit, Chile, showing various zones. *Courtesy: BHP Billiton.*

3.13.1 Target Rings of Alteration

In epigenetic ore deposits, the rocks surrounding them exhibit mineralogical changes in rocks which involve introduction of certain chemical elements and removal of other. Many a times, the chemical changes are minimal but the elements that were present originally merely rearrange themselves into new mineral assemblages. The outer limits of a zone of alteration form the outer rings of a target for exploration. When veins are closely spaced, altered zones merge to form a general halo surrounding the system. The appearance of an altered rock may be the first sign of an approaching ore body. Wall rock alteration is widely used as a guide to hydrothermal ores (Schwartz, 1955; Park and MacDiarmid, 1970). In general, the envelope of alterations around ore are targets of all types of a much larger target than the ore deposit itself and some alteration patterns are easily recognized, particularly when weathering may have leached and/or stained the rock, enhancing the visual effect.

Wall rock alteration has been used most effectively in exploration of hydrothermal, mesothermal districts. Alteration effects are weakly developed in hypothermal zone, difficult to interpret in epithermal districts and subtle in telethermal deposits. Alteration is a result of permeation of hydrothermal solutions into wall rocks and alteration product is influenced by initial rock composition, permeability and structural preparation. In exploration investigations, it is often difficult to be certain of original character of the rock and to separate hydrothermal effects from possible previous regional and thermal metamorphism and possible near-surface weathering effects, past and present.

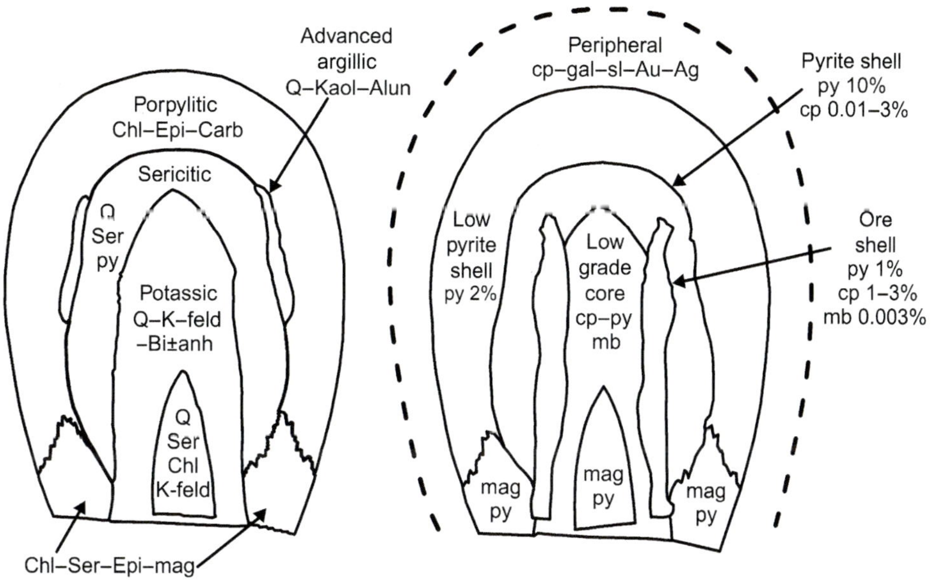

Explanation:
Chl, chlorite; Epi, epidote; Carb, carbonate; Q, quartz; Ser, sericite; K-feld, potassium feldspar; Bi, biotite; Au, gold; Ag, silver; anh, anhydrite; py, pyrite; mag, magnetite; Kaol, kaolinite; Alun, alunite; cp, chalcopyrite; gal, galena; sl, sphalerite; mb, molybdenite

FIGURE 3.6 Hydrothermal alteration and zones, minerals and ores in a porphyry copper deposit.

Application of hydrothermal alteration as a guide to ore requires rather special skill and knowledge because interpretation requires microscopic mineralogical investigation. The simple mineral recognition of the more obvious alteration effects in hand specimens is invaluable in evaluation of new areas and in interpretation of drilling results of blind penetrations of postmineral cover. The alteration effects in porphyry copper deposits are shown in Fig. 3.6. Lowell and Guilbert (1970) and Barnes (1997) have summarized very well the alteration effects of porphyry copper of different areas.

Systematic mapping of altered zones as exposed on the surface and in underground workings serves to delimit the ore-bearing area, discouraging too extensive a program of exploration outside it and calling attention to parts of it that have not been explored and to places where its limits have not been reached. Alteration of a type which habitually sheathes individual ore bodies and veins has a number of practical uses. For example, if a crosscut passes through an altered zone without finding ore, the possibility that the altered zone is part of an outer halo surrounding an ore body may suggest exploration on one side or the other of the crosscut. In a drill hole, altered rock may be the first sign of the approach to an ore body and may serve as a signal for redoubled care in supervising the ensuing drilling. Such a hole, whether or not it encounters ore, should be continued until it has passed completely through the altered rock.

Despite the impressive degree of similarity in the nature of rock alteration associated with similar ore deposits throughout a region, the presence of alteration of a given type is not in itself a guarantee that commercial ore is to be found, although the more consistent the association has proved to be in a given district the more confidence it inspires. Conversely, the absence of strong alteration, although by and large a discouraging sign, is not necessarily a sufficient reason for condemning a vein. Schwartz (1955) discussed more applications and lists the following as guides to ore:

- Intense hydrothermal alteration of rocks on a considerable area is a favorable sign in searching for epigenetic ores in new areas.
- Zones of hydrothermally altered rocks furnish a guide to the direction to follow in searching for ore.
- In limestone formations, areas of dolomite, mangano-calcite are favorable signs; recrystallized limestone or dolomite and bleaching of the formation; intense carbonization of igneous rocks, such as diorite, occurs in connection with many gold deposits.
- Areas of intense silicification, particularly in limestone formations, are commonly a good sign, but are barren probably more often than is the case with some other varieties of mineralization.
- Areas, zones, or layers of intense sericitization are a favorable sign, particularly in rocks of the monzonite and diorite clans.
- Bleaching of rocks is a guide to alteration, and the large number of deposits in or bordered by, leached rocks makes it useful in helping localize more intense prospecting activity. Localized red oxidized zones indicate fomer presence of pyrite and perhaps other sulfides. Any unusual color change merits study, because some hydrothermal minerals (like Chlorite) may cause darker zones than in unaltered rocks.
- Any intense development of a possible hydrothermal mineral always warrants study as a possible guide to mineralization. In addition to more common hydrothermal minerals, such as sericite, biotite, chlorite, quartz, carbonates, clay minerals, feldspar, epidote, zoisite, alunite, tourmaline and pyrite, less common minerals like hematite (specularite), magnetite, mariposite, and talc may be significant; also so-called pneumatolytic products such as topaz, fluorite, apatite, and garnet.
- Alteration along fracture zones, even if restricted, may be an evidence of leakage from more intense alteration below.
- Alteration may give clues to the underlying structure, and this may be helpful, chiefly where relatively barren rocks overlie favorable host rocks, particularly limestone.

3.13.2 Hypogene Zoning as a Guide

All the mineralogical variations from the source channel to the surface from the central axis constitute the hypogene zoning. Noticeable mineralogical changes might take place laterally within a mining district in deposits of epithermal and mesothermal types. If primary mineralization is discovered at depth, it is likely that in higher workings, one may find a zone of enrichment and will serve as a guide.

District-scale zoning is widely recognized as a guide to the development of partially explored mining districts (Parks and MacDiarmid, 1970; Guilbert and Park, Jr, 1986; Sillitoe, 2005). If structural and stratigraphic complexities do not distort simple patterns, a given mining district may show a definite sequence of mineral deposits from the central source of the ore fluids outward. Utilizing district-scale zoning in mineral exploration may be a simple sorting of the various prospects and mines of a district to decipher if unexplored metallization might lie at depth or laterally beneath the cover. The zonal study of a district may emphasize visual identification of minerals themselves. The rock chip geochemical study will be very useful. It is possible that hypogene metal patterns can be considerably distorted because of the varying stabilities of the metal ions under near-surface weathering condition. For example, zinc can be cleanly leached from surface while some of other metals form relatively stable sulfates, carbonates, oxides, and silicates. Many of the secondary minerals are not easily identified visually, when present in trace amounts and mixed with hydrous iron oxides, etc. It should be remembered that the host rock units and the various facies of wall rock alteration developed within them may exert influence on the minerals originally deposited and the supergene mobility of the metal upon weathering. For example, copper is leached readily under acid porous conditions characteristic of a thoroughly sericitized porphyry, while stable copper carbonate might remain near surface in an unaltered calcareous matrix.

Mineralogical Guides to Solution Paths: Hydrothermal ore body must have an inlet and outlet. By the plumbing system, the solutions moved toward the surface through openings of various shapes in permeable horizons.

Zone of Oxidation: Sulfides decompose, sulfur converted to sulfate and their metal content carried away in solution or else fixed as stable compounds (oxides, carbonates, silicates). Limonite and their box works.

3.13.3 Oxidation Products

Since ages, oxidation products of an ore body (leached outcrop, various types of gossans, etc.) were used as a guide by the prospecting geologists and miners. With the better understanding of the geology and chemistry of oxidation products, the use of them has become even more effective. The downward leakage of the waste products of oxidation or enrichment may serve as a guide in UG development. Oxidized ore body may be rich enough to make a mine in its own right (nickel, zinc, lead).

3.13.4 Leached Outcrops

The interpretation of gossans and leached cappings is one of the earliest recognized guide to ore and led to the discovery of many of the important mining districts of the world. Sulfide ore minerals are relatively unstable in surface weathering environment and the familiar term "limonite" is applied to several hydrous iron oxides, often the predominate mineral remaining in the outcrop of the ore deposit. Under favorable conditions, color texture and structure of limonite can furnish valuable clues to the nature of unweathered mineralization. Most of the showing of the limonite, even in remote regions, has been

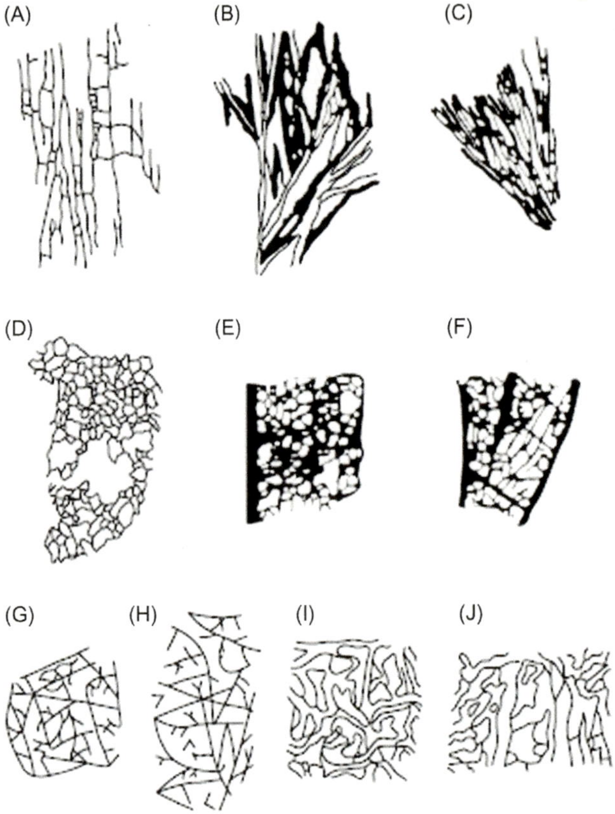

FIGURE 3.7 Boxwork structures developed after the primary ore minerals: (A–C) galena; (D and E) sphalerite; (F) chalcopyrite; (G and H) bornite; (I and J) tetrahedrite. *Source: After Blanchard and Boswell (1934).*

explored by many. It is possible that most readily impressive surface gossans have been explored to some extent but there are many local situations that require recognition of many subtleties in weathered mineral outcrops. The boxwork structures developed after the primary ore minerals are shown in Fig. 3.7.

In exploring through cover, where oxidation and enrichment may have preceded concealment, one is faced not only with drilling blind but also correctly interpreting the possible buried weathered zone first penetrated by the drill. Conditions near ideal for the sulfide minerals usually constitute a relatively small percentage of the total bulk of the rock, and when leached from typical disseminated ore, the rock matrix remains strong enough to support itself and preserve the delicate texture and structures of the limonite for inspection in outcrop. In contrast, many sulfide vein, bedded, pipe deposits contain such a high percentage of sulfides that thorough weathering may remove the bulk of the ore deposit and the unsupported walls collapse against spongy, chaotic mass of limonite, kaolinized wall rock, and corroded fragments of gangue, etc., which may not crop out boldly.

Excellent short summaries are given by McKinstry (1962). Most important outcrops have been stripped over the years and the waste had been piled upon peripheral cropping with each passing year. Unless one has demonstrated his ability to interpret leached outcrops, one should not rely too much on limonite cappings. In preliminary evaluation, exploration geologist should rely on "wall rock alteration and other guides." An estimate should be made of total sulfide content and that of metal sulfide/pyrite ratios. The character of hypogene minerals and evidence of supergene enrichment should be carefully noted. Wall rock alteration and weathering effects upon wall rock commonly neglected, although a thorough understanding of them may permit an inference of more favorable conditions at depth or laterally beneath cover.

3.13.5 Metals in the Oxidized Zones

In oxidized zones, native gold grains, being the most resistant metal, accumulate in fractures and in near-surface rocks resulting in yielding high assay values (residual enrichment). Besides, leaching of near-surface rocks further aid to concentrate gold values. Hence, high gold values in the upper section of the oxidized zone are to be taken with a caution. In arid climates, the generation of requisite chemicals (chlorine, sulfuric acid, manganese, etc.) is possible in oxidizing deposits that will make gold, soluble as gold chloride. Instances have been noticed that gold has migrated in the presence of manganese, providing small pockets/patches of zone of enrichment. Metals of varying proportion found in the oxidized zone include tin, silver, copper, lead, nickel, cobalt, molybdenum, zinc, chromium, etc.

3.14 STRATIGRAPHIC AND LITHOLOGIC GUIDES

If a sedimentary bed exclusively hosts ore, then that bed constitutes an ideal stratigraphic guide. Of a group of beds contain ore bodies; still, they qualify to call them as stratigraphic guide, though other sedimentary horizons may not be entirely barren. On the contrary, if an intrusive body or a volcanic flow hosts the ore, instead of "stratigraphic," the term "lithologic" is used. The ore deposition may be syngenetic (contemporaneous with the rocks in which they occur) or it may be epigenetic (introduced later into the rock) (eg, Bushveld complex, S. Africa and Stillwater complex, Montana).

In Syngenetic Deposits: Syngenetic deposits are formed coeval as the rocks in which they occur. They have formed by the same process and at the same time of the geological time frame as the enclosing rocks. They are sometimes part of the succession on the earth's surface in the form of a sedimentary layer (iron-rich sedimentary horizon). These are two types: (1) Igneous: magmatic-layered mafic intrusion during the crystallization of magma, usually mafic, ultramafic heavy, metal-rich liquids settle and accumulate at specific sites, often at the base, within the intrusion (Platinum group of metals; chromite; Ni–Cu (PGM) and (2) Sedimentary (deposits occur as beds/bed-like matter and are conformable with underlying and overlying rocks). They include many sedimentary mineral deposits: limestone, oil shale, bauxite, coal, phosphorite, iron and manganese ores, etc. Certain deposits of chromium, uranium, and vanadium belong to this type. The width and thickness of these deposits range from a few meters to a few hundred meters and strike length running to km.

In Epigenetic Deposits: In epigenetic deposits, the ore solutions are introduced later into the rocks, in which they occur. They have been introduced into the pre-existing country rocks after their formation via metal-bearing fluids. A mineral vein in a host rock is a good example. Fracturing and breaking of the rock along weak plane (or fault) at a depth ranging from surface to several kilometers below surface is a prerequisite. Hydrothermal solutions pass through open spaces/fractures and deposit the minerals. These deposits have various forms like fissure veins and sheet-like form, etc. Chlorites, slates, and phyllites host many gold deposits which are also in basic to intermediate igneous rocks than in rhyolites, quartzites, limestones, etc. When ore is introduced into the already existing rocks (epigenetic), like many minerals in pegmatite, shows strong partiality to certain favorable formations. Some examples of epigenetic deposits are: (1) Porphyry type deposits, that are large, low-grade deposits associated with porphyritic intrusive body; (2) Skarn-type mineral deposits formed by replacement of limestone by ore and calc-silicate minerals adjacent to a felsic or granitic intrusive body; and (3) Fracture-filling vein type deposits which have both lateral and depth extension but are narrow (hypothermal; mesothermal; epithermal).

Competent Versus Incompetent Formations: The more hospitable hosts to ore deposition are competent rocks. Since they are strong, they break when fail like any brittle material. Quartzites, conglomerates, and fresh igneous rocks are competent formations under normal conditions. On the other hand, the incompetent rocks are weak and have a tendency to plastically deform or follow flowage. Shales, schists, limestones, and igneous rocks that have been altered to sericite, chlorite, or serpentine are incompetent formations.

3.15 STRUCTURAL GUIDES

3.15.1 Fracture Patterns as Guides

Fractures which are the result of failure in a rock mass are seminal for the formation of ore deposit in solid rock. Generally fractures facilitate as channel ways for the entry of mineral laden solutions, the receptacles for ore deposition, and the loci for replacement (Fig. 3.8). It is possible that metallic and other elements can pass through massive rocks by diffusion, by capillary or subcapillary openings. Usually the local transportation from the main zone through diffusion is a slow process. Formation of vein deposits is predominantly due to fractures present which have noteworthy influence on the form of the ore bodies. The shapes of fractures and fracture systems, generally, reflect in the structure of many ore deposits.

Parallel or nearly parallel veins, arranged in an "en echelon" form, are the simplest pattern. In many places, the veins take a pattern shape resembling "Y." Some veins could a "braided system," wherein two veins run parallel for some distance and then join, to separate again and become parallel resembling the pattern of chicken-wire mesh. Multitude of characteristic shapes is formed by individual veins or vein groups. But the most common ones are fractures arranged "en echelon" (Fig. 3.9).

3.15.2 Contacts as Guides

Owing to contrast in strength between massive intrusive body and weaker sediments, the "contacts" are the most favored place for ore deposition. Contacts are also quite

FIGURE 3.8 Hot fluids that pass through fractures in deep rock can crystallize and fill the fracture to form mineral veins (fracture filled gold veins; Photo: Gandhi).

FIGURE 3.9 (Left) Conjucate pair of two extension fractures in quartz; (right) crysotile asbestos veining and complex branching in microfractures. *Photo: S. Jacob.*

vulnerable places for fracturing during regional deformation and during adjustments accompanying the emplacement of an intrusive. Hence, they are of interest both as broad loci of regional scale and as short range guides to the positions of veins. It is possible for a vein to follow a contact for long distance or part of it, depending upon various conditions. The place in which a vein crosses a contact is quite likely to be the site of an ore shoot, and such a condition provides a very simple X-like target for ore search. The contact between igneous rock masses and the intruded rocks is favorable horizons for occurrence even though the ore is not genetically related to the intrusive mass. In many cases, chemical as well as physical influences may be potent. When solutions ascend through fractures in the intrusive, the contact may be the first place where they encounter hospitable rocks. Thus the contact and the fracture may constitute intersecting loci.

Most pyrometasomatic deposits are localized by faults, fractures, or breccia zones, although some preserve no evidence of fracturing of the wall rock. Selective replacement of beds is very common. Thus contact deposits, although they are associated with a special

FIGURE 3.10 (Left) The presence of disseminated selective replacement of sulfides in shales, filling fractures; (right) magmatic chromite ore seams in the Bushveld layered mafic intrusion, Dwars River area, S. Africa. *Photos: S. Jacob and John Friske.*

group of earlier silicate minerals (their only distinctive characteristic), are not different morphologically from other types of selective replacement deposits (Fig. 3.10, left). Although some deposits are found along a contact or making outward from it, they are subject to the same types of structural control as other replacement deposits. In some districts, contacts are useful guides. "Pyrometasomatic" deposits are commonly found in the general vicinity of igneous contacts (older term "contact metamorphic"), although they are not restricted to such positions. It was generally believed that contact-metamorphic ore was deposited from solutions or emanations that carne from the magma during its intrusion (Fig. 3.10, right). The ore in pyrometasomatic deposits is not uniformly distributed around the contact and, in fact, is usually less abundant in the immediate vicinity of the intrusive than farther away. Furthermore, the ore minerals, particularly those of copper, lead, and zinc, are later than silicate minerals and appear to belong to a younger hydrothermal stage.

3.15.3 Folds

Depending upon a variety of circumstances, the ore is found in different parts of the folds, viz., crusts of anticline, troughs of syncline, or on the intervening limbs. The date of folding with respect to ore deposition will be a determining factor. Before folding, if the ore was present, it will naturally get deformed along with the enclosing rocks (Fig. 3.11). If the ore itself forms a sedimentary bed, it will have the same general shape as the beds above and below it. If the folding is gentle, the ore usually will have uniform thickness throughout the fold. Alternatively, if there are variations in thickness, they will reflect differences that were present before folding and will have no relation to the shape of the folds. The distribution of ore may get modified if the events took place after the folding. It is possible that sufficient erosion might have planed off the anticlinal part of the ore bed leaving ore in the synclines.

FIGURE 3.11 The accordion-folded and squeezed mineralized rocks, Rajpura-Dariba Mine, India. *Photo: Gandhi.*

The ore invariably reflects the folded structure, if the ore has been introduced into already folded rocks. The imposition of the effects of the structure of rocks on the ore deposition can take place through (1) influence of folded beds on fracturing, (2) influence of the folds on the flow of ore-bearing solutions, and (3) the shapes of replaced beds. Thinning and thickening, drag folding, flowage folding on the flexure are the common features found in tight folds.

3.16 GEOCHEMICAL GUIDES

Detection of abnormal concentrations of chemical/metallic ions in surficial materials, such as soils, stream sediments, rocks, soil, water, plants, called "geochemical anomaly" can be used as guide to mineralization. The anomalous values might indicate the presence of mineralization below earth. A map showing distribution of metal ions may help to locate target rings surrounding a possible mineralization. The groundwater in a mineralized area (under oxidizing conditions) contains metals/sulfates in amounts ranging from traces to so much that the water is not potable (eg, water in copper deposits of Zambia). It is likely that some of the metals (like Cu, Zn, Pb, Ni, Co, Mo, Sb, Bi, etc.) mobilized from the source and present in groundwater are likely to give clues about the possible mineralization.

3.16.1 Biogeochemical and Geobotanical Guides

Biogeochemical analysis is very useful as a guide to the underlying geology in areas of transported cover when the signature in vegetation can be better than soil. Normally, biogeochemical surveys are of two types: (1) utilizes the trace element content of plants to outline dispersion haloes, trains, fans related to ore mineralization and (2) specific plants or the deleterious effects of an excess of element in soils/plants as guide to mineralization (geobotany). Geobotany has been used as an aid to locate mineral deposits in many parts of the world. The presence of indicator plants signals the existence of a particular element in the soil in which they grow. Those plants that point to the presence of specific element are called "universal indicators." Metal-rich soil and associated vegetation can also be located by remote sensing using multispectral techniques.

3.17 ANIMAL ACTIVITY

Small rock fragments, soil, etc. are brought up to surface by the burrowing animals, rabbits, termites, etc., which might give a clue regarding what lies below. In some parts of the world, dogs were trained to sniff mineral-bearing boulders (with SO_2 other gases associated with oxidizing sulfides) to trace back their source. In some aquatic creatures (like fishes, mollusks insects, etc.), the abnormal concentrations of copper, lead, zinc in parts of their bodies have given clues for mineralization, in the vicinity of their habitat. It is noted that animals of some areas suffer from specific diseases due to excess intake or deficiency of some elements concentrated in soils/sediments/water, around their habitat.

References

Barnett, C.T., Williams, P.M., 2006. Mineral exploration using modern data mining techniques. Society of Economic Geologists, Special Publication, 12, pp. 295–310.

Barnes, H.L. (Ed.), 1997. Geochemistry of Hydrothermal Ore Deposits. #rd ed. John Wiley & Sons, New York.

Barnes, J., Lisle, R., 2004. Basic Geological Mapping. John Wiley & Sons, New York.

Blanchard, R., Boswell, P.F., 1934. Additional limonite types of galena and sphalerite derivation. Econ. Geol. 29, 671–690.

Brimhall, G.H., Dilles, T.H., Proffett, J.M., 2006. The role of geological mapping in mineral exploration. Society of Economic Geologists, Special Publication, 12, pp. 221–241.

Chaussier, J.-B., Morer, J., 1987. Mineral Prospecting Manual. Elsevier, Amsterdam.

Goodchild, M.F., 1992. Geographic information science. Int J Geogr Inform System 6 (1), 31–45.

Guilbert, J.M., Park Jr., C.F., 1986. The Geology of Ore Deposits. W. H. Freeman and Company, New York, p. 985.

Kramer, J.H., 2000, Digital mapping systems for field data collection. In: Soller, D.R., ed., Proceedings of a Workshop on Digital Mapping Techniques. U.S. Geological Survey, Open-file Report 00-325, pp. 13–19. http://pubs.usgs.gov/openfile/of00-325/kramer.html.

Kreiter, V.M., 1968. Geological Prospecting and Exploration. Mir Publishers.

Li, Z., Zhu, Q., Gold, C., 2005. Digital Terrain Modeling: Principles and Methodology. CRC Press, Boca Raton.

Lindley, I.D., 1987. The discovery and exploration of the Wild Dog gold-silver deposit. East New Britain, PNG. Pacific Rim Congress '87. AIMM, Melbourne, pp. 283–286.

Lowell, J.D., Guilbert, J.M., 1970. Lateral and vertical alteration mineralization zoning in porphyry ore deposits. Econ. Geol 65 (4).

McKinstry, H.E., 1962. Mining Geology, second ed. Asia Publishing House, p. 680.

Park Jr., C.F., MacDiarmid, R.A., 1970. Ore Deposits, second ed. W.H. Freeman, San Fransisco, p. 169.

Schwartz, G.M., 1955. Hydrothermal alteration as a guide to ore. Econ. Geol., 50th Anniv. Vol., Pt. 1, pp. 330–323.

Sillitoe, R.H., 1999. Field skills in exploration. Society of Economic Geologists Newsletter no. 38, p. 4.

Sillitoe, R.H., 2005. Supergene Oxidized and Enriched Porphyry Copper and Related Deposits: Economic Geology, 100th Anniversary Volume, pp. 723–768.

CHAPTER

4

Remote Sensing Techniques

OUTLINE

4.1 INTRODUCTION

Remote sensing has been utilized as a device for finding potential exploration targets from the initial stage of photogeology. The quality of remotely detected information has expanded as the level of technology has improved. Aerial photographs were used in the early days for wide area topographic survey. The technique steadily progressed to be substantially more refined and it was only post World War II, obtaining of ground geological information was initiated. Using stereoscopes it was possible to interpret geological

Essentials of Mineral Exploration and Evaluation
DOI: http://dx.doi.org/10.1016/B978-0-12-805329-4.00011-9

structures from the aerial photographs. The essential utilization of remotely sensed information was relative. If mining was being carried out for a specific sort of deposit in an area, use of aerial photographs would be made to find comparable geological elements somewhere else in the same area.

Relative pattern of aerial photo use went ahead till the satellite period at which point of time commercial availability of satellite imageries became prevalent. To assess a large area in greater detail, geoscientists started using multispectral, radar, and infrared (IR) imaging in a range of combinations. Furthermore, view of a prospect could be made from various angles in various seasons owing to multiple flyovers. As the visits to the prospect for reassessment were not required on a repeated basis, the cost of regional exploration enormously decreased. The capability for acquisition of information through radar imaging for cloud and surface cover is an added advantage. In addition, the data on areas (tropics and arid regions) which were impossible to large-scale regional field exploration could be possible. With the advent of computer capabilities, the data imageries could be digitally enhanced to highlight specific features. The identification of specific minerals from space could now be possible by spectral studies.

4.2 REMOTE SENSING

Remote sensing is a process to acquire, prepare, and decipher information of spectral and spatio-temporal nature on objects, phenomenon or areas under investigation without being in direct physical contact. Transfer of information is carried out using "electromagnetic radiation" (EMR) in remote sensing. Sabins (1997) stated "EMR is a form of energy that reveals its presence by the observable effects that are produced when it strikes the matter."

Electromagnetic energy, when incident on a certain feature of earth surface, the energy can be reflected, absorbed, or transmitted, which will vary in proportion depending upon the material type and conditions of different earth features. The distinctions allow to recognize diverse elements on a satellite image. At different wavelengths, extent of three basic energy interactions would differ even within a given feature type. In one spectral range two features may be distinguishable but altogether different on other band of wavelength. The resultant optical effect due to spectral differences within the visible portion of the spectrum is called "color." To discriminate among various objects, spectral differences in the extent of reflected energy are utilized by human eyes.

Different types of resolution of remotely sensed imageries are:

- **Spectral resolution** speaks of the limits of individual spectral wavelength ranges in defining fine wavelength intervals. Finer spectral resolution is more closely associated with narrower range of wavelength for a specific band.
- **Spatial resolution** speaks of the detectable subtle element in the satellite image. Finer spatial resolution is required in detailed mapping of wetlands than regional mapping of physiographic areas. Pixel size determines spatial resolution, while the radiance quantization affects radiometric resolution (Schowengerdt, 1983).

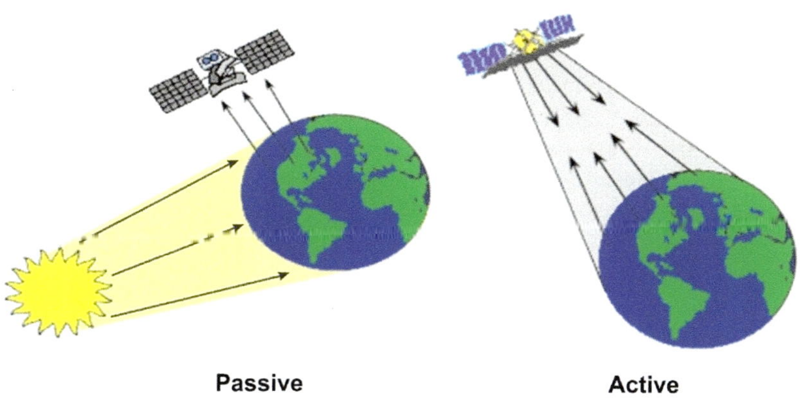

FIGURE 4.1 Remote sensing data gathering system. *Courtesy: Porwal.*

- **Radiometric resolution** speaks of the capability to distinguish slightest variances in energy. Better the radiometric resolution of a sensor, further sensitive it will be for sensing minor variances in reflected or emitted energy.
- **Temporal resolution** speaks of the time gap amid images. Many uses require monitoring information on recurrent basis.

Remote sensing data gathering systems are divided into two different types, viz., (1) Passive Remote Sensing and (2) Active Remote Sensing (Fig. 4.1). Levin (1999) states "A passive Remote Sensing system records the energy naturally radiated or reflected from an object. An active Remote Sensing system supplies its own source of energy, which is directed at the object in order to measure the returned energy. Flash photography is active Remote Sensing in contrast to available light photography, which is passive. Another common form of active Remote Sensing is radar, which provides its own source of Electromagnetic energy in the microwave region. Airborne laser scanning is a relatively later form of active Remote Sensing, operating in the visible and near-infrared wavelength bands."

Numerous remote detecting stages are intended for following a north–south orbit. This when combined with the west–east rotation of the earth permits to scan the majority of the surface of the earth for a specific timespan. Orbits are called "near polar orbits" owing to the orbit's angle with respect to a linear pass through the two poles, viz., north and south. The resulting orbits are sun-synchronous because of their passing through individual world zone at a uniform instant known as local run time. Within the same season, the portion of the sun in the sky will be at a given altitude, as the satellite passes overhead. Uniform brightness is an essential requirement while obtaining images in a particular period of time for consecutive years. It is an important condition for change detection or for combining contiguous images together called "mosaicking" (Levin, 1999).

Present day satellites used for the purpose of remote sensing are typically in "near polar" orbits. In the first half of its orbit, the satellite moves towards north and then moves towards south in the second half of its orbit. This phenomena is known as "ascending and descending passes." Ascending pass most likely falls on the darker side if orbit is sun-synchronous whereas descending pass will be on the brighter sunlit side. Image of the earth's surface is obtained on a descending pass by the sensors recording reflected solar energy only during availability of solar illumination. Both active and passive sensors would be able to image the earth's surface on ascending pass.

The sensor scans some part of the surface of the earth when a satellite revolves around the Earth. The region on the earth's surface imaged is referred to as the "swath." Sapce-borne sensors image swaths that are widely extensive. During the orbiting of a satellite, the position of east–west of it will remain unchanged. Both the rotation of earth and satellite orbits jointly function to permit complete passing over the Earth's surface on completing one orbit cycle.

The technique of measuring various ground parameters from an orbiting earth satellite is called "satellite remote sensing," the product of which is called "satellite imagery." Satellite imageries have brought a revolution in sophistication, strategies, and procedures in exploration and mapping. The satellite imageries with ever-improving high accuracy facilitate efficiently to map the geological formations, structure, drainage pattern, weathering pattern, etc. These imageries have been the basic input for wide-ranging applications, viz., agriculture, forestry, environment, mineral exploration geodesy, etc.

4.3 WHY REMOTE SENSING

In order to develop regional scale geological maps for use in small-scale survey, scheduling ground activity for large-scale surveying and to study the field geological features together with their geographical locations and association of various geological units on surface, remote sensing has been in use extensively. In order to have a comprehensive understanding of lithostratigraphy, multiple data sources are required to be integrated.

Three-dimensional view of the local relief will be provided by stereo imagery to facilitate delineation and identification of units. For field analysis and ground truth, aerial photographs and satellite imageries are carried along with for use in the form of base maps in the field. Aerial photographs generally provide high-resolution information (like weathering, drainage patterns, etc.) for site-specific analysis. Large coverage area and moderate resolution are required for regional overview. In geological exploration, since the various elements of concern are not dynamic, frequency of imaging has been never an issue.

One of the biggest problems faced in the application of remote sensing are the thick vegetation canopy. For such cases, the technique of remote sensing is of great aid to geologists for indicating rock type on the basis of the growth of the vegetation type on it (geobotany). The growth of particular type of vegetation is controlled by specific type of mineral and rock constituents on which it grows. A study involving natural association of various geological units with the growth of specific type of vegetation on them acts as good indicators and is of much use in geological mapping. By integrating different source of image data (optical, radar) at an appropriate scale, remote sensing is optimally used. Worldwide data are readily available in imageries captured through satellites and are of reasonable price (price per km).

4.4 MAJOR REMOTE SENSING SATELLITE SYSTEMS

Most of the satellites orbit at an altitude between 700 and 920 km. The earth rotates and the satellites scan 185 km length of its swath. Taranik (2009) has summarized the available remote sensing systems and their capabilities in the world.

LANDSAT (Introduced by US Government)

First generation (Landsats 1, 2, and 3 of 1972), Landsats 4 and 5 of 1984, Landsat MSS (repeat coverage: 16 days), Landsat Thematic mapper (TM) (repeat coverage: 16 days), and Landsat 6, Landsat enhanced TM, launched in 1999 (repeat coverage: 16 days).

SPOT (Introduced by French Government)

Multispectral scanner (XS) multispectral mode acquires three bands of data green, red, reflected IR wavelength with spatial resolution of 20 m.

Panchromatic (Pan) acquires a single band of data, primarily green and red wavelengths with spatial resolution of 10 m; both image modes cover 60 × 60 km of terrain (repeat coverage: 26 days).

AVIRIS (Advanced Visible/Infrared Imaging Spectrometer):

Conventional multispectral scanning systems, such as Landsat TM, SPOT XS, record up to 10 spectral bands and bandwidths of 0.10 μm. Hyperspectral scanners are a special type of multispectral scanner that record many tens of bands with bandwidths on the order of 0.01 μm. At visible wavelengths and at reflected IR wavelengths, many minerals have distinctive spectral reflectance patterns. Many minerals may be identified on suitably processed hyperspectral data. AVIRIS image strips are 10.5 km wide and several tens of kilometers long (Sabins, 1999).

ASTER (Advanced Space borne Thermal Emission and Reflectance Radiometer); **ATLAS** (Airborne Terrestrial Applications Sensor) JV US & Japanese Govts, launched in 1999

Hyperspectral scanners

VNR: 3 bands, 15 m resolution; SWIR: 6 bands, 30 m resolution; TIR: 5 bands, 90 m resolution.

ASTER is a multispectral sensor with 14 "geoscience-tuned" spectral bands which provide geological information far superior to that available from Landsat TM but at lower accuracy and mineralogical detail compared with hyperspectral systems, such as the 126 channel airborne HyMap sensor (http://c3dmm.csiro.au/ASTER%20Map%20of%20Australia%20EOI%20flyer.pdf).

Thailand Launch ***Theos***: 4 bands R, G, B, IR; 2 m; Panchromatic, 15 m, color

Taiwan ***FormoSAT-2***: Launched in 2008; 4 bands; R, G, B, IR0; 2 m; Panchromatic, 8 m, color; guaranteed acquisition with tasking, 3-day revisit capability.

4.4.1 High-Resolution Satellites

IKONOS: GeoEye (US Private Co.), launched 1999

Panchromatic (nadir): 1 band, 0.82–1 m resolution; Multispectral: 4 bands, 3.28 m resolution

IKONOS, Panchromatic, 1 m; IRS-1D Panchromatic, 5.8 m; SPOT Panchromatic, 12 m. Satellite imageries of millennium wheel, London, using IKONOS (left), IRS-1D (middle), and SPOT systems (right), depicting the pixel qualities, are shown in Fig. 4.2.

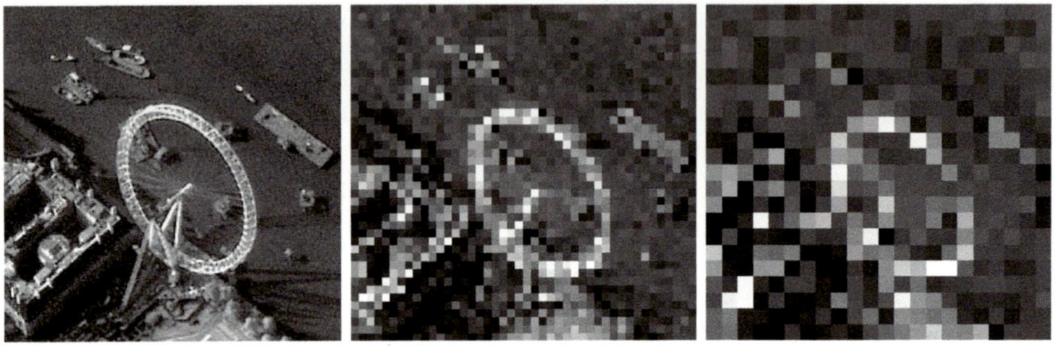

FIGURE 4.2 Satellite imageries of millennium wheel, London, using IKONOS (left), IRS-1D (middle), and SPOT systems (right). *Source: From Taranik (2009).*

QUICKBIRD: Digital Globe (US Private Co.), launched in 2001

Panchromatic (at nadir): 1 band, 0.6 m resolution; Multispectral: 4 bands, 2.4 m resolution.

SPOT 5 (French Govt.), launched in 2002

Panchromatic: 1 band, 2.5 m resolution; Multispectral: 4 bands, 10 m resolution

ALOS (Japanese Govt.): Advanced *Land Observing Satellite*, launched in 2006

Panchromatic: 1 band, 2.5 m resolution; Multispectral: 4 bands, 10 m resolution

5 channels VNIR—natural color, IR, Pan; Swath 70 km, DEM 10 m.

WorldView 1: Digital Globe, US Private Co., launched in 2007

Panchromatic Nadir: 1 band only, 0.5 m resolution; 1:2000 scale hardcopy; 60 × 110 km swath; 2.0 m R, G, B, IR; coastal, yellow, red edge, and near-infrared; DEM accuracy ±1 m.

GeoEye: *GeoEye (US Private co.)*, launched in 2008; 5 channels VNIR—natural color, IR, Panchromatic, Geolocation accuracy <3 m; Panchromatic: 1 band, 0.41 m resolution

Multispectral: 4 bands, resolution 1.64 m

Tasking—1–4 weeks in good weather; Cloud Cover rating 0–15%; Minimum area 100 km^2; Collection capacity 5–7 times of QB/IK.

WorldView 2 Digital Globe (US Private Co.), launched in 2009

Panchromatic (at Nadir): 1 band, resolution 0.46 m

Multispectral: 8 bands, resolution 1.8 m

German ***RapidEye***: Launched in 2008; Five Rapideye satellites orbit at an altitude of 630 km.

Major advantage for rapid repeat coverage; 5 bands—R, G, B, Red-Edge, NIR

Constellation of 5 satellites, 5 m pixel size, 48 × 48 km scene size.

4.5 RADAR AND THERMAL INFRARED SENSORS

Radar provides its own source of electromagnetic energy to illuminate the terrain hence it is an active form of remote sensing. In tropical regions, radar energy is measured in wavelengths of centimeters and has an advantage of penetrating rain and clouds. Radar images may also be acquired at a low depression angle that enhances the subtle topographic features, which are commonly the expression of faults, fractures, and lithology. In vegetated regions, radar images record vegetation surface only and not the underlying terrain (Sabins, 1999).

Early Systems—SEASAT (1978)—ocean floor bathymetry

First Generation Systems

ERS-1, ERS-2, JERS (early 1990s); RadarSat-1 (1995); ENVISat (2002)

Structural mapping, geology based on roughness

Second Generation Systems, launched in the recent past

- ALOS—L-band PALSAR system 10 m resolution—great for digital elevation
- Radarsat-2—C-band system, 3 m resolution
- Cosmo Skymed—X-band constellation, 1 m resolution
- TerraSAR-X—X-band system, 1 m resolution

High spatial resolution, flexible swath widths, look angles, and multipolarization interferometry and mine wall/slope stability, subsidence in historical mine areas.

RADARSAT-2 Now Fully Operational (http://www.radarsat2.info)

Canadian Synthetic Aperature Radar (SAR) Launched in 2008

Multi-Polarization—Single *(HH, VV, HV, VH)*—Dual *(HH + VV)*

Quad-Polarization *(HH + VV + HV + VH)*

- 3 m Ultrafine Single Pole
- 8 m Fine Quad Pole
- 25 m Standard Quad Pole

Satellite imageries of the same area using ASTER and ALOS systems depicting the pixel clarity are shown in Fig. 4.3.

FIGURE 4.3 Satellite imagery (JAXA, Geoimages) of an area in West Pakistan, using ASTER (left), ALOS systems (middle), and interpreted geological map (right). *Source: From Taranik (2009).*

Thermal Infrared Hyperspectral Sensors

Six specialized sensors to meet most remote sensing needs.

New 1800 pixel Broadband **TABI,** launched in 2009.

Thermal Infrared (TIR) (http://www.itres.com)

ITRES Thermal Airborne Broadband Imager **(TASI)**

Canadian airborne instrument flying late 2006; 32 channels located in the 8–12 μm (TIR); Airborne spectral resolution typically 50 cm to 3 m; Application to Ni/PGE's, skarns Zn/Ag

The current resolutions of various satellites launched in different periods of time are listed below:

Satellite With Year of Launch	Resolutions (m)	Satellite With Year of Launch	Resolutions (m)
GeoEye-1 (2008)	0.4/1.6	Worldview-1 (2007)	0.5/2.0
Quickbird (2001)	0.7	Ikonos (1999)	1/4
Orbview (2014)	1/4	THEOS (2008)	2/15
ALOS (2006)	2.5/10	RapidEye (2008)	5/5
RadarSat-2 (2007)	3/8/25	SPOT (2002)	2.5

4.6 DIGITAL IMAGE PROCESSING

Sabins (1997, 1999) and Drury (1986, 2001) have explained the methods of digital image processing in detail and how they can be applied to geological remote sensing.

- **Image restoration**: With an objective of ensuring recorded image look the same on the ground, the image errors, noise, genetic disorders incorporated while imaging, storing and playback activities are compensated. Some of the defects routinely corrected are: replacing lost data (dropped scan lines; bad pixels), filtering atmospheric noise, and geometrical corrections.
- **Image enhancement**: To better the information substance of the image, the visible effect which the image may be having on the interpretation with an object is altered. Some of the routines used to enhance the images are: (1) contrast enhancement—a simple linear transformation, called contrast stretch, is used to enhance the contrast of a displayed image by expanding the original grey level range (Drury, 2001); (2) spatial filtering—to enhance naturally occurring features such as fractures, faults, joints; (3) density slicing—continuous grey tone range is converted into a sequence of density ranges, individually conforming to a particular digital interval. Each slice is given an individual color; and (4) false color composite images—of three bands (MSS bands 4, 5, and 7), increase the amount of information available for interpretation.
- **Information extraction**: Computer is used to relate among diverse features of information sets.

4.7 APPLICATION OF REMOTE SENSING

Remote sensing has revolutionized mineral or petroleum exploration in myriad ways. Apart from its ability to cover large and inaccessible areas rapidly, the volume of data acquired enables us to use as a powerful tool for mineral targeting, besides many other applications. In different fields, there can be many applications of remote sensing. Remote sensing has specific demands for each application such as spectral resolution, spatial resolution, and temporal resolution. Remote sensing satellite imageries are widely used now in mineral exploration, structural investigations, subsurface information, and in hydrogeology. It is often combined with other data sources providing complementary measurements (Goetz and Rowan, 1981; Deutsch et al., 1981; Peters, 1983; Drury, 1986, 2001; Bultman and Getting, 1991). According to Sabins (1999), "some systems are deployed only on satellites (Landsat, SPOT). Other systems are currently deployed only on aircraft (hyperspectral) systems. Radar systems are deployed on both satellites and aircraft."

4.7.1 Mapping of Geology and Fracture Patterns at Regional and Local Scales

For any mineral exploration program, the geological mapping provides the basic ground. Data from other sources combined with remote sensing provide complementary measurements. Expression of surface topography and roughness is provided by "Radar." A host of geological applications in remote sensing includes: (1) surficial/lithological and structural mapping, (2) mineral/hydrocarbon exploration, (3) geo-environmental and geohazard mapping, (4) baseline infrastructure, (5) sedimentation mapping and monitoring, (6) geobotany, (7) sand & gravel exploration and exploitation, among others.

Geological exploration is the most rudimentary operation in remote detecting when aerial photographs are used utilized to recognize topographic surface components which might suggest subsurface features. At the point when searching for alike mineral deposits in a specific locale, surface components, eg, differential weathering, pattern of drainage, folds/faults, can be distinguished that can be contrasted with exploration targets elsewhere (Fig. 4.4). The importance of regional and local fracture patterns as controls of ore deposits has been recognized for long time by prospectors and mining geologists. Regional and local fracture patterns localize many ore deposits. Fracture patterns acted as conduits for ore forming solutions to penetrate host rocks and make excellent targets for future investigation. To map such fracture patterns, Landsat and Radar are often used. In order to interpret both structure and hydrothermal alteration, Landsat Thematic mapper (TM) and "satellite" images are widely used. Two assemblages of hydrothermal alteration minerals (iron minerals, clays plus alunite) can be identified by "Digitally processed TM ratio images." TM ratio images defined the prospects that are now major copper deposits (Collahuasi, Ujina) in northern Chile (Sabins, 1999).

FIGURE 4.4 (Left) A fault trace near Moab, UT, USA, that would be difficult to detect on the ground is easily seen in an aerial photograph. (Right) The Ray Rock gold prospect, North West Territories, Canada. A large-scale linear feature related to an ophiolite sequence developed during Precambrian tectonism. *Source: http://www.bdrg.esci.keele.ac.uk/Staff/Images/court.jpg (left) and http://www.ersi.ca/ (right).*

4.7.2 Hydrothermally Altered Rocks and Associated Mineral Deposits

The spectra utilized for remote sensing can readily differentiate clays and oxides. Without the need of extensive soil sampling program, potentially valuable ores may be distinguished through correlation of altered form to original constituents. Differentiating various types of vegetation is another valuable component of spectral analysis. The assemblages and alteration minerals that occur in hydrothermally altered rocks are well recognizable by the spectral bands of Landsat TM. Hydrothermally altered rocks associated with many ore deposits have distinctive special features that are recognizable in digitally processed TM images (Fig. 4.5). Hyperspectral imaging systems can identify individual species of iron and clay minerals, which can provide details of hydrothermal zoning. Silicification, which is an important indicator of hydrothermal alteration, is not recognizable on TM and hyperspectral images (Sabins, 1999).

The surface temperature of the earth, measured as "thermal infrared" (IR) is also of great interest to an exploration geologist. Multispectral IR (in the range of 8–14 mm) enables identification of rock types by their silica content. The emission spectra of silicate minerals contain a broad absorption "trough" caused by Si–O bonding. With an increase in silica in the rock, the absorption features shift to shorter wavelengths. This shift is a potential means of distinguishing various rock types. Besides this, the technique can be used for mapping the ground moisture, geological structure, geothermal reservoirs, underground fires in mines, etc. In spectral remote sensing, it is not only possible to map the individual mineral species but also chemical variation within the molecular structure of the crystal lattice (Lipton, 1997).

FIGURE 4.5 (Left) Satellite imagery of La Escondida Cu, Au, and Ag deposits, Chile. (Right) 3, 2, 1 RGB composite SWIR bands and 4, 6, 8 in RGB showing lithology and alteration differences. *Source: www.zonu.com; www.news.satimagingcorp.com*

4.8 ADVANTAGES OF SATELLITE IMAGERIES

A percentage of the preferences in utilizing remote detecting satellite images include: (1) remote detecting gives the outline required to build quick regional scale geological maps for use in small-scale survey, scheduling ground activity for large-scale surveying; (2) geographical locations and association of various geological units on surface; (3) a brief perspective of regional scale provides a superior viewpoint as compared to small ground perception for mapping of structures; (4) permits a geoscientist to look at other reference secondary information at the same time; (5) recognizing geomorphic units and deciding rock lithostratigraphy outcrop units (colored digital images facilitate better to differentiate litho units); (6) very cost effective; since many organizations offer the free source of landsat data; (7) computer processing enables discrination and detection of specified rocks/areas; and (8) commercially available global coverage, without any restrctions or conditions, thererby empowering information procurement from remote zones.

4.9 REMOTE SENSING AND GEOGRAPHIC INFORMATION SYSTEM

Bultman and Getting (1991) and many others have demonstrated the value of integrating remote sensing and Geographic Information System (GIS). Recent advances in computer technology "Digital image data" now permit inferencing of images in a completely digital managing setting. In the time to come, computerized data handling will take advantage of the advancements in information technology for GIS (Bonham-Carter, 1997). Remote detecting items are appealing for GIS database improvement. Coverage of wide area can be made economically in digital form within GIS environment. Data collected in raster format in remote sensing can be conveniently transformed to vector format for the purpose of GIS modeling applications. Resolution is the major problem with the satellite imagery data. Through the recently launched satellites, the spatial resolution of images has improved a lot.

In order to detect surface manifestation of structural features, use of a variety of sensors and resolutions has been studied. Rowan and Bowers (1995) states "delineation of structures that controlled tertiary precious metal mineralization in Nevada and California was made possible from the combined spatial resolution of Landsat TM and side looking airborne radar (SLAR) images."

4.10 REMOTE SENSING VERSUS AERIAL PHOTOGRAPHY/PHOTOGRAMMETRY

By measuring the EMR from airborne system, remote sensing and aerial photography acquire information on Earth's upper surface. However there are differences between them as given below:

Remote Sensing	Aerial Photography/Photogrammetry
1. Images are taken from satellites	Usually taken from air planes
2. Images gathered by a digital CCD camera	Photos obtained using analog camera, converted to digital mode through scanning
3. In CCD, radiation reaching the sensor is measured quantitatively	Film has high resolution
4. Linewise image generated, therefore geometrical correction is complicated	Whole picture is taken at one instance, hence is a central projection
5. Devised for gauging radiation in electromagnetic spectrum	Gathers data in the visible spectrum
6. Images are affected by absorption	Air photos are affected from haze
7. Using atmospheric corrections, analysis of inbound electromagnetic spectrum	Accurate 3D model generation for boundary location plot of objects and generation of Digital Elevation Model
8. Images problematic to process and require skilled people	Air photos can be interpreted more easily
9. Quite beneficial for following happenings on worldwide scale. Satellites cover an extensive stretch and acquire all time images and revisit the same place at fixed interval	----------------------

4.11 REMOTE SENSING AND MULTISPECTRAL IMAGING

With higher resolution imageries available every year, remote sensing has been used as a standard first step for a well-planned exploration program. Depending upon the terrain constituents, satellite imagery maps only the superficial ground cover whereas higher resolution aeromagnetic maps unseen or subsurface lithology, actual depth, etc. Hyperspectral remote sensing uses reflected shortwave IR light to map alteration of rocks even seeing through the cover of trees. Scanners can map large areas (>3000 sq. km per day) at a resolution of 10 pixels, thereby reducing unit costs.

Satellite images have been used by geoscientists to serve as databases from which many things can be done as given below:

- Rock units, structural details (folds/faults) can be picked up
- Mapping subregional surface geology; advantage of creating large-scale area maps allowing examining in single scene or in mosaic
- Study of landforms and evaluation of dynamic changes in natural events (floods, volcanoes, etc.)
- Many a times, imageries function as a visual base on which geologic map is drawn either directly or on the overlay
- Valuable data for the planning an exploration program can be created by using a well-collated and structured "database," integrated into a powerful GIS
- To plan, manage, and monitor mineral exploration program, remote sensed data and GIS are essential for accurate information in a region.

4.12 REMOTE SENSING VERSUS SONAR

Examining the depth of the ocean floor by SONAR can likewise be considered as remote sensing. SONAR uses sound waves and not EMR hence it is an active type of remote sensing. Sonar and remote sensing frameworks transmit waves through a meddling medium (water, air) that adds noise to the information. Hence the raw data collected need corrections. Radar is thought to be climate reliant and environmental turbulences influence passive remote detecting. Both the systems depend on calibration from field data to make necessary corrections (viz., salinity, temperature, pressure measured by ship, or measurements of atmospheric profile using "radiosonde"). Bathymetry of the sea is produced by using SONAR while remote sensing focuses more on identification of material properties.

Point vector data containing XYZ are created both by Eco-sounders (single, multibeam) and Airborne Laser Scanning. This data are handled with a specific end goal to eliminate noise. While managing bathymetry, an added complexity is for the tide correction. The aftereffects of the remote detecting investigation can be contrasted effortlessly with the field (aerial photos, maps, field measurements) though in Sonar, the underlying bottom of the sea is unseen.

4.13 REMOTE SENSING INDUSTRY—PRESENT TRENDS AND OUTLOOK

Main uses of remote sensing are for

- regional Exploration targeting and logistics, structural mapping, change detection, community, and environmental baseline mapping, R&D

- Increasing demand for high-resolution imagery, for time series, changes monitoring, Exploration
- Main data used are IKONOS, Quickbird, ASTER, Landsat, SPOT
- Main image processing software tools used are ENVI/IDL, ERDAS Imagine, PCI Geomatics, ERMapper, IDRIS, Ecognition, TNTMIPS, Adobe Photoshop, plus a variety of custom in-house software.

Remote sensing would increasingly play its continuous role for the management of natural resources. The technical capabilities of sensors, space platforms, data communication systems, GPSs, digital image processing systems, and GISs are improving on almost a daily basis. At the same time, we are witnessing an evolution of various remote sensing procedures from being purely research activities to being commercially available services. Most importantly, we are becoming increasingly aware of how interrelated and fragile the elements of our global resource base really are and of the role remote sensing can play in inventorying, monitoring, and managing earth resources and in modeling and helping us understand the global ecosystem.

Various tools of remote sensing exist as on date to an exploration geologist. With the help of GIS technology, suitable combination of different layers of images, computer data handling and management, spectral study, mapping using remote sensing techniques, ground survey, analysis, modeling and presentation are integrated to provide a powerful decision-making tool. In mineral exploration, search for mineral deposits is based on certain scientific principles and methods with an integrated involvement of geological, geochemical, and geophysical techniques for an understanding of favorable setting within which a mineral deposit is likely to occur. GIS with its flexibility of experimentation and capability to extract topological attributes from maps works as unique tool to derive mineral potential maps and assist in fixing exploration priorities.

Demand for natural resources continues to increase with the ever increasing population and industrialization of many nations. Spread of population centers and increasing pressure for environmental sustainability have forced the search for economic mineral deposits into more remote regions. Above issues compel for growing demand of remote sensing in geological investigation. With the improvement of technology, it is hoped that the thought power of geologist to acquire further comprehensive evidence on application of remote sensing will also advance in the days to come.

References

Bonham-Carter, G.F., 1997. GIS methods for integrating exploration data sets. In: Proc. Exploration'97, Geophysics and Geochemistry of the Millennium. Geol. Surv. of Canada, pp. 59–64.

Bultman, M.W., Getting, M.E., 1991. Integration and visualization of geoscience data applied to the search for concealed mineral deposits. In: Proceedings of the Eighth Thematic Conference of Geologic Remote Sensing, Environmental Research Institute of Michigan, Ann Arbor, Michigan, vol. 2, pp. 1323–1324.

Deutsch, M., Wiesnet, D.R., Rango, A. (Eds.), 1981. Satellite Hydrogeology. American Water Resources Association, Minneapolis, MN.

Drury, S.A., 1986. Remote sensing of geological structure in European agricultural terrains. Geol. Mag. 123, 113–121.

Drury, S.A., 2001. Image interpretation, Geology, 3rd edn Allen & Unwin, London.

Goetz, A.F.H., Rowan, L.C., 1981. Geologic remote sensing. Science 211, 781–791.

Levin, N., 1999. Fundamentals of Remote Sensing. Remote Sensing Laboratory, Geography Department, Tel Aviv University, Israel, p. 225.

Lipton, G., 1997. Spectral and microwave remote sensing: an evolution from small scale regional studies to mineral mapping and ore deposit targeting. In: Proc. Exploration'97, Geophysics and Geochemistry of the Millennium. Geol. Surv. of Canada, pp. 43–58.

Peters, E.R., 1983. The use of multispectral satellite imagery in the exploration for petroleum and minerals. Phil. Trans. R. Soc. Lond. A 309, 243–255.

Rowan, L.C., Bowers, T.L., 1995. Analysis of linear features mapped in Landsat Thematic Mapper and Side-Looking Airborne Radar Images of the Reno 1 by 1 quadrangle, Nevada and California: Implications for mineral resource studies. Photogr. Eng. Remote Sens. 61 (6), 749–759.

Sabins, F.F., 1997. Remote sensing. Principles and Interpretation, 5th Edition Freeman, New York.

Sabins, F.F., 1999. Remote sensing for mineral exploration. Ore Geol. Rev. 14, 157–183.

Schowengerdt, R.A., 1983. Techniques for Image Processing and Classification in Remote Sensing. Academic Press, New York.

Taranik, D., 2009. Remote Sensing for Mineral Exploration. Irish Assn. for Econ. Geol. and Joly Geol. Soc., Trinity College, Dublin, p. 75.

CHAPTER

5

Geophysical Exploration

OUTLINE

Essentials of Mineral Exploration and Evaluation
DOI: http://dx.doi.org/10.1016/B978-0-12-805329-4.00012-0

5.1 INTRODUCTION

The primary concern of geophysical exploration is associated with detecting changes in geological conditions (from a distance) that may be directly or indirectly related to economic mineral deposits. To make this possible, there must be a change in some physical properties directly associated with the change in geological conditions. The "volume" exhibiting physical property change (contrast) must possess a size, shape, and location in some coordinate system with respect to the system of measurement, ie, "geometry." The volume exhibiting a physical property contrast is acted upon by some natural or artificial "field" and, if the physical property contrast and the geometry are adequate, an observable "anomaly" (a statistically significant departure from normal values) is caused in the applied field and the volume can be detected (Rogers, 1973).

The geophysical methods, the basic physical property or properties used, and the type of applied field are given below:

Methods	Physical Properties	Types of Fields
Gravity	Density	Earth's natural gravity field
Magnetic	Magnetic susceptibility; natural remnant magnetism	Earth's natural magnetic field
Electrical	Electrical conductivity	Earth's natural telluric currents
Self-potential	Electrical/thermal conductivity	
Resistivity	Electrical conductivity	Earth's resistance
Eletromagnetic	Electrical conductivity and inductance	Response to applied electric or electromagnetic field
Induced polarization	Electrical capacitance	Polarizability voltage or frequency-dependent ground
Radar	Dielectric constant	Travel times of reflected radar pulses
Seismic	Seismic wave velocity (elastic moduli and density)	Artificially created seismic waves
Radiometric	Radioactive decay	Natural radioactive radiation
Thermal	Thermal conductivity	Earth's natural thermal conductivity

The final result of geophysical measurements usually is a map or profile, portraying the variation of the measured field in space. Geophysical measurements are taken at points (stations) whose location must be known with sufficient accuracy (1) to indicate how the measured quantity changes with location, (2) to determine certain corrections, and (3) to enable anomalous areas to be relocated at any time for further exploration work. Stations can either be in a regular grid or an irregular network and the station spacing is determined by the problem at hand, ie, the expected size of the sought after feature or condition. In airborne surveys, data are continuously recorded along profiles.

TABLE 5.1 Geophysical Methods and the Corresponding Targets

Geophysical Method	Direct and Indirect Targets
Gravity	Mapping of sedimentary basins, geological structures like buried channel, folds, faults, cavities, voids, etc.; high-density ores and minerals like chromite, manganese, barites, sulfides, etc., and low-density minerals like coal, lignite, etc.
Magnetic	Mapping of igneous, sedimentary, and metamorphic units, geological structures; detection of buried pipelines, caves, tunnels, groundwater, etc.
Electromagnetic	Sulfides. Some oxides, mapping of conductive formations, geological structure, groundwater, engineering, environmental problems
Self-potential	Sulfides, graphites, formational contacts, seepage, etc.
Induced Polarization	Sulfides, some oxides, disseminated ore (porphyry type), groundwater, and environmental problems
Seismics	Mapping of sedimentary formation (coal beds), structures, bedrock and basement rock, geotechnical and environmental problems
Radiometric	Radioactive minerals, geological formations, structures, environmental problems
Thermal	Sulfides, geological formation mapping, geothermal springs, environmental problems

5.2 GEOPHYSICAL METHODS AND TARGETS

The efficacy of geophysical techniques in mineral exploration depends not only on the contrasts of physical property but also on local geological environment, topography, etc. Each area tends to have its own distinct geophysical identity necessitating the adoption of appropriate geophysical strategy. Usually, more than one technique is employed in one survey, to reduce ambiguity in interpreting the nature of unknown resources in the subsurface. Geophysical targets include ore bodies, bedrock top, saturated zone, etc. With the advancement of technology in geophysical instrumentation, data processing, etc., the geophysical methods are used in diverse fields, viz., engineering, groundwater, environment, etc. The geophysical methods and the corresponding targets for which they are employed, either directly or indirectly, are listed in Table 5.1 (Ramakrishna, 2006).

5.3 CHOICE OF A TECHNIQUE

Rapid advances in geophysical instruments in recent decades have transformed exploration, making it possible to identify potential ore bodies without the long process of ground-based exploration that may take many months. A number of techniques are available for geophysical surveys for quick appraisal of airborne anomalies and also for prospecting and detailed exploration work. Some of the important methods are Induced Polarization (IP), both Frequency & Time Domain (all with dipole–dipole arrays), Transient Electromagnetics (TEM) and Gravity. While TEM has been successfully used for major discoveries of massive sulfides, the IP technique is most applicable in disseminated copper sulfides.

The geology and the size of the project area, the likely depth and size, accessibility to the terrain, the survey purpose, and available budget dictate the choice of the geophysical tools. The above-mentioned details will help to create a geological model to estimate the target response and consideration of economics of various survey methods. Invariably more than one technique is used. Integrated interpretation using multiple survey method is often used since finding the deposits is becoming more difficult.

Various Geophysical Exploration survey methods (both airborne and ground) are summarized later.

5.4 GRAVITY TECHNIQUES—GRAVITY GRADIOMETRY, GEODESY, MICROGRAVITY SURVEYS

In the earth, gravity measurements define anomalous densities which are measured at different points by the ground-based gravimeters. Source body gravity variations are found by gravitational anomalies calculated by deducting the regional gravity field from the measured field. Shallow high-density bodies are associated with positive gravity anomalies whereas shallow low-density bodies are associated with gravity lows (Fig. 5.1). Gravity highs are yielded by deposits of high density (chromite, hematite, barite, etc.) whereas gravity lows are yielded by deposits of low density (halite, weathered kimberlite, etc.). This technique facilitates to predict the total anomalous mass (ore tonnage) responsible for anomaly. Gravity methods detect lateral contrasts in density. Vertical as well as lateral contrasts of resistivity and velocity could be detected by seismic and electrical methods.

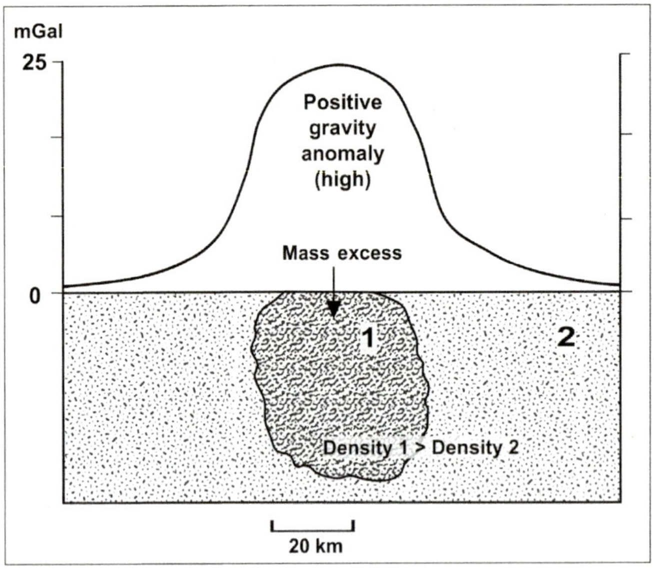

FIGURE 5.1 Sketch showing an excess mass in the crust locally enhances the gravity field producing gravity high (positive anomaly).

Several types of gravity surveys are available and the choice depends upon the purpose. In order to map deep geological structures, airborne gravity surveys are more appropriate. More details of the geological structures closer to the surface are provided by gravity gradiometry surveys. The targets that have been detected by airborne data can be further delineated in detail by ground gravity surveys. For larger survey areas, high-resolution gravity information in remote areas can be acquired by "Full Tensor Gravity (FTG) gradiometry" which has almost 10 times much higher efficiency than ground surveys. Identification of lithologies, structures, and ore bodies themselves is possible by applying gravity methods. Underground workings, unless at shallow depth, are not easily detected by gravity surveys.

The structures and rock types serving as spatial and temporal controls of mineralization usually give rise to fairly distinctive geophysical responses. Horsts and grabens are marked by diagnostic gravity highs and lows. Greenstone belts give rise to positive gravity anomalies compared to surrounding rocks. Magmatic bodies and faults respond to both gravity and magnetic methods. Fractures, shear zones, and the likes are routinely mappable by magnetic and EM methods, either from ground or from air. As they are zones of weakness, they are recognizable as electrical conductors in EM survey because of the presence of electrolytes, clays, and graphite. In many cases, the airborne EM survey brings out a swarm of geological conductors posing a serious problem of distinguishing between the barren shears/fracture/fault zones, and the ones that are most likely to carry economic mineralization. The discrimination is often resolved by the supporting evidence of other geophysical investigations carried out in the same area (Ramakrishna, 2006). Rajasthan state in NW India is an important metallogenic province hosting about 85% of base metal occurrence. The gravity and magnetic signatures on a regional scale (1 km interval) in Pur-Banera, Rajpura-Dariba, Khetri and Rampura-Agucha belt areas are shown in Fig. 5.2. All the above mineral belts are located on the flanks of gravity highs and are associated with sharp magnetic signatures of a few hundred "nT," suggesting a tectonic connection envisaged. A more detailed profile across Rampura-Agucha supergiant deposit is shown in Fig. 5.3, wherein the flank of the gravity high is further resolved. The zinc–lead mineralization is brought out in the form of gravity high of about "half a milligal." The conductors that fall in the vicinity of structurally favorable locales endorsed by gravity and magnetic necessitate a detailed follow-up investigations (Gandhi, 1985).

5.5 MAGNETIC TECHNIQUES

The earth's magnetic field varies from 0.35 oersteds in a horizontal direction at the magnetic equator to 0.65 oersteds in a vertical direction at the magnetic poles. Rocks forming ferromagnetic susceptibility become magnetized in the direction of the earth's field and some rocks exhibit permanent or remnant magnetization that may be independent of the earth's present field. The presence of such rocks distorts the normal magnetic field in their vicinity. Minor variations in magnetic mineralogy (magnetic oxide and iron-titanium oxide minerals) among rocks are exploited in magnetic surveys.

The intensity of earth's magnetic field can be measured to an accuracy of a few gammas by varieties of magnetometers, viz., fluxgate, proton–precession, overhauser that are available for exploration survey. A map/profile is prepared showing how magnetic intensity

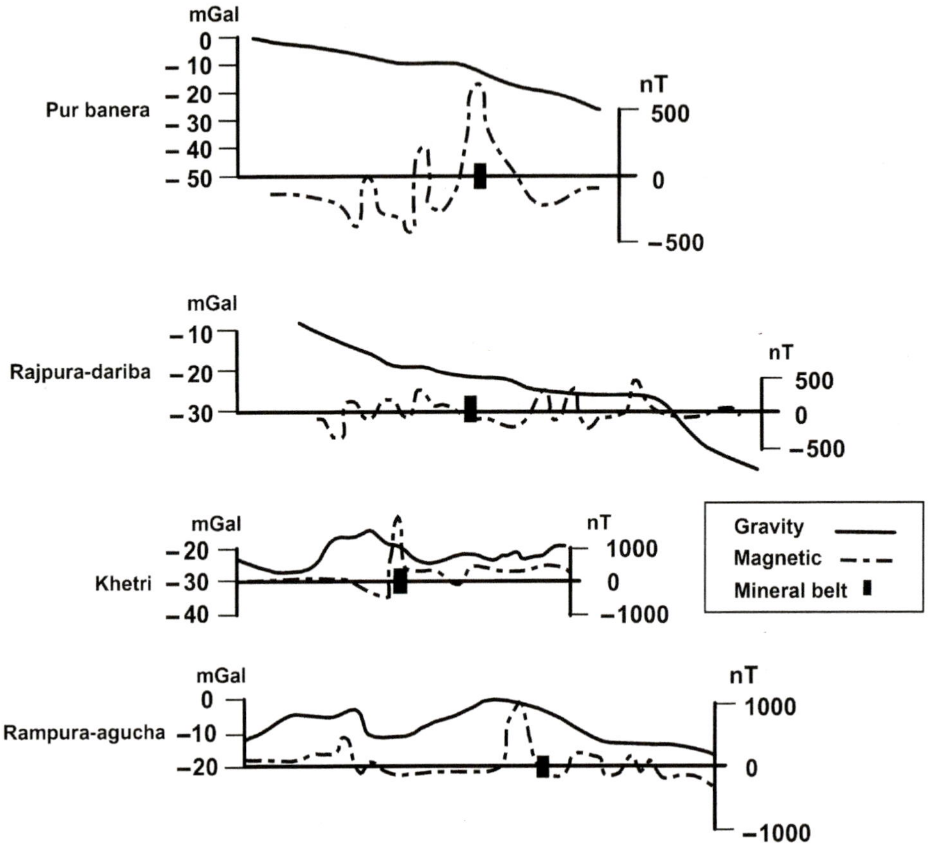

FIGURE 5.2 Gravity and magnetic signatures over known mineral belts in Rajasthan, India. *Source: From Gandhi (1985).*

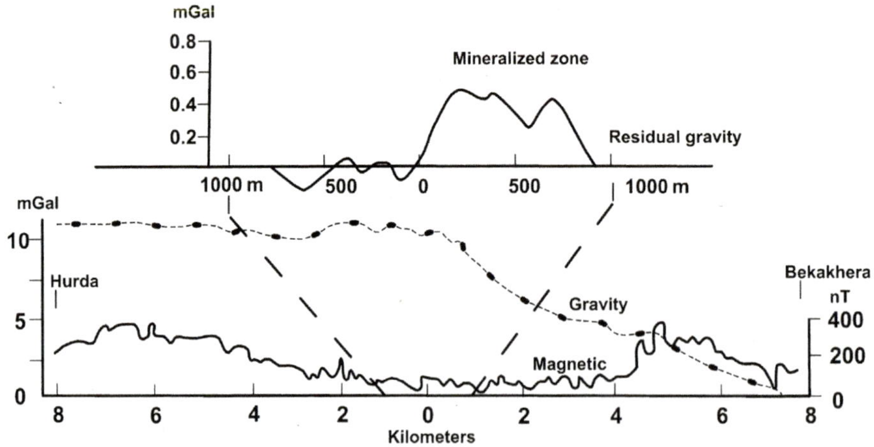

FIGURE 5.3 Gravity and magnetic profiles over Rampura-Agucha mineral belt, India. *Source: From Gandhi (1985).*

varies over the area being surveyed and this is the basis of interpreting the probable distribution of magnetic rocks in earth's upper crust. Earth's primary field is perturbed by rocks that are magnetic and contains various combinations of induced and remnant magnetization (Reynolds, 1997). The quantity, composition, and size of the magnetic mineral grains are dependent upon the magnitudes of both induced and remnant magnetization. Primary igneous/sedimentary processes that establish the magnetic mineralogy have direct bearing on the magnetic anomalies. Magnetic and radiometric profiles over magnetite pitchblende mineralization in Labrador, Canada, and topographic, magnetic, and HeliEM response data from a disseminated sulfide deposit are shown in Fig. 5.4. The secondary effects in rocks that host ore deposits associated with hydrothermal systems are important in mineral exploration and geoenvironmental considerations.

Magnetic surveys detect changes in magnetic concentration of near-surface rocks and are low cost method. They are also used to map the underlying geology and depth of basement of large areas. The magnetic anomaly helps to estimate the size, shape, strike, and depth of the ore body. For geophysical targets that are electrically conductive, EM surveys work well. Mapping of iron-bearing rocks with magnetic method gives a clue of magnetic history and lithology could be differentiated based on remnant direction.

The geophysical signatures (magnetic, gravity, conductivity, and electromagnetic) of Armstrong B massive sulfide deposit in Bathurst, NB, Canada, are shown in Fig. 5.5. In Armstrong deposit, Cu, Pb, and Zn often occur in highly conductive lenses of massive

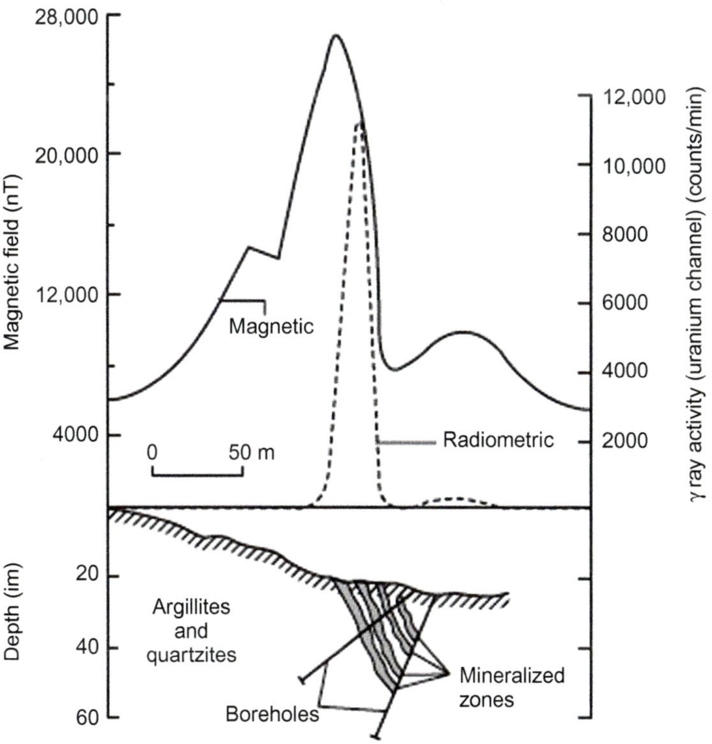

FIGURE 5.4 Radiometric and magnetic profiles over pitchblende–magnetite mineralization in Labrador. *Source: From Telford et al. (1990).*

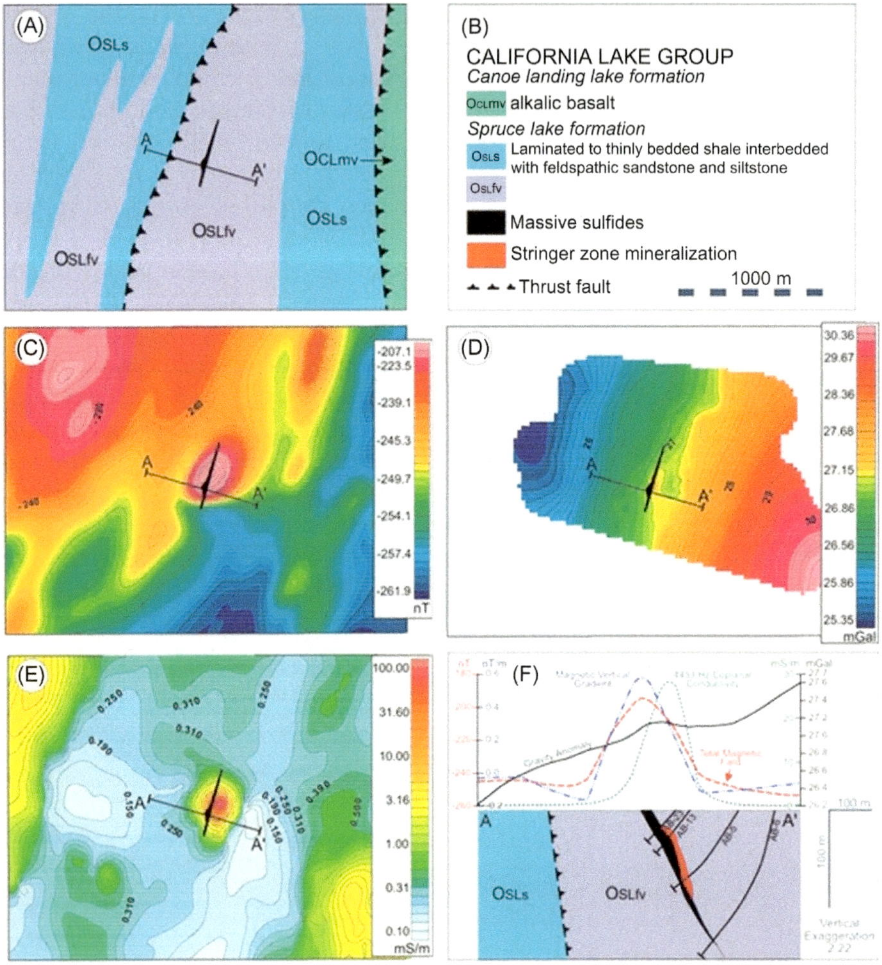

FIGURE 5.5 Armstrong MS Deposit, Bathurst, NB, Canada: (A) geological map, (B) geol legend, (C) magnetic field, (D) gravity anomaly, (E) conductivity, (F) total magnetic gradient, gravity anomaly, conductivity. *Source: From Ford et al. (2007).*

sulfides, which are ideal EM targets. Massive sulfides have a positive density contrast and gravity can be useful in identification. Pyrrhotite is often associated with the deposits resulting in a magnetic anomaly. Pyrrhotite-rich graphite beds have a similar signature, hence volcanogenic massive sulfide (VMS) deposits are not that easy to detect.

Magnetic methods directly detect some iron ore deposits and are also useful for detecting subsurface structure/lithology, favorable tracts for deposits beneath surficial cover. Geoenvironmental applications include identification of magnetic minerals associated with ore/waste rock and their likely release. Mapping of structural features (folds/faults) by the application of magnetic methods is very useful and constitutes an excellent component to any survey plan.

5.6 ELECTROMAGNETIC METHODS

The theory of induction for measuring electrical conductivity of the subsurface is used in electromagnetic induction (EM) surveys. No ground contact is required unlike the conventional resistivity techniques. EM methods provide rapid and easy data collection and are employed as reconnaissance tools to identify anomalies for greater detailing. Electromagnetic methods (EM) utilize inductive coupling supported by resistivity methods. There are many hybrid systems available. In pure EM method, the ground is inductively energized by means of alternate current flowing in a transmitter coil. The signal containing ground response characteristics is deducted inductively by a receiver coil. Frequencies from a few 100 to a few 1000 hertz are used. EM methods are rapid and are widely used when shallow massive sulfides are expected because, unlike resistivity method, inductive methods duly utilize the high conductivity associated with sulfides. The EM signature over Sue A, C and D Uranium Deposit, Athabasca Basin, Canada, is shown in Fig. 5.6.

According to Ford et al. (2007), "Ground time domain EM techniques using large loops can detect conductive targets at depths in excess of 1 km."

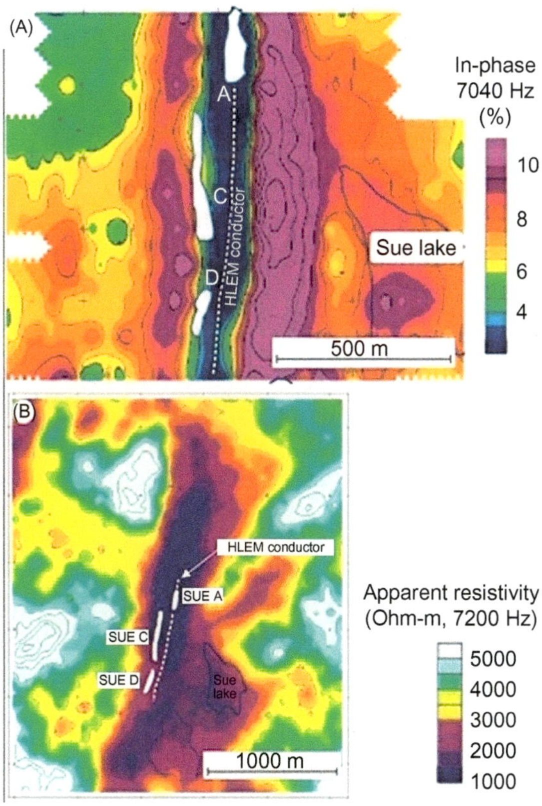

FIGURE 5.6 EM response, Sue A, C and D uranium deposits. Athabasca Basin: (A) MaxMin, 1–10 EM, in-phase and (B) Dighem (Airborne) FD resistivity. *Source: www.geoscan.nrcan.gc.ca/Natural Resources Canada/ Publications and Reports.*

5.6.1 Limitations

Despite being a multipurpose and resourceful technique, the electromagnetic method suffers from a few shortcomings. Besides the anomalies caused by economic resources with a high conductive ore bodies, many noneconomic resources such as graphite, water-filled shear zones, bodies of water, and other man-made features can also yield EM anomalies.

5.7 RADIOMETRIC (GAMMA RAY) METHOD—AERORADIOMETRIC SURVEYS

The technique of measurement of natural radiation in the earth's surface is called "radiometric method" (also known as Gamma ray spectrometry). This is a process in which the unstable atom becomes stable in the process of its nucleus. Energy is released in the form of radiation, viz., alpha, beta, and gamma rays. Radioactive uranium and thorium minerals occur naturally in earth materials. Since they are radioactive, their presence or any anomalous concentration can be detected by radiometric surveys (Fig. 5.3). Although alpha, beta, and gamma particles may be emitted, only gamma particles will penetrate a sufficient distance to be useful in exploration. Gamma rays have the highest energy and will penetrate several hundred feet through atmosphere but only a few inches of earth are sufficient to attenuate natural radiation by one half. Therefore, the radioactive elements or their daughter products must occur in outcrop to be detected. The energy of gamma ray is characteristic of the radioactive element, it came from. Gamma ray can be measured in the ground or by a low flying aircraft.

The concentration, spatial distribution of radioactive elements such as uranium, thorium, and potassium from the upper layer (10–60 cm) of earth's surface having no vegetation, is measured by radiometric surveys. The gamma ray detection indicates the abundance of these elements. The alteration zones mapping can be done by knowing the variation in the element concentration in the host rock. The presence of uranium can be detected directly by radiometric surveys. Greater cost of the radiometric surveying is compensated by the additional information gathered. Wherever the primary mineralization process is related to potash metasomatism (like tin, tungsten, and gold deposits), radiometrics is also used. Higher altitudes (above 300 m) may affect radiometric data negatively. Radiometrics cannot be used offshore or in perma-frost areas since water attenuates the radiometric signal.

In order to recognize the incidence of usual radioactive elements such as uranium, thorium, and potassium, scintillometry (gamma ray method) is used. Multichannel spectrometers are used to detect individual radioelement abundances. The method has application in exploration for uranium since it can detect directly. Among the three radioelements, "thorium" is the most immobile and resembles in behavior to that of zirconium. With increase in alkalinity in rocks (as in felsic rocks), thorium content increases like uranium content. Radiometric surveys can cover large areas and ideally suited for plane areas.

5.8 SEISMIC METHODS

Seismic methods depend upon velocities of acoustical energy in earth materials. Accordingly, they involve the generation of a short pulse of seismic energy and the permanent recording of the arrival of seismic pulse at distant locations, with time intervals after the pulse instant determined to millisecond accuracy. Some types of explosive or the impact of a mass furnishes the energy which is detected by sensitive seismometers operating with electronic amplifiers and a suitable recorder. Wave theory of travel time, reflection, absorption, diffusion, and refraction are applicable to seismic interpretation and quite different principles are involved than the "potential" methods which include gravity, magnetic, and electrical techniques.

Reflection seismology, seismic refraction, seismic tomography, 2D, 3D seismic to survey for deposits from 500 to 2000 m depth below the reach of conventional geophysical techniques are expensive (\$50000–\$70000 per sq. km), but the right tool for areas which have mining infrastructure.

5.9 ELECTRICAL TECHNIQUES

A wide variety of electrical methods have evolved, but all are dependent primarily on the physical property of earth materials known as electrical conductivity. Natural electrical fields occur in the earth due to normal telluric current flow, worldwide thunderstorms activity, and solar related magnetic storms. These energy sources are utilized in telluric, magnetotelluric methods. In "spontaneous polarization" (SP) method, natural electrochemical reactions near the surface of the earth where metallic luster minerals are present, are used. Despite the availability of more sophisticated techniques, SP surveys often form the first step in geophysical exploration and also find application in special problems like investigation for geothermal areas and to estimate the rate of seepage in reservoirs, landslides, and in several environmental and engineering problems. A special conductivity-related phenomenon which is the basis for the IP method occurs when there are mixed methods of conduction—both ionic and electronic—in a volume of earth materials. In such cases, the ground will conduct alternating current better than direct. The electrical conductivity of mineralized ground then will change with the frequency of the applied current, while the conductivity of barren ground will be constant while the frequency is varied.

A multiplicity of separate techniques of electrical methods are available that employ differing instruments and procedures. These techniques have variable exploration depth and lateral resolution. In spite of all the variants, measurements fundamentally are of the earth's electrical impedance or relate to changes in impedance. The electrical methods described hereunder are: (1) direct current resistivity, (2) electromagnetic, (3) mise-a-la-masse, (4) self-potential, and (5) induced polarization.

5.9.1 Direct Current Resistivity Method

Electric conduction in minerals takes place in three ways, viz., electronic conduction, dielectric conduction, and electrolytic conduction. Electronic conduction takes place in metals, dielectric conductance in insulators, and electrolytic conductance in aqueous solutions. A small percentage of conducting minerals in a rock matrix considerably reduces its overall resistivity, provided there is continuity among mineral grains. Bulk resistivity of a rock is determined primarily by the electrolytic fluids contained in it. Hence it is the "porosity," the amount of water saturation and nature of concentration of salts in these fluids that govern the resistivity of a rock formation. The resistance that is offered to the flow of current between the opposite faces of a rock of a cubic meter is called resistivity or specific resistance of that rock.

5.9.2 Electromagnetic Method

To induce measurable current in earth, alternating magnetic fields are used. EM methods have been traditionally used in exploration of sulfide mineral deposits with low resistivity (high conductive). Very large areas can be screened by airborne EM methods to identify potential areas. EM methods are extensively used to map rock types and geological structures.

5.9.3 Mise-a-la-Masse Method

To assess the extension and geometry of a known ore body, the "mise-a-la-masse" method can be very useful electrical method. The method involves connecting a current electrode to an ore body intersected in a drill hole and another current electrode is placed at infinity on the surface. Similarly, one of the potential electrodes is also sent to infinity in opposite direction to current electrode while the other potential electrode is run along a rectangular or square grid lay out on the ground, covering the borehole and the adjoining area.

As in the case of resistivity method, the self-potential has to be bucked out before sending a current into the ground. As the current is sent into the ground, a proportionate voltage develops across the potential electrodes, depending on the subsurface conductivity. As the current may vary from station to station due to variable contact resistance, the readings are normalized by calculating the ratio of voltage to current. The resulting potential distribution when contoured broadly indicates the geometry of the ore body.

5.9.4 Self-Potential Method

Quantifiable self-potentials (SPs) are generated by several possible natural sources. There are four types of SPs, viz., (1) electrokinetic potential across a boundary between two rock media with differing streaming potential coefficients; (2) diffusion potentials generated due to differences in concentration of electrolytes in the ground; (3) adsorption potentials which arise due to the adsorption of positive and negative ions on the

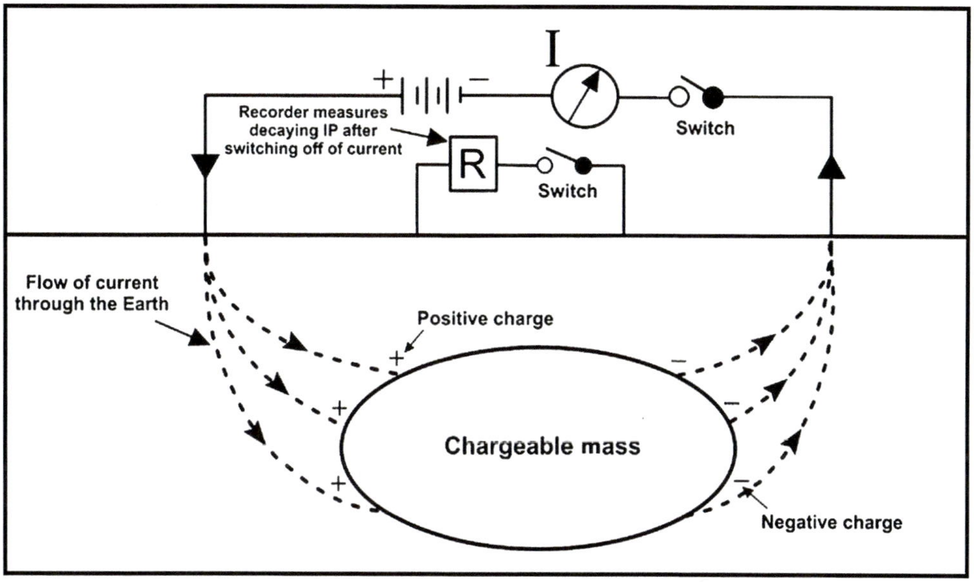

FIGURE 5.7 Schematic sketch map how Induced Polarization technique works.

surface of ores, veins, clays, etc.; and (4) mineralization potentials arising due to conducting minerals like sulfides, graphites, magnetite, etc. Gently or shallow dipping ore bodies may give rise to dipolar anomalies because of the proximity of lower end of the ore body to ground surface. In mineral exploration, closed SP anomalies of a few tens to a couple of 100 millivolts or those extending linearly over a limited strike length are generally considered favorable for follow-up, by other methods. The strike extent, width, and symmetry of the anomaly provide suggestive information on the ore body geometry and its disposition. One should be cautious because when closely spaced multiple bodies give a response which is similar to that of large and deeper ore body (Ramakrishna, 2006).

5.9.5 Induced Polarization Method

The IP method was developed to aid in exploration of disseminated electronically conducting mineralization that may not be detected by resistivity or electromagnetic methods. There are two measures in IP. In "time domain" (TD), the measure of polarizability is called chargeability which may be defined as the polarization voltage across a unit cube of material energized by unit current. The measurement of IP in TD involves charging the ground with a specific current for a specific period and recording decay voltage after the stoppage of the current. In "frequency domain" (FD), the resistivity of a rock measured with a high frequency is always lower than that measured with a lower frequency or DC, because due to dispersion, the magnitude of current increases with increase of frequency (Fig. 5.7).

5.10 THERMAL METHODS

Multispectral thermal infrared data enables identification of rock types by their silica content. The emission spectra of silicate minerals contain broad absorption trough caused by Si–O bonding. The interaction of "Si" with "O" determines the position and shape of this trough. With the increase in silica content in rocks, the absorption features shift to shorter wavelengths. This shift is a potential means of distinguishing various rock types. Thermal methods can be used for mapping the ground moisture, geological structure, geothermal reservoirs, underground fires in mines, etc.

5.11 REMOTE SENSING METHODS

The remote sensing methods make utilization of images acquired in the ultraviolet visible, near infrared bands of the electromagnetic spectrum. Thermal infrared observations are also part of remote sensing. Digital form of the remote sensing data can be processed conveniently. Iron hydroxide minerals, silica, clay alteration, etc. can be demarcated over large areas by comparing the known spectral responses of minerals/mineral groups.

5.12 BOREHOLE GEOPHYSICS (GEOPHYSICAL LOGGING)

The geophysical responses from targets buried in the earth decrease with increasing depth of burial of the body. If we can send the sensors close to the target, the response is enhanced. This is the basis of borehole geophysics, which provides a means of obtaining number of physical properties of subsurface rocks, ore bodies, sedimentary fluids and analyzing them for a much better understanding of the subsurface. Geophysical logging is essential in petroleum, groundwater, coal, potash exploration, where drilling is usually noncoring type. In mineral exploration, these logs are very useful as they can provide valuable information, otherwise lost in zones of poor core recovery and also where boreholes narrowly miss an ore body.

The basic principles involved in geophysical logging are the same as their surface counterpart. Probes/sensors that measure different physical properties are lowered through the borehole to collect data continuously or at regular intervals, that is displayed digitally and/or graphically as a geophysical log. Electrical logging cannot be done in the absence of borehole fluid and iron-cased holes cannot be logged by electrical, electromagnetic, and magnetic methods. In PVC cased holes, magnetic, EM logging is possible.

According to Dyck (1983), "Well logging is performed in boreholes drilled for the oil and gas, groundwater, mineral and geothermal exploration, as well as part of environmental and geotechnical studies" (www.finance.newsdemocrat.com). A summary of geophysical logging techniques is given below:

Types of Logs	Parameter	Remarks
Gravity	Density	Accurate *in situ* determination of densities
Magnetic	Magnetic field	Magnetic zonation
ELECTRICAL		
Resistance	Relative resistance	Simple measurement, qualitative analysis only true resistivities can be calculated; porosity, saturation, pore fluid resistivity can be determined. Resistive beds appear thinner
Resistivity	Apparent resistivity	
SP	Natural potential	Shale, nonshale discrimination, qualitative
IP	Chargeability	Can pick up disseminated ore mineralization, distinguish between clay zone and brackish water
EM	Conductivity	Used in dry holes, where resistivity fails; cannot pick up disseminated zones
RADIOACTIVE		
Natural gamma	Natural radioactivity	Clay shale zones easily distinguished, used in dry and cased holes also
Gamma gamma	Density	*In situ* density can be used only in dry holes

5.13 LITHOLOGY LOGS

Lithology logs can be obtained by many techniques. A sonic log delivers a formation interval time, which depends on with lithology and rock type and texture. Internal transit time is recorded as the time consumed by the sound waves to propagate through distance fixed between piezoelectric transmitter and receiver. In order to distinguish between sands and shales in a siliciclastic environment, natural radioactivity log (gamma log) of the formation along the borehole will be useful. Sandstones are composed of usually nonradioactive quartz whereas shales are naturally radioactive due to potassium isotopes in clays, and adsorbed uranium and thorium.

5.14 GROUND PENETRATING RADAR SURVEYS

Ground Penetrating Radar (GPR) is used for identifying the exact location of concealed mineral bodies and also to map the stratigraphy. Cross-sectional measurement of shallow subsurface is provided by this method. These surveys can attain a resolution of a fraction of a meter and effectively map fractures, cavities, sink holes, and also underground pipelines, cables, etc. In areas where GPR is effective, there is a better understanding of the geological structures as well as the physical properties that control the penetration and reflection of radio waves. Under favorable geological settings, GPR is unparalleled in the wealth of information it can provide (Annan and Davis, 1997).

This method has a great similarity with seismic and may be considered as a mini reflection seismic survey, using EM waves instead of seismic waves. Two suitable antennae when travel into the subsurface, get reflected from the objects in the subsurface and return to the receiving antenna. If the time taken for the pulse from transmitter to receiver is known, the depth of the reflector can easily be calculated.

5.15 VERY LOW FREQUENCY SURVEYS

Very Low Frequency (VLF) electromagnetic technology is used to examine the geoenvironmental problems in shallow, low conductivity sedimentary layers, in groundwater exploration and in locating the position of anomalous source bodies beneath the surface. VLF electromagnetic technology is used in the frequency range of 15–30 kHz. VLF utilizes the carrier waves of distant powerful communications transmission of military organizations also. A VLF transmitter when traversed across the earth's surface will receive high signals that are carried over the resistive portion of the crystalline rock mass. VLF methodology can also detect any linear conductive body besides water-bearing fractures. VLF techniques are particularly popular in their airborne version. This method can also be used successfully for searching of both conductive and resistive targets (McNeill and Lobson, 1991).

5.16 OTHER METHODS

Several traditional geophysical techniques are used in exploration for a range of mineral deposit types. Traditional geophysical techniques normally perform the detection of medium to small ore bodies at shallow to moderate depth. Some of such techniques include the piezoelectric method, ultraviolet laser-induced fluorescence (the luminex method, Seigel and Robbins, 1983), airborne sniffing (used for mercury exploration), and radon sensing.

Modern explorationists are working in an especially challenging time. Technology is expanding very quickly than ever. Occurrence of ore bodies is generally found associated with particular type of stratigraphic, structural, and tectonic environments in a geological setting. A careful study of the geological controls and associated geographic distribution spanning over large areas would be of aid in predicting deep-seated ore bodies. New discoveries thus require constant advancements and innovations in exploration together with development of new tools.

5.17 GEOPHYSICAL INVERSION TECHNIQUE

Geophysics is becoming an increasingly important discovery tool as exploration moves deeper into subsurface. The ability to rapidly iterate and combine data and constrain models within project timelines is essential for making geophysical modeling a more practical and reliable aid. As explorationists are compelled to look deeper in to the subsurface for minerals, geophysical inversion is emerging as a powerful tool for improving exploration success. Over the decade passed by, the technique has evolved from a complex, time-consuming process into a rapid and integrated approach for visualizing what lies beneath the surface. Today inversion is being applied at every stage of the exploration programme, from prospect generation to resource definition.

Inversion is a process whereby explorationists use geophysical data, what they know about a region or project and other existing data, to construct the most likely distribution of physical properties under the surface. The technique can extract more insight from geophysical data than traditional approaches, because geophysical measurements are converted into 3D images of the subsurface that can be integrated with geological observations. The process of 3D geophysical inversion aims to produce the most likely distribution of physical rock properties that validates what is observed. Advances in cloud computing and software technology over the years have made the technique accessible to any explorationist with an understanding of geophysics.

Geosoft has developed cloud-based geological inversion software ("VOXI Earth Modeling") that helps to generate 3D voxel models from ground or airborne gravity and magnetic data. VOXI has made large, multiparameter geophysical inversion modeling faster, more responsive and effective as an exploration tool to assist with predictive modeling for project generation and target delineation, as well as more advanced exploration. As exploration progresses to the advanced stage, the inversion process provides an increasingly accurate picture of subsurface geology and structural controls on mineralization. Whether generating new projects or finding new ore deposits around existing operations, geophysical inversion can save time and money by enhancing then 3D image of the subsurface and providing more insight into potential targets for follow-up and drilling. Inversion also helps to prove a geological concept that can be taken to another area that appears to have the same geological setting. It is a valuable tool for the geologist to have at every stage of exploration from grassroots to resource definition.

5.18 EMERGING GEOPHYSICAL TECHNIQUE

As most of the near-surface mineral discoveries have already been made, the future discoveries are likely to be from depths greater than the present day deposits and might probably covered by thick layer of overburden. It is necessary that new sophisticated exploration technology is required to find these deposits than the present day conventional methods. Recent innovations in airborne and downhole EM, use of audio magnetotellurics (AMT), and adoption of 3D-Seismics are the steps in this direction. The earth's electrical resistivity structure from depths of a few 100 m to several 100 km could be imaged by using MT survey, which has been found to be cost-effective, rapid, and accurate for identification of mineral zones. MT surveys are employed for (1) large area target potential evaluation, (2) targeting of Titan surveys, (3) basin mapping (depth of cover), (4) shallow crustal scale evaluation for hydrocarbon and geothermal exploration, (5) deep crustal regional scale structural mapping, and (6) resistivity logs for lithological determinations.

5.18.1 Magnetotelluric Technique

The MT technique is a development of the telluric method which exploits certain natural earth currents (telluric currents) that propagate as sheets over vast areas on the earth. Their cause is attributed to several factors like the rotation of the earth, ionospheric currents, tropical storms, etc. These currents are interrupted and modified by large-scale natural electrical discontinuities like major geological contacts, folds, faults, shear zones, etc. MT has become one of the most important tools in deep Earth research. MT method can resolve geoelectric structure from depths of tens of meters to depths of tens of kilometers depending upon signal frequency and resistivity of material being studied. Hence, depth interpretation based on MT data is much more definitive than that based on gravity or magnetic data (Vozoff, 1972).

The low frequencies employed in MT enable larger depth penetration of several kilometers and also the resistivity of the ground obtained will be close to DC resistivity. With the fast development of technology of low noise sensors and signal processing acquisition of data up to 500 Hz is now possible. Hence its use in exploration of deep-seated mineral deposits becomes viable as well as attractive. MT has been found by many operators to be a cost-effective way to enhance an exploration program.

5.18.1.1 Controlled Source Magnetotelluric (CSMT)

An electrical dipole (an artificial source) connecting directly to the ground at a certain distance from the receivers is used in CSMT. This creates a situation similar to an EM plane wave traveling downward from the sky, in which the measurements at a certain frequency are related to the skin depth of the EM field (www.detectation.com). Natural electromagnetic waves that are generated in the earth's atmosphere by a range of physical mechanisms are being measured in CSMT. Natural electromagnetic waves decay at a rate dependent upon their wavelengths, as they travel into earth's interior (Yamashita, 1984) (www.geoservices.co.id). CSMT allows rapid reconnaissance of areas while detecting conductive zones up to 2000 m deeper. MT surveys are also employed for large area target potential evaluation and shallow crustal as well as deep crustal regional scale structural mapping (data inverse modeled to even 25 km depth). For locating future mineral deposits, combined use of 3D geophysical inversion and 3D geological models is essential.

5.18.1.2 Audio Magnetotelluric (AMT)

Instead of expensive traditional diamond drilling and borehole geophysics, AMT, a shallow penetrating EM technique, can be used as an alternative. In searching for new deposits of both brownfield and greenfield exploration, MT and AMT are being used increasingly by major and junior companies. This technique has logistic simplicity (portable by backpack to anywhere) which reduces cost and increases productivity. The installation of this system is environmentally benign and the technique is practical anywhere. Unlike seismic and IP survey methods, meaningful profiles could be constructed due to flexible site location and offline sensitivity. The direction and relative strength of offline rigid grid of conductors are indicated by "induction vectors" and are useful especially where the surface is resistive or frozen. The great advantage of MT/AMT survey is that it can penetrate zones of thick conductive clay that covers many prospective sites. Other airborne or surface technique cannot penetrate thick conductive clay zones (Quantec Geoscience, 2011).

5.18.1.3 Deep Rapid Reconnaissance and Detailed Follow-up

MT allows rapid reconnaissance of large areas, while detecting conductive zones from near surface down to 2000 m and more. Continuous picture of subsurface resistivity structure and high lateral resolution data may be redundant by placing close-spaced stations along lines. In order to keep the cost minimum, a "two-pass" methodology can be followed wherein station and line spacing are as wide as possible in the "first pass." A "second pass" with more stations at closer spacing can be done to increase resolution, once the areas of interest have been identified. This way, cost-effective identification of conductive mineralized zones could be done rapidly and accurately. This trend of two AMT sites adopted by INCO in Sudbury, Ontario, Canada, sensed a 1750-m-deep nickel deposit. Following similar method, Falconbridge also could locate two Ni–Cu mineralization zones (at ~800 and ~1350 m) (www.phoenix-geophysics.com).

5.18.1.4 Advantages of MT Surveys

MT surveys have many advantages, viz., (1) suitable for both Greenfield and Brownfield exploration; (2) exploration from surface to kilometers deep (better than other surface or airborne techniques); (3) practical over geological composition from highly resistive to highly conductive cover areas; (4) light weight portable equipment with negligible environmental impact; and (5) exploration in any season.

In various countries in the world (Canada, United States, Australia, Chile, Argentina, PNG, Russia, Mexico, etc.) MT/AMT surveys are used for base metals exploration. In many parts, exploration for precious metals and diamond (kimberlite mapping) is also done. Titan 24 magnetotelluric survey carried out in Half Mile Zn, Cu Deposit, NB, Canada, and Kidd Creek deposit area, Ontario, Canada, are shown in Fig. 5.8. In shield areas, significant MT surveys have also been done. Surveys that can be done using this technique are (1) advanced deep penetrating Titan 24 technology for MT and IP and (2)

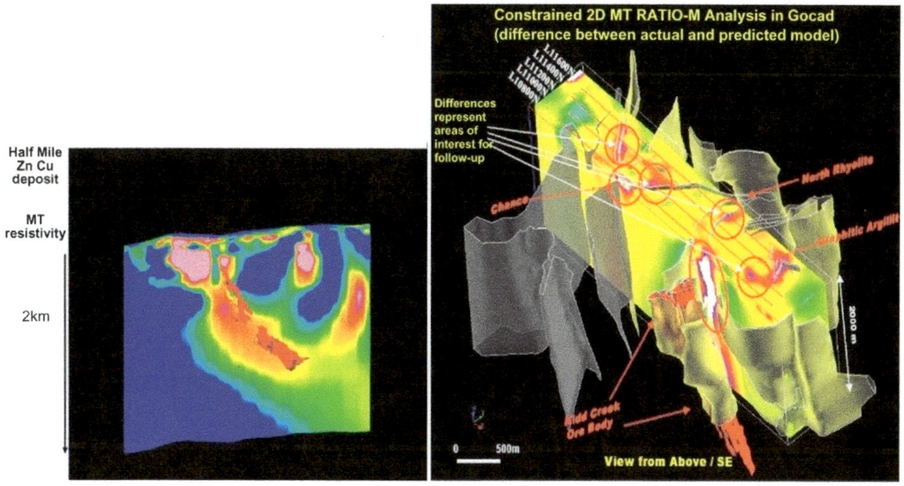

FIGURE 5.8 Titan 24 Magnetotelluric survey: (left) Half Mile Zn, Cu Deposit, NB, Canada and (right) Kidd Creek deposit area, Ontario, Canada. *Source: From Hollyer (2012).*

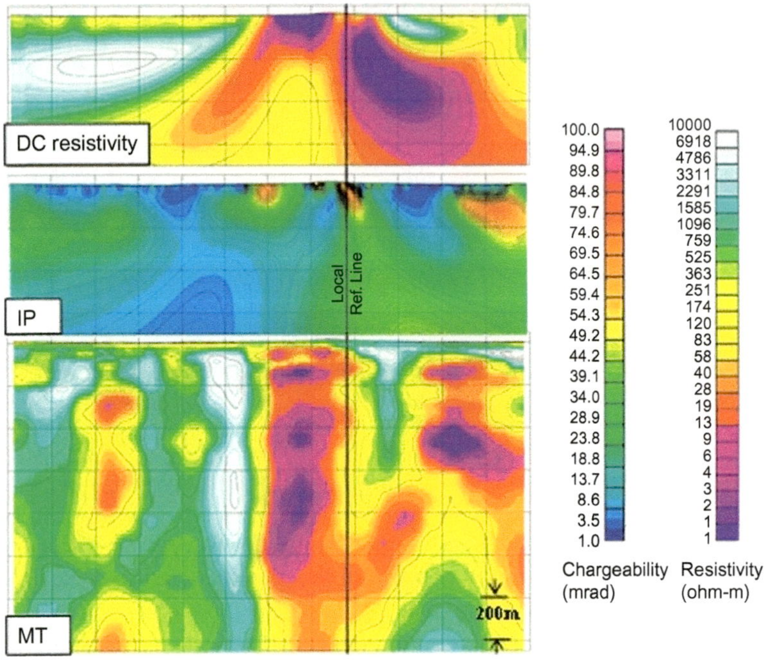

FIGURE 5.9 Inverted sections of DC resistivity, IP and MT along Line-5 in Pur-Dariba area (Titan 24 Survey, MT Section), Rajasthan, India. *Source: From Kavdia et al. (2015).*

Spartan MT surveys for regional and terrain scale exploration to depths of 30 km. MT surveys have been successful in locating deep metallic mineral deposits and elusive kimberlite pipes undetectable with other EM methods. MT survey (Titan 24) carried out in Pur-Banera Prospect, Rajasthan, India, sensed the buried copper ore bodies, which were drill-proved later (Kavdia et al., 2015). The Titan 24 survey MT section and delineation of buried ore bodies along with the drill section are given in Figs. 5.9 and 5.10.

Titan 24 IP & MT surveys look deep, map mineralization, alteration, structure and lithologies in three dimensions, enable explorationists to aim their drilling programs more effectively, and find the big one early. These surveys aid in deep crustal studies (deep transect over 300 km traverse, basin and range setting, northwest Nevada; Soundings distributed at 3–4 km intervals; Data inverse modeled to 25 km depth; Basin and range geological setting). Titan 24, EM/electrical system provides DC resistivity and IP chargeability sections up to a depth of 750 m. Titan 24 is a deep earth imaging system, maps the depth, and focuses on drill programs. Titan 24 IP & MT and Spartan MT surveys are effective in the discovery of VMS-NiS, Porphyry Cu, U, and Au deposits and enable explorationists to aim their drilling program more effectively. These deep penetrating surveys help for regional and terrain scale exploration up to depths of 30 km and also enable exploration to detect positive targets at depth and let you find the big one early.

Recent innovations in airborne and downhole EM, use of AMT, adoption of 3D-Seismics are the steps in this direction. AMT and 3D-Seismic systems are being increasingly used for locating deep-seated conductive mineralized bodies (+500 m depth) in already explored and developed mining districts. Recently, 3D Seismic system located a 15-Mt massive sulfide deposit at a depth of 1100 m in New Brunswick.

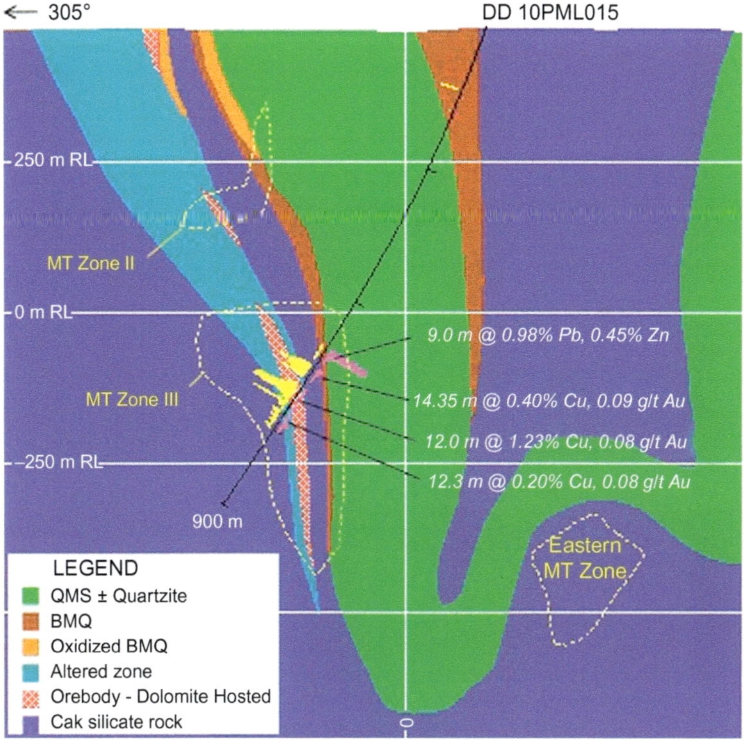

FIGURE 5.10 Interpreted Geological Section along Line 4 showing conductive MT zones and drill hole through Zone III, Pur-Dariba prospect, Rajasthan Histoplots: Yellow—Cu (Au) and Magenta—Pb + Zn. *Source: From Kavdia et al. (2015).*

5.19 AIRBORNE GEOPHYSICAL SURVEY

Airborne geophysical survey is an obvious choice for rapidly scanning regional target areas in a cost-effective manner. Airborne surveying is used by many exploration companies in order to ascertain the geophysical aspects of potential resource area. Depending upon the location and geology, there are a variety of ore deposits that can have similar or unique geophysical characteristics. Information on areas conducive to mineralization that may be difficult to access from the ground can be provided by aerial surveying (fixed wing aircraft/helicopter) in rapid manner (Fig. 5.11). Aeromagnetic, electromagnetic, radiometric, or gravimetric techniques may be used and each method has its own advantages. Certain methods may be better suited than others since different methods measure different physical properties. It also depends upon the geological conditions and the objective of the survey. In order to gather complete information set on one area, it is necessary to conduct more than one type of survey. In an overall exploration cycle, airborne geophysical survey is an important and expensive one.

The techniques which find application from an airborne platform include magnetic, radiometric, and electromagnetic. Gravity surveys are in an advanced stage of development and would find commercial application in the near future. Aeromagnetic and radiometric maps apart from being direct detection tools for specific targets are also useful in a

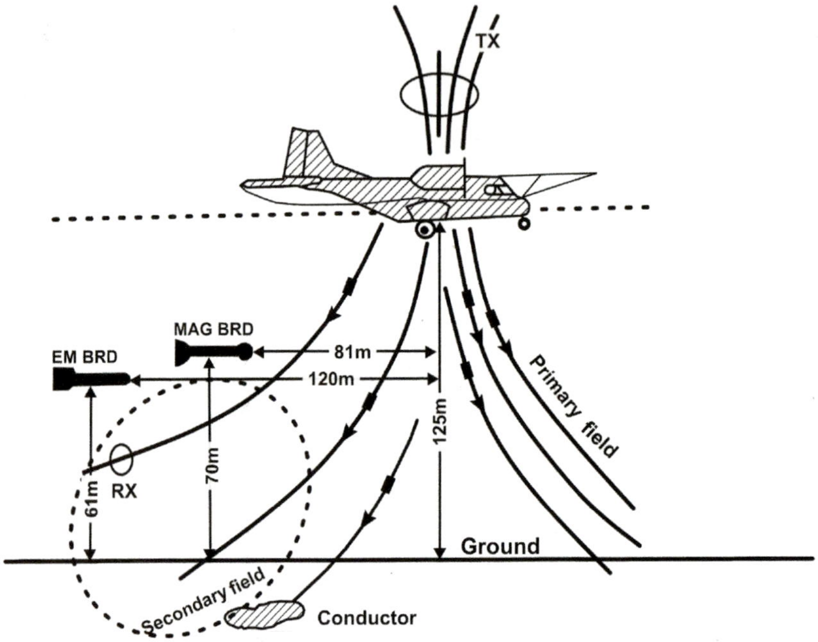

FIGURE 5.11 Sketch of a typical airborne EM survey system (time domain).

variety of exploration scenarios. They help to delineate lithology, distribution pattern of intrusive, major faults/folds and many other geological features, which cannot be covered by conventional geological mapping. High-resolution surveys at 25-m flight-line spacing with 25-m average ground clearance may effectively replace the ground magnetic surveys in future.

The choice of geophysical tool in airborne surveys is dictated by many factors, viz., the size of the prospect area, local geology, the physical characteristics of the target, and the surrounding rock types, the size and orientation of the target, purpose of the survey, location and access to the survey area, terrain, weather conditions and the budget, etc. Data resolution will be controlled by the height above the terrain, line spacing, flight speed, and sampling rate will be controlled by the data resolution. Different survey techniques will be used for different types of mineral deposits. For example, EM survey is better suited to VMS targets than porphyry targets. One should be cautious when using more than one type of survey technique since that can affect the acquired data.

The primary objective of airborne electromagnetic (AEM) survey is to explore the presence of conductive mineralization. EM survey penetrates deep down to detect mineralized bands, sulfide bodies, and other geological features depending upon the surface resistance. The minerals sought are usually massive accumulation of metallic sulfides having a direct relationship with the EM response. Indirectly, graphitic horizons or hydrothermal alteration zones, because of their relationship with mineralization as also conductive clays that result from weathering processes over kimberlites, are also potential targets for such

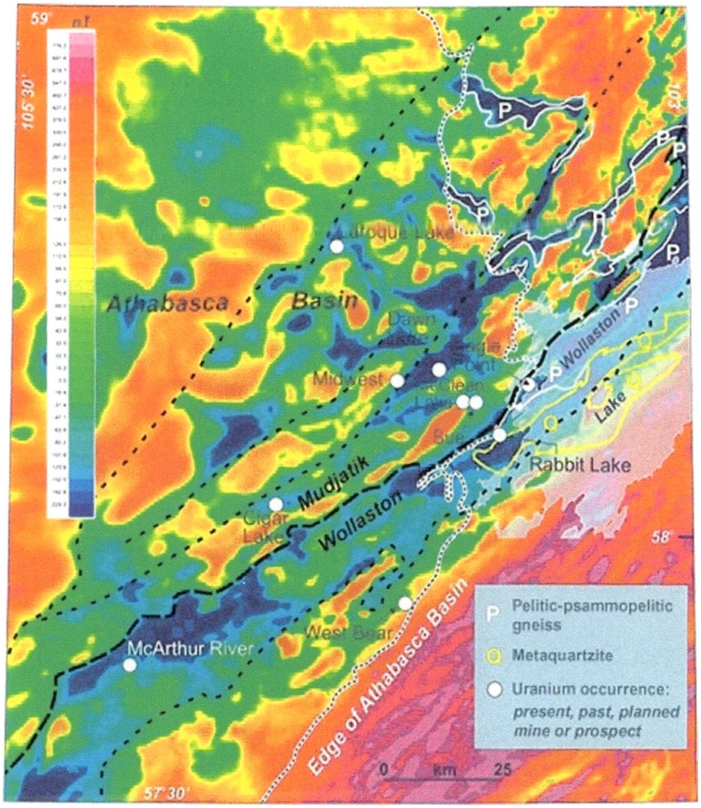

FIGURE 5.12 Aeromagnetic map (Elev. 305 m; LS 805 m) South margin of Athabasca Basin in Canadian Shield. Low magnetic field is in blue shades. Desired targets of "U" exploration, many deposits were discovered in the flanks of magnetic lows. (For interpretation of the references to color in this figure legend see the color plate.) *Source: From Ford et al. (2007).*

surveys. Some of the systems commercially available are GEOTEM, MEGATEM, SPECTRUM/QUESTEM TEMPEST, etc. These techniques have recently been extensively utilized for outlining deep-seated targets in Argentina, Australia, Brazil, Canada, Chile, Finland, Namibia, etc. An aeromagnetic map of south margin of Athabasca Basin (Canadian Shield) showing geophysical signatures is given in Fig. 5.12. In the Athabasca basin, in the 1970s, airborne spectrometry and radiometric boulder prospecting discovered several deposits in the shallower part of the basin. Subsequent recognition of relationship between mineralization and the basement rocks/faults, etc., led to discovery of basement conductors using airborne and ground EM surveys. Aeromagnetic surveys have mapped basement, critically outlining politic metasedimentary rocks, coinciding with a belt of magnetic lows (Ford et al., 2007).

Among other EM methods, time domain electromagnetic system (TDEM) has greater depth penetration. TDEM can refer to either surface or airborne EM system. The two helicopter systems of TDEM used in surveys are called SkyTEM and VTEM. Despite their differences in technical specifications, they both operate and produce data in a similar manner. EM transmitter and receiver loops are attached beneath a helicopter. It has been found that Heliborne TDEM could detect conductive material buried in the ground.

5.19.1 Advantages

Considerable expertise in system designs, maintenance, and operation are required for airborne geophysical technology which is expensive and complex. Some of the leading companies that provide services are: Bell Geospace, Gugro Geospatial, Goldak Airborne surveys, MPX Geophysics, New Sense Geophysics, Precision Geosurveys, Sander Geophysics, etc. Since many potential service providers are competing, a company that developed innovations in the survey technique or instrumentation gets an edge over others. For airborne surveying, multinational mining companies have maintained their own systems. But the companies which specialize only in "data acquisition" remain independent and work for multiple companies.

Airborne geophysical techniques have various advantages, viz., (1) they are cost effective to cover large areas rapidly and efficiently; (2) can cover areas inaccessible to ground-based work; (3) can provide evenly spaced data which helps in interpretation and minimizes environmental impact on the ground; (4) they are economical when presented with the large unexplored areas, to narrow down the area to be covered by detailed ground studies/drilling, etc.; and (5) information on regional structure/geology can be obtained to locate sites for detailed exploration.

5.20 HIGH-DEFINITION AIRBORNE GRAVITY GRADIOMETRY

"Gedex" has recently developed a new technology of High-Definition Airborne Gravity Gradiometry (HD-AGG), a new way of imaging Earth's subsurface. Image changes in the Earth's geology based on density. An elegant solution to density-based imaging is to measure minor variations in the gravitational field of the Earth from an aircraft, understanding of top layers of Earth by imaging down to a depth of 15 km by Gravity Gradiometer, and differentiate by density (since density is the prime indicator of commercial opportunity). HD-AGG, with its precision, speed, and depth of search, will significantly and efficiently find, delineate, and validate subsurface anomalies and potential resources. It has a high sampling rate (one image, every 60 m), builds up a detailed map essential for database management. HD-AGG technique brings in a new high precision data and ability to create insightful images of subsurface geology that can aid to integrated detailed exploration (Mining Magazine, 2013).

5.21 UNMANNED AERIAL VEHICLES

In the days to come, the unmanned aerial vehicles (UAVs) will be used more. The higher accuracy than conventional aerial photos is possible because of lower altitude of UAV, compared to aircraft. Despite this, for foreseeable future, most geophysical surveys will continue to be carried out by manned aircraft. In order to obtain a comprehensive interpretation of geological structures and targets, developing better methods to integrate multiple datasets will be the biggest challenge of the industry.

5.22 FUTURE TRENDS

Except for accuracy and improved resolution, it is not anticipated any great revolution in the industry. The Laws of physics are better known, hence, it is well known that no new technique could be developed despite the claims by some parties that they have developed something new which was not previously possible. Although there is a lot of opportunity to improve EM and gravity surveys, very less advancement can be possible in aeromagnetic and radiometric systems.

5.23 MARINE GEOPHYSICAL EXPLORATION SURVEY

Extensive source of living and nonliving resources are in the oceans which cover 71% of Earth's surface. Chemical elements like Na, Mg, K, Ba, Cl, Sr, and B, besides "salt" are important mineral resource in the sea water. Commercially, only Na, Cl, Mg, and Br are extracted from sea water in large scale. The interaction between the sea water with sediment deposition, biological activity, and magmatism related to specific geological environments controls the mineral resources in the oceans. The processes like precipitation, sedimentation, biological metabolism, diagenesis, and volcanic activity facilitate to enrich mineral deposits and ore formation in the oceans.

Marine geophysical surveys may be classified into (1) coastal surveys related to myriad engineering and environmental problems, (2) surveys to investigate structures and locate natural resources like oil and natural gas in the shelf areas, and (3) deep sea research projects. In the course of offshore exploration, mostly seismic methods are used. Besides, a wide spectrum of other geophysical methods was also developed within the last few years. The range of exploration extends from engineering problems to scientific investigation in the deep oceans. There are many limiting conditions between shallow and deep water geophysical methods. The selection of a vessel is one of the criteria, for near shore surveys small boats/ships are sufficient, but for open sea explorations, fix-mounted equipment, eco-sounders are needed. Marine exploration at sea starts with a topographic survey of sea floor. Depending upon the problem of investigation, geophysical and geological methods are available, which either can be applied parallel to the topographic survey (seismic, magnetic, sub-bottom profiling, etc.).

5.23.1 Exploration for New Resources

There are no straightforward methods available to predict the existence of natural resources below sea floor from magnetic anomalies. A chance for a direct proof may exist in case of placer deposits, where heavy mineral assemblage is often accompanied by magnetite and other magnetic minerals. Even in these cases, it is important to delineate source rocks magnetically. The hydrocarbon potential of an area is asserted only indirectly. Magnetic measurements give important clues to the assessment of hydrothermal deposits because they are clearly connected to the exhalation of igneous rocks. A direct connection between magnetic anomalies and the places where hot brines rise out of sea bed does not exist. Mn crusts/nodules cannot be found by measurements. There is no relation to the magnetic signature and composition of Mn nodules/crusts.

Depending upon the changes in the international scenario, marine deposits represent important resources for medium and long terms. Besides oil exploration, only a few other marine deposits like marine placers, gravel, sand, shells, evaporates, and phosphates are economically mined. Strategic marine mineral resources/reserves of deep sea deposits (polymetallic sulfides and nodules) demand a thorough investigation because they mainly occur in international waters. Sooner or later, the countries have to deal with and establish their policies on exploitation and mining these mineral resources.

5.24 SATELLITE GEOPHYSICS

The data on earth's gravity, magnetic field, surface temperature (thermal infrared) are of great interest to an exploration geophysicist, besides the multiparametric data obtained from satellites. Satellite-based sensing surveys may be applied during "site screening" and characteristic investigation to provide important information about surficial geology, geomorphology, water systems, biosphere, etc. It is also possible to interpret the locations of faults and other structural features. The satellite gravity and magnetic maps are very useful in the context of mapping deep crustal features and the thermal imageries have applications in problems related to volcanology, mining, environment, and integrated mineral exploration projects.

In satellite-based systems, the data are acquired digitally with specific radiometric characteristics which can then be utilized in other image analyses. Satellite images can now provide excellent resolution with high spectral accuracies. The weakness being that the image acquisition times can vary depending on satellite orbitability and environmental conditions in the project area.

References

Annan, A.P., Davis, J.L., 1997. Ground Penetrating Radar—coming of Age at last!. Exploration'97, Geol. Surv. Canada Publn., pp. 515–522.

Brock, J.S., 1973. Geophysical exploration leading to the discovery of the Faro deposit. CIM Bulletin, pp. 97–116.

Dyck, A., 1983. Drillhole electromagnetic methods. In: Nabighian, M.N.. Ed., Electromagnetic methods in Applied Geophysics—Applications. Part A & B, Soc. of Expln. Geophysics, Publ., pp. 881–930.

Ford, K., Keating, P., Thomas, M.D., 2007. Overview of geophysical signatures associated with Canadian Ore deposits. In: W.D. Goodfellow, Ed., Mineral Deposits of Canada, a synthesis of Major deposit types, District metallogeny, The evolution of geological provinces and Exploration methods. Geological Association of Canada Special Publication 5, 2007, pp. 1–21.

Gandhi, S.M., 1985. Geophysical and geochemical investigations carried out by national/other agencies in and around Rampura–Agucha deposit. Unpublished HZL report, 32 p.

Hollyer, G., 2012. In the shadow of the Head frame: working with Titan 24 at Mine sites. Quantec Geoscience, ppt presentation, 43p.

Kavdia, N.K., Nath, D., Sharma, S., 2015. Exploration success in the quest for hidden mineralization—A case study from Pur-Dariba Copper Prospect, Bhilwara Dist., Rajasthan. In: P.R. Golani, Ed., Recent Developments in Metallogeny and Mineral Exploration in Rajasthan. GSI Special Publication no. 101, pp. 29–42.

McNeill, J.D., Lobson, V., 1991. Geological mapping using VLF radio waves. In: Nabighian, M.N., Ed. Electromagnetic methods in Applied Geophysics—theory. Soc. of Expln. Geophys., Tulsa, 972 p.

Mining Magazine, 2013. Imaginative imaging: interview with Gedex. Min. Mag. 20–23, Nov. 2013.

Quantec Geoscience, 2011. <http://www.quantecgeoscience.com/>. Leaders in Deep Exploration.

Ramakrishna, T.S., 2006. Geophysical practice in mineral exploration and mapping. Geol. Soc. India, Mem. No. 62, 382.

Reynolds, J.M., 1997. An Introduction to Applied and Environmental Geophysics. Wiley, Chichester.

Rogers, G.R., 1973: Geophysical surveys. In Exploration for Mineral Deposits (Ed) A.L. Payne, in Mining Engg. Handbook (Gen Eds. A.B. Cummings and I.A. Given) SME, vol. 1, pp. 5–24.

Seigel, H.S., Robbins, J.C., 1983. Exploring with Luminex: Mining Engineering v. 35, 729–732.

Telford, W.M., Geldart, L.P., Sheriff, R.E., 1990. Applied Geophysics, 2nd edn Cambridge University Press, Cambridge.

Vozoff, K., 1972. The magnetotelluric method in the exploration of sedimentary basins. Geophysics, 37, pp. 98–141.

Yamashita, M., 1984. Controlled source Audio Magnetotelluric. Phoenix Geophysics Ltd., Ont. Canada.

CHAPTER

6

Geochemical Exploration

OUTLINE

Essentials of Mineral Exploration and Evaluation
DOI: http://dx.doi.org/10.1016/B978-0-12-805329-4.00013-2

6.1 INTRODUCTION

A geochemical exploration campaign aims at locating economic mineral deposits through recognition of unusual concentrations of chemical components in surficial materials such as soils, stream sediments, rocks, water, plants, and air. The object of the search is detecting certain distribution of chemical components at levels necessarily above usual to be termed as a geochemical anomaly. An anomaly might indicate the presence of mineralization below the earth. Geochemical surveys warrant for systematic collection and chemical analysis of any of a wide variety of naturally occurring materials such as rocks, soils, stream sediments, surface/ground water, vegetation, and air. Geochemistry detects trace quantities of metals or associated elements shed from a mineral resource into the natural setting over a considerable time period. Rocks are broken down into constituent minerals at or near the earth's surface. Some minerals are soluble and are rapidly dispersed by rainwater, eg, salt. Others are chemically stable, physically resistant, and can remain for many years in the weathering environment, and are commonly concentrated in streams and rivers, eg, diamonds, tin, and gold. Different techniques can be used to detect variations in the minute levels of the elements, and these anomalies lead the prospector to a source. Samples can be collected from streams, soils, gravels left by ice sheets, desert sands, vegetation, and water bores. This phase of exploration results in the definition of anomalies, which are unusual coincidences of geophysical, geochemical, and geological features that could signal the presence of mineralization.

Exploration geochemistry relies on certain basic geochemical concepts. The general principles underlying element distribution in the natural environment imply that elements tend to be associated with one or another of these major phases, or the atmospheric and biologic phases, which had subsequently developed. Based on these features, elements can be classified as:

1. Siderophile (ie, affinity for iron)
2. Chalcophile (ie, affinity for sulfur)
3. Lithophile (ie, affinity for silicates)
4. Atmophile—present as gas in the atmosphere
5. Biophile—occurring as biological material.

Thus element distributions in the natural environment are primarily influenced by their fundamental chemical properties which in turn depend on the element's electronic constitution and hence position in the periodic table.

6.2 THE GEOCHEMICAL CYCLE

A rock may be considered as a chemical system in which chemical changes can be brought about by various agencies (Clarke, 1942). According to Rose et al. (1979), the earth is a dynamic system in which materials are moved from place to place and changed in form and composition by a wide variety of geological and geochemical processes. The geochemical environment, as defined by pressure, temperature, and availability of most abundant chemical components, determines the stability of mineral and fluid phases at any given point. The geochemical environments of the earth are classified into two broad groups, viz., primary (deep-seated) and secondary (surficial), which are distinguished by gross differences in pressure, temperature, and chemistry. The movement of earth materials from one environment to another can be conceptualized in the form of a partly closed cycle called the Geochemical Cycle. There are two parts of the cycle, viz., one comprising the deep-seated processes of metamorphism and igneous differentiation and the other involving surficial processes of weathering, transportation, and sedimentation. A simplified diagram of a typical geochemical cycle is presented in Fig. 6.1.

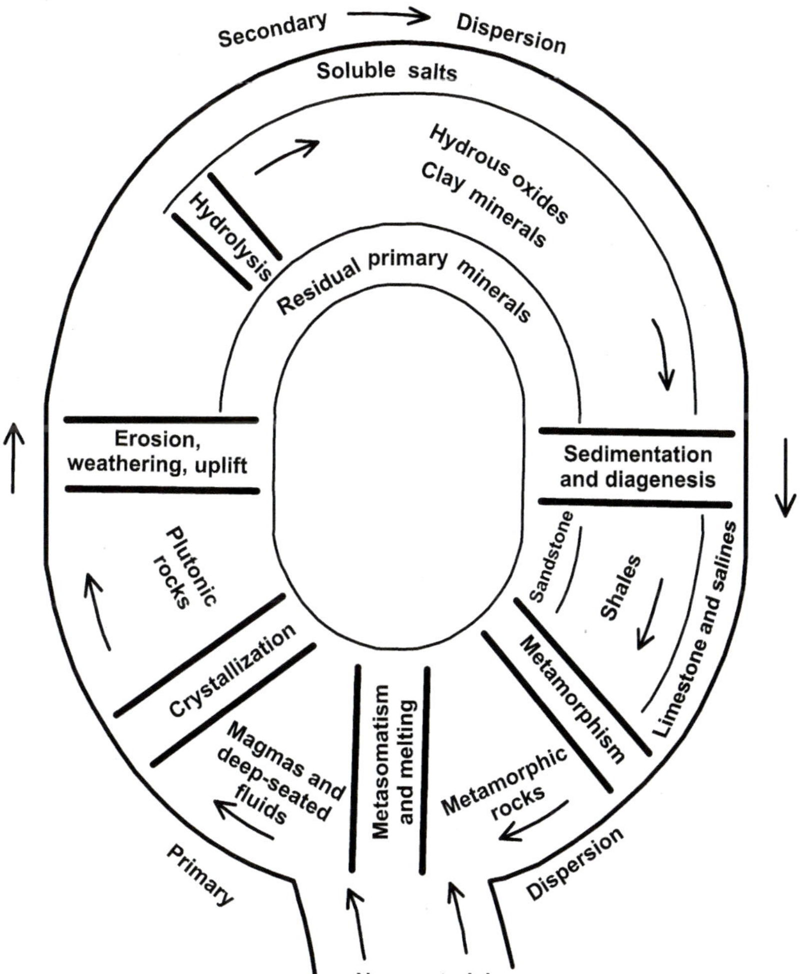

FIGURE 6.1 The geochemical cycle.

6.3 GENERAL PRINCIPLES

Exhaustive summary of the details of the processes with regard to exploration geochemistry is given by Hawkes and Webb (1962), Beus and Grigoriyan (1975), Rose et al. (1979), Levinson et al. (1980), Horsnail (2001) and Lett (2009). However, a brief summary is given below.

Mineral deposits characterize abnormal concentrations of particular chemical components, for the most part inside of a moderately restricted volume of the earth's crust. Majority of mineral deposits incorporate a zone at its center, in which the significant chemical components or minerals are concentrated to a quantity adequate to allow mining. The chemical components encompassing the center, by and large, diminish outward in concentrations so as to be measured in parts per million (ppm) or parts per billion (ppb), which apparently surpasses the usual background level of the host/country rocks. These zones, called as geochemical anomalies, help in detecting and tracing mineral resources. Geochemical anomalies are geochemical highlights not quite the same as what is viewed as ordinary. They can be the aftereffect of rare or unprecedented processes concentrating specific chemical components (eg, a mineral forming process, weathering and element dispersion from an abnormal chemical component concentration, eg, an ore body) or component collection or concentration by normal procedures acting over long time (eg, searching and concentration of specific chemical components by ironstones, ferruginous regolith, or manganese oxides), or nonnatural contamination of locales or samples.

A geochemical anomaly is any high or low variation in chemical component, not described by a characteristic geochemical variation. Determining the natural variation or geochemical background in respect of different chemical components provides a means of interpreting a geochemical survey. The higher bound of geochemical background is called threshold above which any value is thought to be anomalous (Fig. 6.2). Closely related to establishing threshold is selecting sampling and analytical strategies that would give maximum anomaly contrast (ie, peak value to threshold ratio). Identification of the mechanism controlling movement of elements in a region can aid in adequate planning a geochemical survey in order to obtain the best anomaly contrast.

The spread and intensity of geochemical anomaly are influenced by the following factors:

- Mobility of elements in the physical and chemical milieu
- Depth of the upper altered tier of the deposit
- Shape and size of the deposit including the attitude
- Permeability, porosity, and mineralogical composition of host rock
- Local topography
- Hydrologic regime
- Nature of weathered profile
- Tectonic movements
- Sampling media and density
- Precision and accuracy of the analytical procedure.

When a mass of earth material (rocks or minerals) passes through major alterations of a geochemical cycle, it tends to be redistributed, fractionated, and mixed with other masses

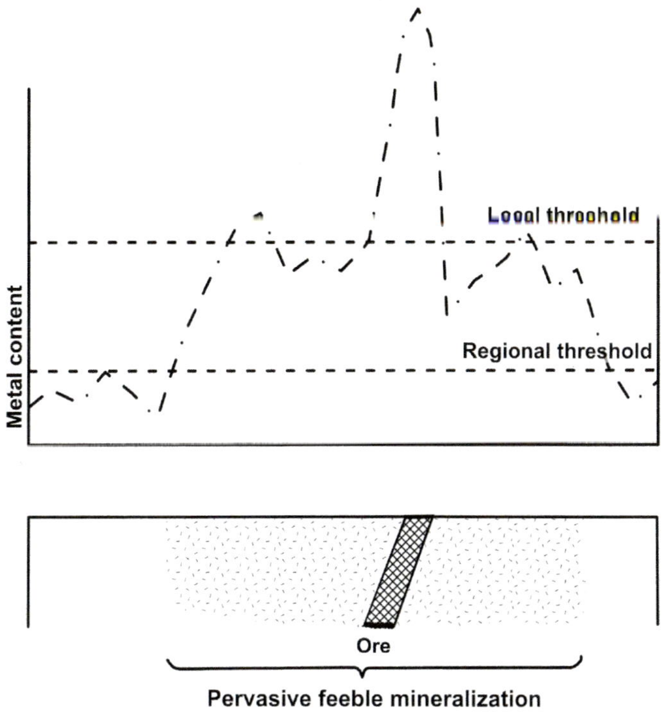

FIGURE 6.2 Diagram illustrating local and regional threshold values.

of materials. This process in which grains and atoms move to another geochemical environment is called geochemical dispersion. There are two main processes that cause geochemical dispersion of elements.

6.3.1 Physical Dispersion

This process is the effect of purely mechanical agencies such as the movement of material by stream or glacial action. Excepting for alluvial sorting of clay and sand, purely mechanical processes generally include product mixing but not separation into specialized fractions. Anomalies develop at places of mineral concentration (eg, heavy mineral concentration in high-energy streams, mineral concentration by moss, transport of rock fragments in glacial till).

6.3.2 Chemical Dispersion

This process generates fractions of widely differing chemical composition. In this process, chemical weathering prevails and elements get sorted and dispersed according to their differing mobility. Temperature, pressure, pH, and Eh are the fundamental variables determining dispersion of chemical elements. Anomalies develop in response to differing pH, Eh, temperature, and pressure.

6.3.3 Primary and Secondary Environments

The nearly equidimensional zone enclosing the source material (eg, ore deposit) is called primary dispersion halo that signifies the distribution of elements in a systematic and distinct pattern generated as a consequence of primary dispersion. Size and shape of resulting dispersion halos differ significantly as an aftereffect of the various physical and chemical variables that influence passage of fluid in rocks. A few halos can be recognized at separations of several meters from their source bodies, while others are closely located. Few of the controlling factors for the formation of primary halos include existence of fractures or otherwise in enclosing rock, porosity, permeability of enclosing rock, reactivity of ore fluid with enclosing rock among others. The primary halos can be either syngenetic or epigenetic depending upon their formation along with or after the formation of enclosing rocks. When the rock mass is altered by hydrothermal process, epigenetic halos can have wide areal coverage or can also be confined to channel courses through which ore fluids have passed through open spaces. They are much of the time developed in rocks overlying a mineral body and thus can be of great importance in geochemical prospecting.

Secondary dispersion halos are at times described as dispersion trains, geometry, and coverage of which rest on various factors among which physiography and ground water movement are by far the most important. Ground water very often aids in dissolving and transporting certain constituents of ore bodies for significant distances prior to ultimately meeting in streams. Further scattering might result in stream sediments when soil or weathering trash that has abnormal metal substance gets to be consolidated through erosion in stream sediment. Examination of the fine sand and silt of stream sediments can be a useful technique for identification of ore bodies inside of the zone drained by the stream.

6.3.4 Geochemical Mobility

Geochemical mobility is of much importance in geochemical prospecting. It determines the nature of dispersion patterns. In the secondary environment, it relies on upon certain intrinsic properties of the elements being referred to and the main features of the site. To this reference, critical properties of elements include electronic arrangement, ionic potential, stability relations with change in pH and Eh, tendency to form complexes with organic substance, and propensity to be coprecipitated or absorbed with iron or manganese hydroxides. Characteristics of the site that either increase or limit geochemical mobility are physiography, soil type, pH and Eh of ground water, vicinity or nonappearance of decomposing vegetation in soils, calcium carbonate precipitates in soils, and iron and manganese hydroxide precipitates in stream channels.

6.3.5 Pathfinders

Pathfinder elements/minerals provide good clues in the search for hidden ore bodies because they generally form large haloes. Pathfinders, used in both primary and secondary environments, are defined as relatively mobile element (or gases) clearly associated with the element being sought, but more easily located either because they form a broader halo or because they are easier to detect by current analytical procedures. The types of deposit and their pathfinder elements as per Rose et al. (1979) are given in the table.

Pathfinder Elements	Types of Deposits
As	Au–Ag vein type; Au–Ag–Cu–Co–Zn complex sulfides
B	W–Be–Zn–Mo–Cu–Pb skarns; Sn–W–Bo veins or gneisses
Hg	Pb–Zn–Ag complex sulfide deposits
Mo	W–Sn contact metamorphic deposits
Mn	Ba–Ag vein deposit, porphyry copper
Se, V, Mo	U, sedimentary type
Mo, Te, Au	Porphyry copper
Pd–Cr–Cu–Ni–Co	Pt in ultramafic rocks
Rn	U, all types
Zn–Cu	Ag–Pb–Zn; Cu–Pb–Zn sulfide deposits
SO_4	Sulfide deposits of all types

Besides, Pd-isotope analysis of regional samples of lead in surface gossans/ore minerals may indicate a mineral deposit and the probable point of origin of the mineralization.

6.4 GEOCHEMICAL EXPLORATION SURVEYS

These are most amenable to sulfide ores of Cu, Zn, Pb, Ni, Mo and to a lesser extent, to the ores of U, W, Sn, Hg, Ag, and Au. These surveys are not useful in looking for indicator minerals, gemstones, the ores of Mn, Cr, Ti, bauxite, iron ores, etc. Geochemical methods are most rewarding, where (1) older conventional methods were ineffective, particularly in the primary reconnaissance of very large tracts of remote terrain never thoroughly prospected and (2) outcrops are rare due to forest cover, deep weathering, and glacial overburden.

Exploration situations where geochemical surveys have been particularly useful include:

- In reconnaissance for ore types only recently of economic value
- In reconnaissance for low-grade deposits where ore minerals/weathered products are inconspicuous, hence previously not recognized (low-grade porphyry deposits of Canada, United States)
- In detailed survey to locate additional deposits hidden under a cover of glacial till/soil/vegetation/forest litter
- As a method of complementing ground checking, airborne, or ground geophysical anomalies
- Along with geological and geophysical surveys in appraising known but poorly exposed deposits.

Geochemical surveys are *not to be depended for pin-pointing drill targets* in exploratory drilling. It can only indicate areas generally favorable for mineralization, but drill targets are to be decided on the basis of integrated geological, geophysical, and other observations.

6.4.1 Types of Geochemical Surveys

Geochemical surveys are carried out under two broad stages: reconnaissance and detailed. The key purpose of a reconnaissance survey is preliminary assessment of a very large area using a small scale of mapping. Such surveys are normally done by sampling stream sediments. Detailed surveys use closely spaced samples over an area of a few square kilometers, and individual samples may be as close as few meters apart. The purpose of a detailed survey is identification and delineation of a geochemical anomaly that would aid in locating a deposit of particular mineral or the probable contiguity of an already identified deposit.

6.5 VARIOUS GEOCHEMICAL EXPLORATION SURVEYS

Various geochemical exploration surveys include:

- Pedogeochemical surveys
- Stream sediment surveys
- Lithogeochemical surveys
- Biogeochemical and geobotanical surveys
- Hydrogeochemical surveys
- Atmochemical surveys
- Electrogeochemical surveys.

6.5.1 Pedogeochemical Survey

The interpretation of data of soil survey will be different depending upon the material sampled: (1) residual soil—developed by the weathering of bedrock without much lateral movement; (2) glacial moraine/outwash that has been transported laterally; and (3) soil/air with constituents which may have percolated upward from depth in vapor form.

Residual soil:

- Temperate climate—Residual soil anomaly occurs as a thin blanket of loose material overlying the bedrock
- Tropical climate (as in Africa, S. America), it can be 20–40 m thick, completely decomposed from which identity of parent rock cannot be ascertained.

Residual soils generally carry chemical memory of the composition of the parent rock. Residual soil derived from mineralized rock will always be anomalous.

Residual soil surveys are:

- best applicable in areas where fresh outcrops are rare and all surface materials look alike;

- used effectively in primary "blue sky" reconnaissance of very large area where poorly developed system of surface drainage channels (central and west Africa);
- used extensively in all topographic and climatic conditions in the follow-up or confirming anomalies uncovered by geochemical sediment survey or geophysical survey;
- useful in conjunction with other methods in exploring immediate vicinity of known deposits;
- not useful in Mountain terrains where outcrops are abundant and where conventional prospecting methods are cheaper and more effective.

6.5.1.1 Collection of Sample

Normally about 100 g collected at fixed intervals along straight traverse lines, along topographic, cultural features, or ridge tops/roads. The spacing should be close enough to permit the samples fall within anomalous feature being sought (if the target is of narrow vein type—15 m wide × 150 m long, then the sample spacing should be about 8 m along lines and 80 m apart). Proper identification of soil horizons and care should be taken in soil sampling.

6.5.1.2 Soils

Soils are the upper layered part of the regolith. Soils are characteristically organized into layering differing from each other and from the underlying parent material in their properties and composition. Apart from color and texture which aid in recognition in the field, the properties of greatest significance that affect geochemical dispersion of elements are pH, organic matter, clay mineral type and assemblage, amount of sesquioxides, etc. The individual layers are the soil horizons (few centimeters to several meters in thickness). The horizon comprises the soil profile (Fig. 6.3) which is the result of vertical movement of material in solution and suspension accompanied by complex series of chemical reactions,

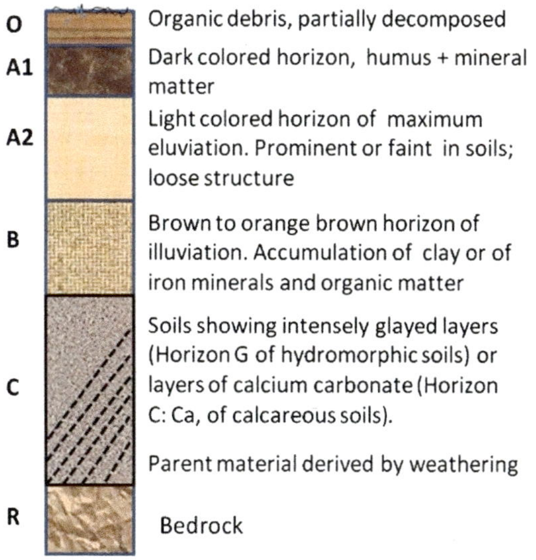

FIGURE 6.3 Idealized soil profile.

many of which are organic in origin. Water is the essential medium in which the transfer and reconstitution take place.

According to genetic and geographic environments, soil profiles vary in the make up within wide limits. Most profiles from surface downwards comprise three principal horizons and are identified by codes A, B, and C. The entire sequence need not always be represented. Incomplete profiles may lack some horizons or erosion might prompt trimmed profiles to the degree of uncovering C horizon. Metals indigenous to parent material vary in their response during the development of soil horizon. Soluble metals, absorbed, incorporated in clays, colloid are liable to be removed from "A" horizon (resistant primary minerals enriched in that region).

6.5.1.2.1 FACTORS AFFECTING SOIL FORMATION

Parent material: Permeable parent rocks are readily decomposed—soil develops rapidly.

Climate: Temperature and rainfall decompose rock; soils of humid regions—thoroughly leached and to possess Fe-rich "B" horizon; Arid climate—calcareous soils (in semi-arid and warm regions also). Precipitation to evaporation when high supports the gravity-driven downward movement of ground water through soil and when low supports rise of soil moisture from the underlying water table.

Biological activity: It is a function of vegetation and the role of microorganisms helping in decomposing plant debris. These are in turn related to climate.

Relief: Formation of soil is controlled by topographic relief through its association with the level of ground water, drainage, and erosion.

The accumulation of parent material by weathering takes longer than its differentiation into soil horizon. Moderate humidity and free drainage help to develop a faint "A" horizon in some decades. Development of "B" horizon needs much greater length of time, in 100s and 1000s of years.

Study of soil and glacial till has been utilized widely as a part of geochemical survey and has led to finding of several mineralized bodies. By and large, such studies are of a comprehensive nature and are carried out in a densely spread grid. The design and interpretation of various soil surveys have been dealt in detail by Fletcher et al. (1987). Few of the geochemical dispersion processes accountable for development of sediment and soil anomalies are condensed in a conceptual three-dimensional model. This demonstrates the downward varying degree in the content of metal in a classic soil profile on adequately drained and inadequately drained soils altered by glacial transport of the original overburden (Fig. 6.4). According to the model proposal put forward by Cameron et al. (2004) for the formation of soil anomalies through saturated transported cover at the Lake Cross deposit, Canada, reduced metal ions are oxidized at and above the water table by infiltrating oxygen that produce H^+ ions and in turn dissolve soil carbonates (Fig. 6.5).

6.5.2 Stream Sediment Survey

Transported soils used in geochemical exploration include stream sediments, lake sediments, and glacial debris. Stream sediment sampling is the most widely used and effective

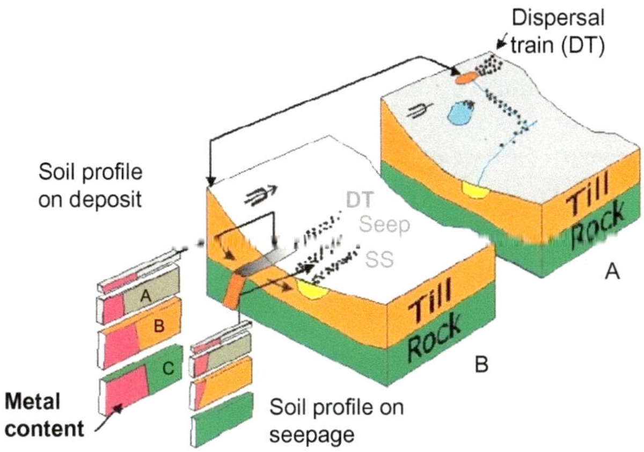

FIGURE 6.4 Sketch map of generalized dispersion and dispersal train model. *Source: From Lett (2007).*

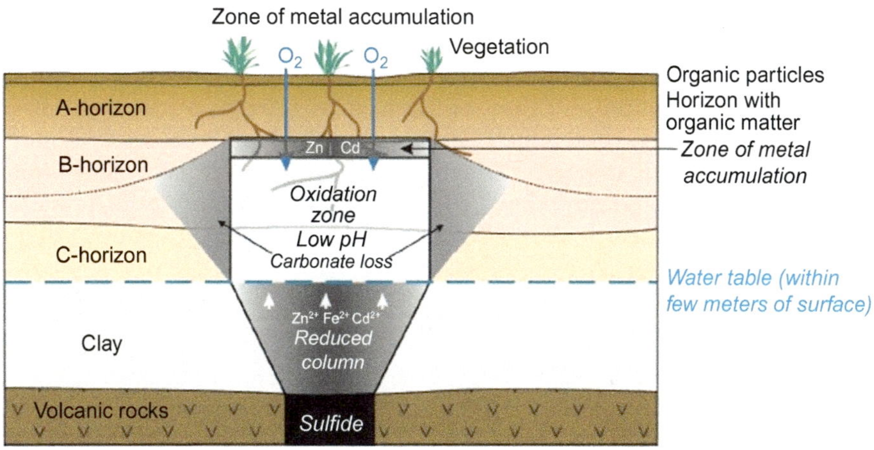

FIGURE 6.5 A model for the development of soil anomalies through saturated transported cover at the Lake Cross deposit, Canada. *Source: From Cameron et al. (2004).*

method of the technique used in exploration. A major problem in stream sediment sampling is the interpretation of the effects, if any, of secondary hydrous iron–manganese scavenging on metallic concentrations in samples. The iron and manganese oxides coprecipitate and concentrate elements such as Zn, Pb, Co, and Mo. Survey involves the sampling and analysis of fine-grained sediments from surface drain channels. The chemical composition of stream sediments reflects, in part, the composition of soil eroded and washed into stream channel and in part, the material chemically precipitated on the surface of the sedimentary particles by stream water. Stream sediment survey is especially suited for rapid and economical reconnaissance of large area. The principal requirement for success is an adequately developed network of surface drainage channels and an area where there is still a chance of undiscovered ore near the surface (Fletcher, 1997).

Ground water that has percolated through ore deposit may carry dissolved metal and later precipitate it on the surface of the granular material of sediment. This chemically precipitated fraction is relatively soluble and can be distinguished from metal held in detrital phase by its ready solubility in weak chemical extractant (as used in cold extractable metal). In either case, the well-chosen and properly prepared sample may indicate the relative likelihood of the occurrence of undiscovered mineralization.

This survey involves collection of samples of fine-grained sand/silt from active channel of stream or from the flood plain near the active channel (Fig. 6.6). Samples are commonly dried, sieved to -80 mesh and the coarse fraction discarded. Samples are analyzed by a suitable analytical method. Cost depends upon the density of sampling and accessibility. Fig. 6.7 shows gold in -80 mesh of regional geochemical survey, stream sediment, and heavy mineral concentrate total gold grain count around Eskay Mine area, BC, Canada. A large number of pristine gold grains suggest a nearby source.

6.5.3 Lake Sediments

If the composition of lake sediments reflects the material within the lake catchment basin, then lake sediment represents a source of data on the presence of mineralization within the basin. Material may be transported into a lake either mechanically in particulate form or chemically in solution and may be derived directly from the weathering of the bedrock or from overburden on the surface. Conventional geochemical exploration based on sampling and analysis of stream sediments, soil, and bedrock has found limited application in areas of glacial overburden with indefinite and disorganized drainage systems (as in large tracts of Canadian Shield). In such areas, lake sediment composition could be

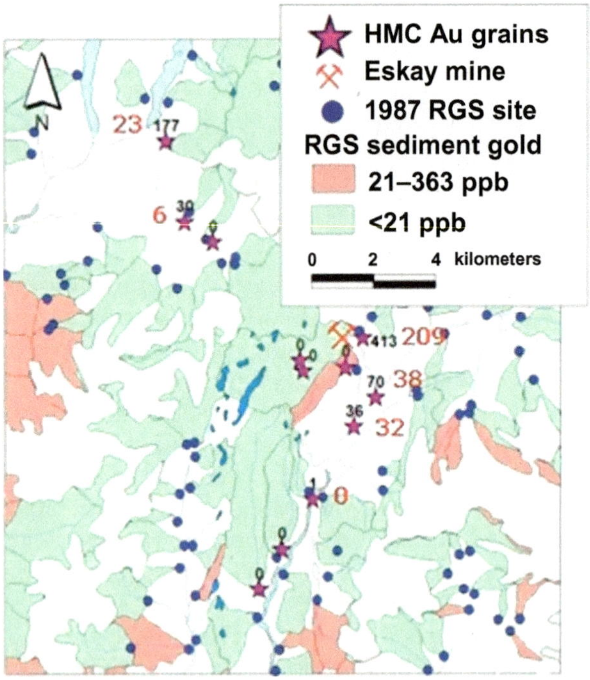

FIGURE 6.6 Gold in -80 mesh of RGS stream sediment and HMC total gold grain count (RGS, Regional Geochemical Survey; HMC, Heavy Mineral Concentrate) around Eskay Mine area, BC, Canada. A large number of pristine gold grains suggest a nearby source. *Source: From Lett (2007).*

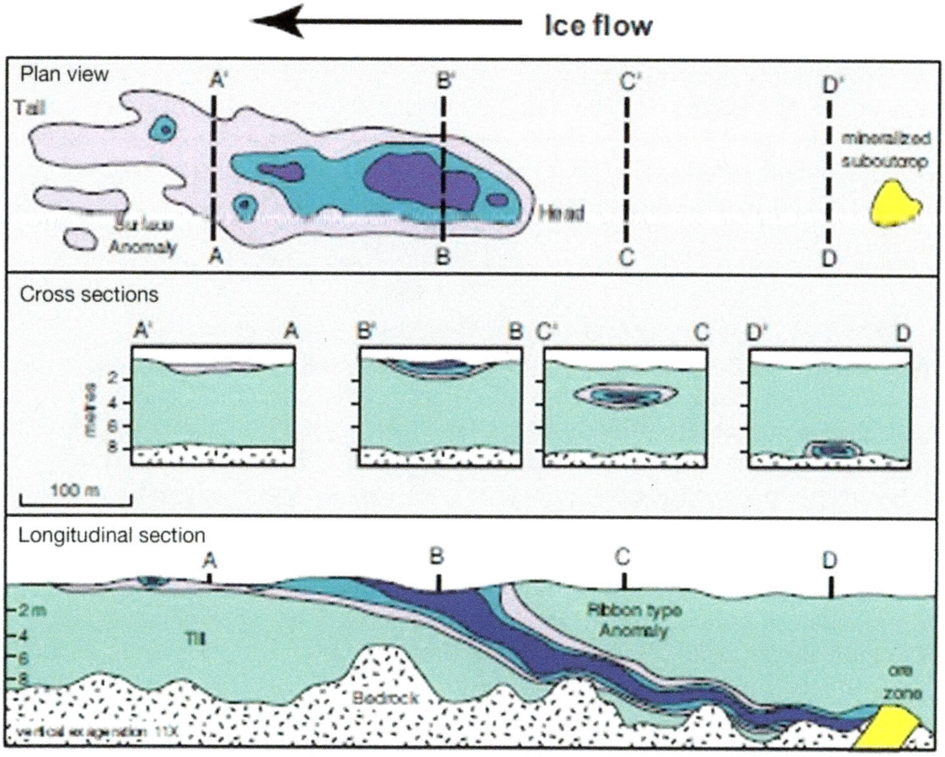

FIGURE 6.7 Idealized model of glacial dispersal train in till in relation to bedrock. *Source: From Miller (1984).*

an indicator of mineralization of the host rock or the mineralization itself may have some diagnostic geochemical characteristic that is transmitted to lake sediments.

6.5.4 Glacial Drift

When combined with conventional techniques, glacial drift methods can be an inexpensive means of locating potentially economic mineralization. This method has been widely used in reconnaissance surveys in Canada, Russia, and northern Europe. The aim of drift exploration is to trace the source of the mineralized material—varying in size from boulders to several microns—that was carried by a glacier and deposited in a moraine when the glacier melted. An indicator train and the density of its indicator elements will depend on the area of outcrop and orientation of the source rock with respect to glacier flow direction, distance of transport of the minerals, erodibility, and the concentration of minerals in the source rock. In areas of thick glacial sediments, dispersal trains may be covered by younger till or by glaciolacustrine or glaciofluvial sediments. Buried trains can only be detected by subsurface till sampling. Because dispersal trains generally rise with increasing distance down ice (Fig. 6.7), anomalies can be expected at any depth within the till unit. Thus, entire till units should be sampled when drilling, not just the lowermost few meters on bedrock, in order to detect the rising tails of dispersal trains or trains within multiple till sequences.

The most useful glacial unit for sampling is basal till, since most of this occurring just above the bedrock, was derived from local sources. Near surface sampling of till combined with boulder tracing is one successful exploration method. Samples of basal till could be obtained by overburden drilling, if the glacial deposits are thick.

6.5.5 Heavy Minerals

In geochemical exploration, the use of heavy minerals has a crucial role in detecting anomalous halos around mineralization. Discoveries of diamond in Canada, South Africa, and Australia among others provide exemplary role of heavy minerals as guides. Heavy mineral concentrates from drainage sediment and glacial till have turned out to be gainful as geochemical and mineralogical guides to gold and diamond exploration. Microprobe analysis of heavy minerals aids in distinction between diamondiferous kimberlite pipes and those devoid of diamonds. It can, much of the time, allow the utilization of surface till sampling, at an impressive reduced cost when contrasted with base of till sampling. Heavy minerals can be used to sort out particular issues, eg, poor or nonlocation of zinc mineralization in carbonate rocks by standard sampling of silt. Numerous Soviet geologists have utilized these techniques to recognize abnormal essential coronas around metal bodies.

Heavy minerals occur in greater abundance near the base of a sequence of alluvial sediments just above the surface of bedrock. Samples are collected by digging pits or sinking holes. The favored method is to wet-sieve the sample (Fig. 6.8) by carefully scooping the sediments into a -20 mesh stainless steel sieve.

FIGURE 6.8 Heavy mineral sediment sample collection and wet-sieving of the sample.

6.5.5.1 *Indicator Minerals*

The use of indicator mineral methods to mineral exploration has extended and grown fundamentally in the course of recent decades. They are currently utilized throughout the world to prospect for a broad range of mineral deposits including diamond-bearing kimberlites, lode gold, magmatic Ni–Cu–PGE, metamorphosed volcanogenic massive sulfides, porphyry Cu, uranium, tin, tungsten, among others (Averill, 2001). Indicator minerals, including ore, accessory, and alteration minerals, are normally sporadically distributed in host rocks. Indicator minerals may be more sporadic in derived sediments, in which case sediment samples must be concentrated allowing to recover and examine those. Most indicator minerals have a moderate to high specific gravity, and thus most processing techniques concentrate indicator minerals using some method of density separation, often in combination with sizing and/or magnetic separations. The presence of specific indicator minerals in unconsolidated sediments provides evidence of a bedrock source in the provenance region and in some cases, the chemical composition of the minerals is associated with the ore grade of the bedrock source. As few as one sand-sized grain of a particular indicator mineral in a 10-kg sample may be significant. To recover such potentially small quantities (equivalent to ppb) of indicator minerals, samples are processed to reduce the volume of material that must be examined. In reducing the volume of material, processing techniques must be able to retain the indicator mineral(s) and do so without contaminating the sample, without losing indicator minerals, and at a reasonable cost. Indicator minerals can be recovered from a variety of sample media, including stream, alluvial, glacial, beach or eolian sediments and residual soils (http://www.vuorimiesyhdistys.fi/). McClenaghan and Kjarsgaard (2007) has reviewed the application of indicator mineral methods.

Thus, indicator mineral strategies can be viewed as a major aspect of a range of clastic-silt based techniques that range from boulder tracing to pebble counts, to indicator minerals, to basic techniques intended to distinguish physically dispersed mineral grains. Representative plan and cross-section perspectives of clastic dispersal and chemical dispersion patterns in different media surrounding a kimberlite in ice terrain are shown in Fig. 6.9.

6.5.6 Lithogeochemical Survey

Lithogeochemical surveys are, by and large, conducted on a local basis and are targeted for the delineation of primary or leakage halos associated with mineral deposits. Primary halos are subdivided into three types, viz. (1) those that cover wide areas, resulting from massive injection of rock by hydrothermal solutions or other fluids percolating from depth; (2) those that are manifested in leakage patterns possessing well-defined channel ways for the dispersement of fluids; and (3) those that form wall rock patterns, where the rock adjacent to these solution channel ways has been hydrothermally altered. Leakage halos vary in dimension, but they are particularly valuable in the search for several types of hidden ore deposits. Gaseous leakage halos are even more important in exploration for deep ore deposits because gases such as mercury, iodine, and bromine move easily through rock and soil pores and fractures.

Lithogeochemcial surveys are effective for reconnaissance surveys and involve sampling of unaltered bedrock with a purpose to demarcate host rocks that are considered to support for

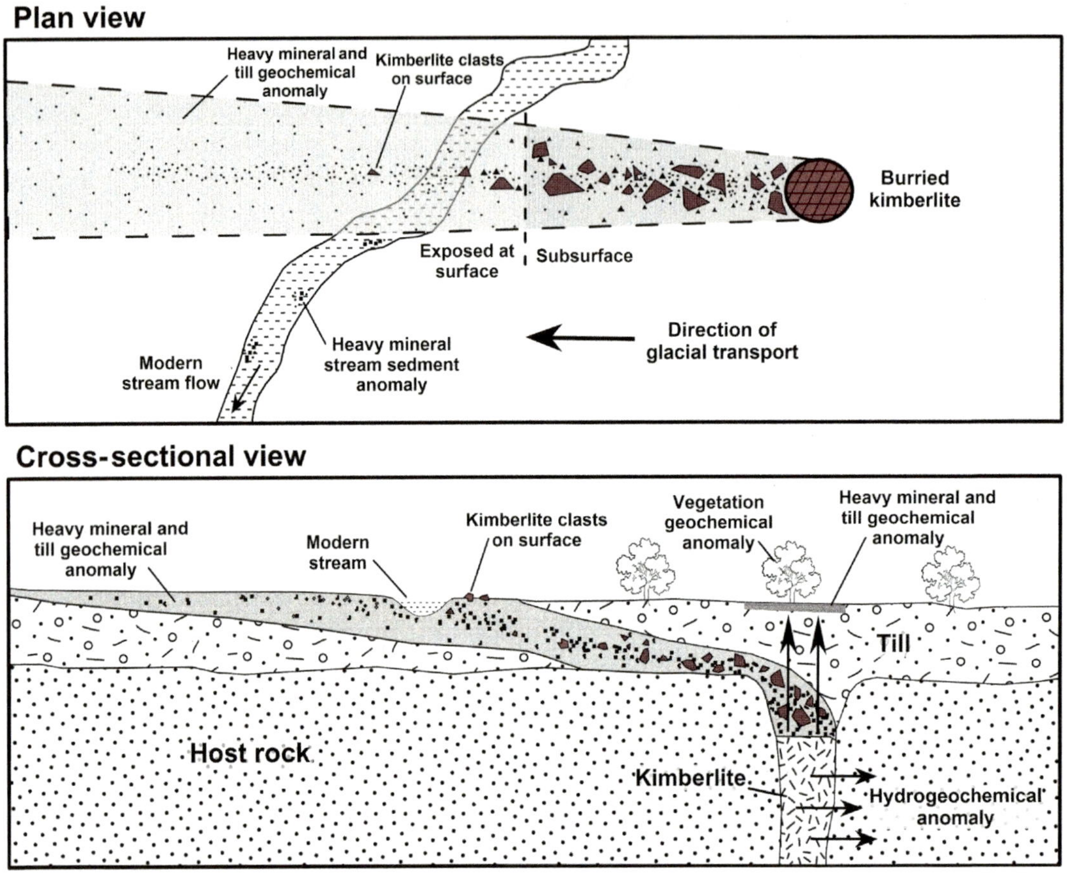

FIGURE 6.9 Schematic plan and cross-sectional views of clastic dispersal and chemical dispersion patterns in various media around a kimberlite in glaciated terrain.

mineralization. Govett (1983, 1989) and Govett et al. (1984) have explained in great detail with apt examples the lithogeochemical surveys carried out in various locations in the world. Suites of samples, strictly comparable as regards rock type, weathering, alteration are collected from mineralized and background areas far removed from mineralization in order to ascertain background values for each type of material sampled. In orientation for reconnaissance rock surveys, one should consider both syngenetic and broad epigenetic anomalies.

- For wall rock aureoles, to take continuous series of channel/composite samples extending away from deposit to unweathered country rock
- In addition to analysis, microscopic studies to be done for mineralogical wall rock alteration
- In orientation surveys, compare the anomalies obtained by rock and soil sampling
- Single element, multiplicative element, and their ratios and distribution with depth to detect zonation, etc. (Figs. 6.10 and 6.11).

Major applications of rock geochemistry are in determining the sense of top and bottom of prospects and alteration, etc., which cannot be otherwise detected easily. Russian geochemists have extensively used this technique by contracting multiplicative indications of younging of variety of deposit types (Govett, 1983, 1989). This technique has been widely applied in the search for VMS-type deposits in Canada to detect alteration zones, enrichment or depletion of metals/metal ratios and their behavior with respect to proximal or distal areas of exhalative mineral deposits.

6.5.6.1 Isotopic Surveys

These techniques are pertinent to elements that are known to be in two or more isotopic forms. The techniques utilize the proportions between isotopes, eg, ^{204}Pb, ^{206}Pb, ^{207}Pb, ^{208}Pb, or ^{32}S and ^{34}S to signpost specific types of ore deposits that might share a common cause (Gulson, 1986). Extensively applied in geochemical exploration and in establishing the age of ore and the source of ore bearing fluids are the lead isotopes. The stable isotopic studies of light elements (viz., H, C, O, S, Cl) and heavy elements (viz., Fe, Cu, Zn, Pb, U) aid in identifying and/or course towards mineralization in hydrothermal and magmatic ore deposits, and to understand controls on migration of elements of environmental importance (eg, Cu, Zn,

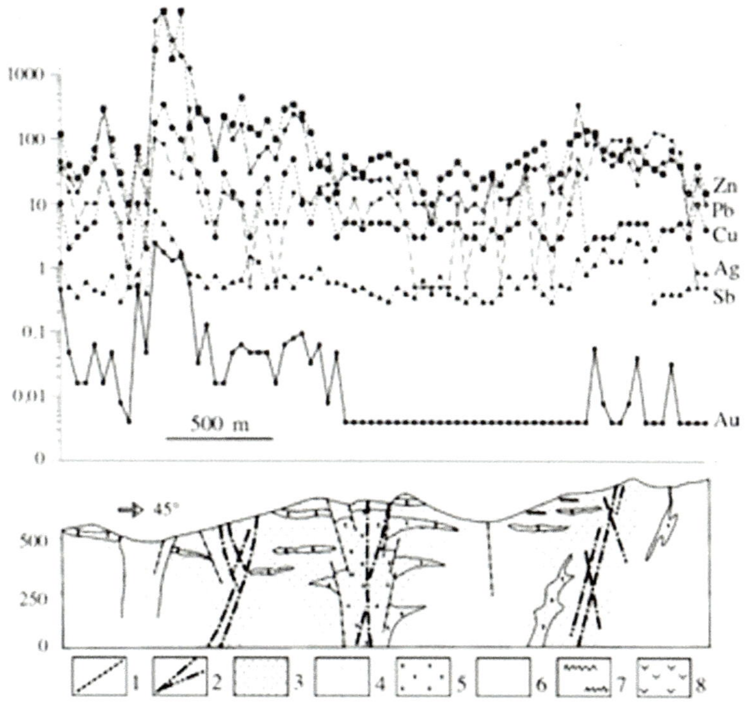

FIGURE 6.10 Element distribution across andesite complex of the ore field Lece-Profile, Rosaraca, Russia. (values in ppm) 1, trace of fracture; 2, quartz–breccia zone; 3, hydrothermal alterations in volcanic rocks (propylitization, sericization, chloritization, kaolinization, silicification); 4, pyroxene andesite; 5, pyroxene-amphibole andesite; 6, pyroclastic rocks (tuff, agglomerate, and breccias); 7, opal chalcedonic volcanic rocks; 8, amphibole–pyroxene andesite. *Source: From Stajevic (2004).*

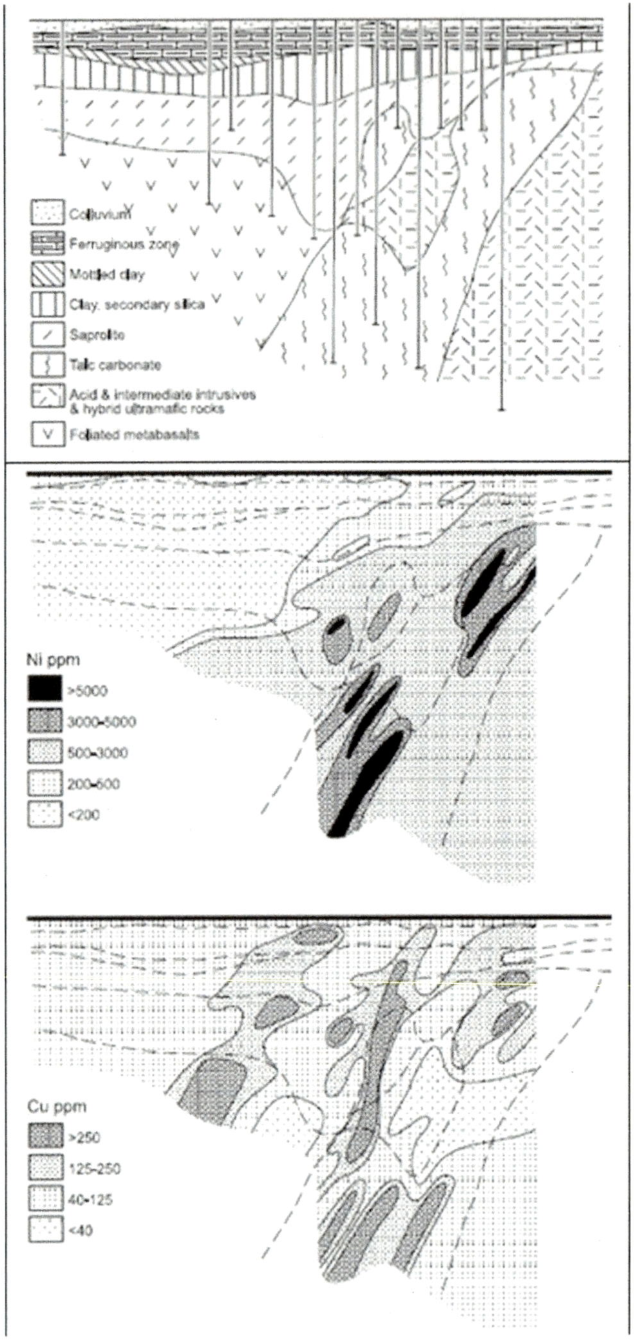

FIGURE 6.11 Geology and isocon contours of Ni and Cu in lateritic terrain, Western Australia. *Source: From Smith (1977).*

etc.). It is found that the content of U and the variation of U/Pb ratio distinctly decrease with increasing depth of lithosphere. Hence the Pb isotopic composition can be utilized as a measure for identifying the location of mineralization in the earth's crust. According to Andrew et al. (1998), isotopic ratios may also be used to assist in elucidating questions of ore formation (www.seas.upenn.edu). The oxygen isotopes ("deuterium" and "tritium" contents) of water are good tracers for the origin of saline geothermal water and the salinity in it.

6.5.7 Hydrogeochemical Survey

Hydrogeochemical survey is rarely competitive with sediment surveys. Samples are more expensive to collect and process. Also most metals are only slightly soluble in water under natural conditions. This survey is focused mainly in lake waters. However, cold and hot spring waters, their precipitates, seeps, streams, and wells have also been sampled. Besides, snow and ice. Rapid analysis of water in streams and surface drainage may help to eliminate certain areas for further consideration for exploration. By eliminating tributaries of low metal ion concentrations and following those of high content, one can trace a mineralized body. The mineralized zone can be defined by combining other geochemical techniques like soil, bio-, and geobotanical. The hydrogeochemical mechanism for dispersion through regolith and also the role of ground water flow in the dispersion trail along its flow from place to place are shown in Fig. 6.12.

Primary metallic ore minerals are stable and are relatively insoluble in water. But under the physical and chemical conditions present in an outcrop zone, most primary minerals are altered to secondary minerals that are more soluble in water. Secondary minerals form sulfurous and sulfuric acids.

Hydrogeochemical exploration for uranium deposits primarily traces uranium, radium, and radon in neutral waters. Under favorable conditions, this survey can locate deposits concealed up to depths of 300–400 m. In environments of extensive oxidation, uranium is leached from rocks. In reducing environments, uranium is precipitated from solution. Water from acid rocks enriched in uranium is more radioactive than water circulating in

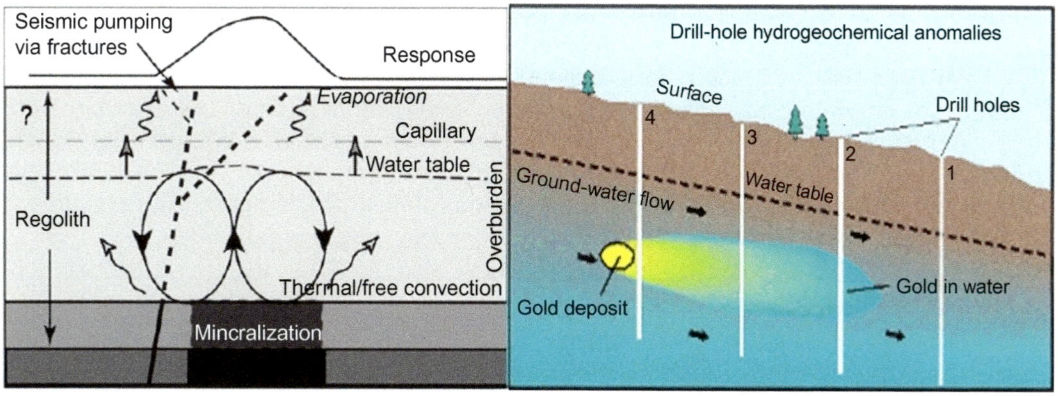

FIGURE 6.12 (Left) Hydrogeochemical mechanism for dispersion through transported regolith. (Right) Sketch indicating Drillhole hydrogeochemical anomalies due to ground water migration along slope. *Source: www.crcleme.org.au/.*

basic rocks. The radioactive equilibrium between uranium, radium, and radon is a useful guide to ore potential. In prospective areas of "U" mineralization, uranium content of water is much more dependable exploration guide than "U" in stream sediments. Uranium can form a very soluble and stable complex ion with carbonate. The uranium concentration in natural waters is typically in the "ppb" range which could be detected by "fission track" technique.

When sediments for one reason or the other are lacking, metal content of water is the only available clue to buried ore (Zn in Wisconsin, USA; Cu in arctic lakes). These surveys involve products of weathering of ore minerals carried from their sources in a buried ore body by natural waters, underground, or surface.

6.5.8 Atmogeochemical Surveys

It has long been known that gaseous halos exist in the atmosphere above various types of mineral deposits. Gas surveys which detect volatile hydrocarbons and "He" have become an integral part of exploration for oil and natural gas.

According to *crcleme.org.au*, "Gas originating from a point source, eg, a mineral body, will diffuse through an overlying porous medium and develop a hemispherical halo inside of the overburden with the peak intensity of the gas species expressed directly over the mineral body. However, the diffusion rate is dependent upon porosity and diffusive nature of individual gases. Regarding the properties of medium, gases on experiencing materials with varying porosity will diffuse along ways of slightest resistance or diffuse through low-porosity materials at rates that can't show a halo of adequate intensity (Hale, 2000). For instance, simulations of gaseous diffusion through porous sand unit overlain by low-porosity clay unit indicate the gas to diffuse freely with sufficient intensity through the sand. Conversely, for gas exuding from a mineral body that is specifically overlain by a low-porosity material, the dispersion might be slow and the resultant halo feeble, and the signature from the source may not be expressed with sufficient intensity to be detectable even if a highly porous material overlies the low-porosity clay. The gases will diffuse rapidly along paths of least resistance such as contacts and structural conduits."

Gaseous halos of radon (Ra) which are present in atmosphere and in soil gases which overly volcanic, metamorphic, or sedimentary rocks containing increased concentration of "U" have been widely used. Atmogeochemical halos from mineral deposits include all components present in the gaseous form which may originate from the ore in mineral deposits or products of supergene destruction of these deposits. The components may be in liquid as well as in solid form. Under atmospheric conditions, they exhibit high vapor pressure which promotes their transition to gaseous state at the temperature prevailing in the lower layer of the atmosphere.

Source of Gaseous Components in Halos	Composition of Gas
Gaseous components of rocks	Br, Cl, F, CO_2
Gaseous products of radioactive decay	Rn
Gaseous products of supergene reactions between ore and gangue mineralization	SO_2, CO_2, F, H_2S
Solid and liquid components of rocks and ores with high volatility	Hg, I

Gaseous halos of "Hg" associated with mercury deposits proper as well as with sulfide deposits containing "Hg" as minute impurities have been widely used in exploration (>20-m-thick Hg halo, detected over glacial moraine over polymetallic sulfide ore). Gas surveys of Hg in regions covered with overburden helped to study fracture pattern of the area. The concentration of Hg in atmospheric air decreases sharply with height. Wind has a direct bearing on Hg halos. Weather (rain, hence dilution) and wind (direction and force) may pose difficulties in interpretation (Australia—positive air anomalies in air over Cu, Ni, Zn deposits; background 1 ppb; over the deposit 12 ppb).

6.5.9 Vegetation

The use of vegetation as a guide to buried ore dates back to early days, when prospectors noticed certain ecological features such as barren areas or peculiar types of plants or trees were associated with certain mineralization (*indicator plants*: Calamine violet—lead to Pb, Zn deposits of West and Eastern Europe; Copper flower—in copper-rich soil in Zambian copper belt).

Prospectors have noticed disease symptoms, the unusual shades of leaves of plants and trees growing over mineralization ground. Presently air photos are widely used. Vegetation rooted in metal-rich environment demonstrated tendency to absorb some of the metal through the root system and concentrate it in the leaves and twigs which can be sampled. Some plants have very deep root enough to tap metal-rich ground water beneath 30–50 m of barren cap rock. Field experiments show that where vegetation is anomalous, the underlying soil also is anomalous, because of the accumulation of metal-rich organic litter.

6.5.9.1 Biogeochemical Surveys

Geochemical prospecting using plant species is of two types, viz., (1) biogeochemical, where the plant is chemically analyzed for trace element concentrations to decipher secondary dispersion patterns, viz., halos, fans, and trains, according to their characteristic shape and relationship to ore deposit and (2) geobotanical, which depends on identification of indicator plants or association and/or morphological color changes in certain plants during their early growth.

The phenomena of selective concentration of certain elements by specific plant species, and more so of their selective parts like leaves, barks, twigs, roots, fruits, among others, provide the main basis for biogeochemical survey. The method offers the most effective means to find concealed/buried ore body, specifically in mountainous desert or rock terrace where plant grows with deep penetration of root.

Root systems extract elements selectively from the substrate, initially by following the lines of least resistance: first elements carried in gaseous phases, then those in solution, then labile phases (elements loosely bound to amorphous oxide coatings on soils), and finally by attacking with their strong chemical and physical powers any further requirements for elements structurally bound in crystal lattices. Thus, they perform the "ultimate selective leach." One needs to unravel the mysteries of the plant kingdom in order to harness the power of plants with respect to their value to assist in geological mapping and exploring for minerals.

6.5.9.2 Geobotanical Surveys

Geobotany is a well-established discipline and has been used since some years, as an aid in mapping geological formations and in locating fresh water, saline aquifers, and mineral deposits. Geobotany requires (1) an understanding of the nature and the distribution of plant communities and indicator plants, (2) recognition of the morphological (form and structural) changes in such plants, and (3) application of this knowledge to botanical information collected through remote sensing (aerial photography). The presence of indicator plants signals the existence of a particular element in the soil in which they grow. Those plants that consistently point to the presence of a specific element are called "universal indicators." Those plants used to locate mineralization only within a specified district are known as local indicators. Metal-rich soil and associated vegetation can be found by remote sensing, using multispectral techniques. In the porphyry copper region of the southwestern United States, studies have shown that abundant populations of "California popy" (*Eschscholtzia mexicana*) grow in close association of copper-rich soils near many porphyry deposits.

The method depends on direct observation of plant morphology and dimension of specific plant species that represents anomalous concentration of one or more elements in supporting soil. It involves: (1) identification of indicator plant; (2) growth or no growth in areas rich in specific elements; and (3) study of the effects of toxic quantities of certain elements on the plant growth, ie, plant morphology. These methods have been successful in deserts and semideserts. Microbiological methods can be used in surface prospecting for petroleum and gas deposits. Aerobic bacteria develop particularly well in soil and sedimentary horizons where a supply of gaseous hydrocarbon is present. In plants, different parts of same plant will show different response. Morphological changes in plants due to toxicity of certain elements are given in the table (Cannon, 1960).

Copper	Dead leaves on lower leaves from tips, purple stems, chloritic leaves with green veins, stunted roots
Cobalt	White dead patches on leaves
Zinc	Chloritic leaves with green veins, white dwarfed forms, roots stunted
Nickel	White dead patches on leaves
Molybdenum	Stunting, yellow orange colored
Manganese	Chloritic leaves, stem and petiole lesions, curling, dead areas on leaf margins
Chromium	Yellow leaves with green veins
Aluminum	Study roots, leaf scorch mottling

Chemical analysis of plants or plant parts can be used as a prospecting tool when the chemical content of the vegetation sampled bears a direct relationship to the chemical content of the soil or bedrock or to the ground water passing through the soil or bedrock. Prospecting with the help of plants has been successful for the following elements: U, Cu, Zn, Mn, Ni, Mo, Cr, V, Ba, Sn, W, and Ba. Biogeochemical sampling is an improvement over soil or rock sampling in regions of widespread but relatively thin postmineralization age overburden, because biogeochemical sampling is based on deeply rooted plants ($\sim$60 m deep). Plant material is sampled and analyzed periodically to minimize the effects of variations in precipitation, temperature, soil Eh, pH, soil type and texture, and plant physiology.

Biological interaction of metals in the ground with the rich organic topsoil has been found to occur. Metals complexed or bound to organic soil materials migrate upward from the soil into air. Barringer Research Ltd. (Canada) developed the "air trace system" which collects air samples from which metals are extracted. Metals directly detected by this system include mercury, copper, lead, zinc, and silver.

6.5.10 Electrogeochemical Specific Ion Surveys

It has been established that an important aspect of metallogenic process is the mobilization of metal ores as metal halide complexes in the mineralizing solution as indicated by numerous fluid inclusions. In most of the base metal mineralization, "flourine" is associated either in the form of a gangue mineral or as inclusions in other gangues. Hence it is logically expected that dispersion of halogen should exist around base metal deposits. Detection of halogens will give a clue to the concealed ore body.

6.5.11 Electrogeochemical Survey

Sulfide ore bodies which are buried deep below the earth act as giant batteries and the constant measurement of specific conductance can help in delineating a target. This is gaining momentum in locating deeply buried sulfide deposits. Electrogeochemistry depends on movement of ion in an electric field. In a mineral body, electrochemical dissolution happens over a huge area of electroactivity. According to Xianrong et al. (2008), "The process leads to elevated concentration of metallic ions in dispersion haloes surrounding a mineral body. Under the activity of geodesic electric field, metallic ions that are not reacted or absorbed will ultimately move upwards into loose sediments/regolith close to the surface and form ion halos in dynamic equilibrium." A schematic diagram showing the profile in rock and regolith and the formation of ICS anomaly is shown in Fig. 6.13.

At the point when an artificially generated electric field is strengthened and the supply time expands, the particles in element balance move upwards one by one and the space abandoned by moved particles will be filled by particles from adjacent minerals in order to keep up dynamic equilibrium. An extensive ionic source (mineral body) at a depth within the range of applied electric field might ceaselessly supply cations to a cathode extractor setup at the surface and will keep on aggregating until another dynamic equilibrium is reached. The particles assembled by the extractor are sourced from both the secondary halos close to the surface and ionic halos of mineral body at depth (Luo et al., 2004, 2010). Geochemical anomalies can be identified for exploration targeting by analyzing ionic concentration in the extractor (www.crcleme.org.au).

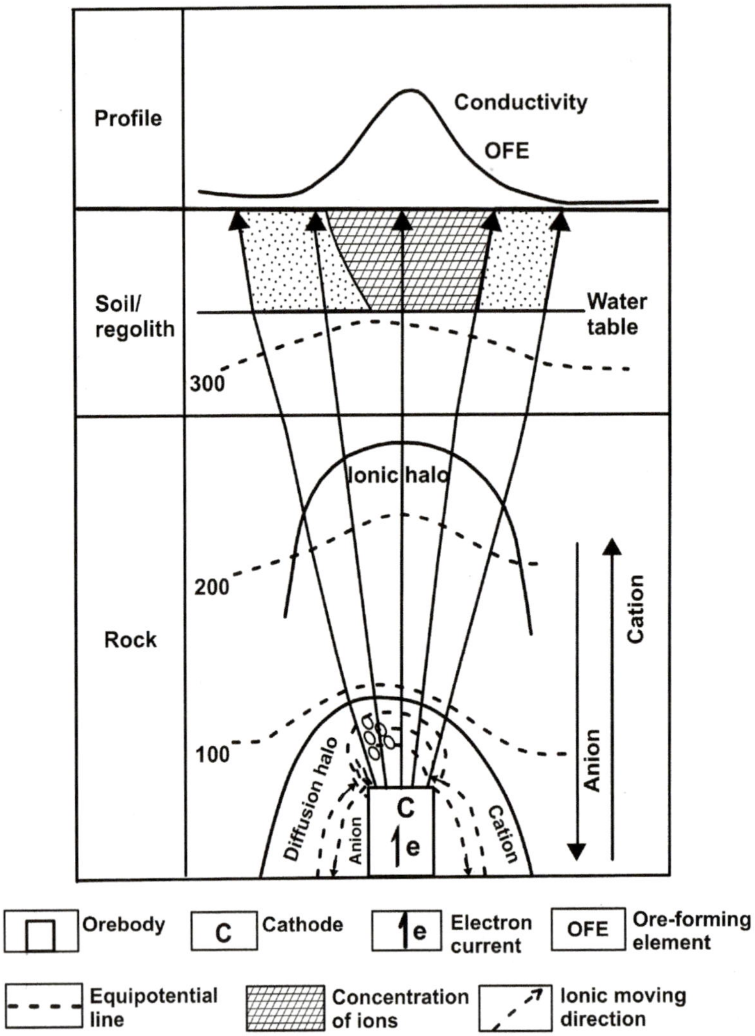

FIGURE 6.13 A schematic diagram showing the formation of ICS anomaly.

6.6 OTHER ADVANCED TECHNIQUES

The use of geochemical methods to recognize concealed ore bodies is confronted with various difficulties, all the more in this way, with regards to exploring landforms underlain by transported and complex regolith. A recent approach is to make utilization of partial and selective weak extractions (SWE), gas examination and biogeochemical studies to depict abnormalities in depositional landforms. Be that as it may, these studies had mixed response, attributable to the inadequately understood and complex mechanism and their usefulness in transferring metals associated with minerals upwards through complex overburden (www.crcleme.org.au). Most of these variants/mechanisms might probably be influenced by complex microbial processes also (Cameron et al., 2004).

As "change is an ongoing process," many new ideas/techniques are ventured by exploration geochemists. Advanced geochemical exploration techniques (Xuequi Wang et al., 1999) that are being practiced in various campaigns for detecting blind mineralization are as follows:

- Electrogeochemical (partial) extraction of mobile elements (CHIM) (Chastichnoe Izvlechennye Metallov). It is an electrogeochemical sampling technique which "allows for detection of buried deposits by surface identification of their secondary epigenetic (superimposed) haloes. CHIM is based on selective *in situ* extraction of electromobile forms of pathfinder elements, with direct electric current, into the specially designed element collectors (ECs), which are embedded in the soil at a number of measurement points along a profile to be explored and are connected to the direct current sources as either anodes or cathodes. A complementary electrode, common to all the ECs, is positioned at infinity" (Levitski et al., 1996).
- Ionic conductivity of soil/overburden (ICS)
- Selective or partial geochemical extraction enzyme leach (SPGE)
- Nanoscale metals in earth gases (NAMEG)
- Mobile forms of metals in overburden (MOMEO).

6.7 DESIGN OF GEOCHEMICAL SURVEY

The level of achievement of a geochemical method in a mineral exploration campaign is an impression of the measure of consideration brought with initial planning and study design. This period of action is alluded to as an orientation survey; its functional significance can't be overemphasized. At the point when a geochemical prospecting study is thought about, four fundamental contemplations must be attended to (1) the nature of ore deposits being looked for; (2) the geochemical properties of the components liable to be available in the target ore deposits; (3) geological components liable to bring about variations in geochemical background; and (4) environmental issue. These components are prone to impact the geochemical manifestations of the target ore deposit. Explanation of these components in an introduction study will allow configuration of a geochemical prospecting overview that is well on the way to demonstrate viable under the overarching conditions (www.lrd.yahooapis.com).

A classic geochemical exploration program consists of:

- Planning
- Sampling
- Chemical analysis
- Interpretation
- Detailed follow-up.

Geochemical exploration surveys can be grouped under two broad categories: (1) strategic or (2) tactical that might be further subgrouped by virtue of the material sampled. Strategic survey suggests a wide area coverage (expressed as thousands of square kilometers) where the essential goal is to recognize locales of improved mineral potential.

Tactical surveys constitute further detailed follow-up to strategic reconnaissance. Usually, the geomorphic area covered by a tactical survey is divided into isolated stretches of significant mineral potential within the normal anomalous provinces (www.seas.upenn.edu).

Geochemical soil surveys, geophysical surveys, and conventional prospecting are necessary before the decision regarding the chances of ores and drilling desirability. Close-spaced sampling to determine the axis of anomaly, pitting and trenching up to bedrock, and determining the metal value distribution is carried out to localize suitable target before initiating drilling. Geochemical surveys are not to be depended for pin-pointing drill targets in exploratory drilling. These surveys can indicate areas favorable for mineralization, but drill targets are to be decided on the integrated geological, geophysical, and other observations. A judicious integration of geological, geophysical, and geochemical exploration techniques principally leads to the greatest success.

6.8 SAMPLING FOR GEOCHEMICAL SURVEYS

Sampling for pedogeochemical, lithogeochemical, and biogeochemical surveys:

1. Random sampling to get a general survey along road, rivers, and in special development areas (granitic area, oxidation, and bleaching zones, etc.)
2. Regular samples: From the profiles perpendicular to the strike of more or less linear bodies.
   a. Distance between profiles—50–200 m
   b. Distance between sampling sites on profiles ~10–25 m
   c. Regular sampling in the form of networks between lines 10 and 250 m.

Soil samples: Three point method (3 samples taken at a distance of 1–2 m from the sampling point and equally spaced about it provide the desired collective sample)

Prospect scale: Profile sampling to follow-up geophysical anomalies (10 times the target width on either side)

Depth: ~10–60 cm, depending largely on the type of soil and the purpose

Amount of sample: ~50–200 g in paper/plastic bags/containers

Till samples: Closer to surface (0.5–1.0 m) in reconnaissance 7 regional surveys; in local surveys, sampling closer to bedrock

Hydrogeochemical sampling: At springs, wells, in rivers especially in the mouth area of tributary rivers

Amount: 0.5–2 litre, in plastic bottles

Pre-enrichment—By ion exchange resins, solvent extraction, filtering after coprecipitation with suitable collector. Concentration by simple evaporation.

Biogeochemical sampling: Twigs, plant, leaves, stem, leaves, etc.

Analysis and evaluation

- For assay of elements transported in dissolved form in water, finest fraction ($<100\ \mu m$) of soil/sediment sample
- Weather resistant minerals (heavy minerals, native elements, barites) fraction of 0.1–1 mm is ideal

- Plant samples are slowly reduced to ash at a temperature of <500°C and the ashes are analyzed either on dry basis or by dissolving them in a suitable acid mixture.

Analysis of samples (methods used)

X-ray Fluorescence (XRF)
Inductively Coupled Plasma Atomic Emission/Absorption Spectroscopy (ICP-AES/AAS)
Inductively Coupled Plasma Mass Spectrometry (ICP-MS)
Radiometry; Neutron Activation Analysis (NAA)
Electron Probe Microanalysis (EPMA).

The desired accuracy of analytical results in exploration geochemical surveys is usually low and is the order of 10–30%. The results are evaluated in the form of relative values.

6.9 GEOCHEMICAL MAPS

In the course of recording and interpreting the data of geochemical prospecting surveys, two different kinds of maps are prepared.

1. Data map—Actual observation and plot profile curves, distribution of metal values along separate lines of sample
2. Interpretation map—Involves grouping of data in range of concentration and plotting of geochemical contours (isogrades). Normally, intervals are selected as factorial multiples of the threshold.

In addition, data on geology, topographic contours, surface drainage, ground water seepage, etc. are used to facilitate interpretation.

6.10 INTERPRETATION OF DATA

6.10.1 Estimation of Background and Threshold

- Geochemical background—Range of variation of local/regional Clarke value of country rock and of water, air, etc.
- Geochemical anomaly—Regional and numerical range of concentration which exceed those of geochemical background
- Kurtosis—(contrast, slope, relief)—of a geochemical anomaly expressed by quotient: maximum/threshold or maximum/background

(A detailed account on geochemical data interpretation is given in the later portion of this chapter).

Homogeneity of geochemical anomaly—Uniform rise in values; variation at short distances will have bearing on the density of sampling

- Autochthonous anomaly—An anomaly about a spring produced by diffusion directly indicates deposit

- Allocthonous anomaly—Mostly produced by mechanical transport of decomposition (weathering) is not found above a deposit/or in direct connection with it.

6.10.2 Distinguishing Between Significant and Nonsignificant Anomalies

- Barren rock types characterized by relatively high background metal content (ultramafic rocks—Ni, Co, Cr)
- Human contamination: Mine dumps, smelter wastes, pattern of agriculture, roads, railways, etc.
- Sampling and analytical errors:
  - Nature of sample—organic matter content, clay, hydrous oxides
  - Depth of soil horizon in relation to sample depth
  - Vegetation and its influence on biogenic metal in soil
  - Local drainage conditions
  - Repeat analysis eliminates analytical errors, erratic high values.

6.10.3 Distinction Between Superjacent and Lateral Anomalies

- Anomaly was assumed to be superjacent, but lateral in reality, due to horizontal movement by ground water or of soil
- Relic pattern in residual soil—primary minerals, low cxMe:Me ratio
- Biogenic—more mobile metals, high cxMe:Me ratio
- Lateral anomaly—close relationship with local land form, changes in ground water level, climate will give false anomaly.

6.10.4 Appraisal of Anomalies

- Magnitude of values, expressed as contrast between peak and threshold
- Size and shape of anomalous area
- Geological setting
- Influence extent of local environment on metal content/pattern.

Procedure for follow-up

1. Revisit area, to confirm the cause whether any evidence for local enhancement or suppression of anomalous values. Recheck anomalous values.
2. Plan follow-up program of work.

Source of lateral anomalies will be sought—upslope, upglacier, updrainage, according to the mechanism of dispersion.

6.11 GEOCHEMICAL DATA PROCESSING

The objective of a geochemical exploration campaign is the detection of geochemical anomalies that have relationship with mineral deposits. In order to identify what is anomalous (ie, abnormal), one should have an adequate understanding of background values (ie, elemental value range occurring under natural conditions) along with the threshold value (ie, the lower limit of the anomalous values) and anomaly contrast (ie, ratio of anomalous to background values). Statistical methods of data processing provide a reliable means to geological understanding related to the geochemical data since it is the geology that governs the process of mineralization. Surveys which involve large number of geochemical data necessitate the use of statistical software packages for data processing and interpretation. Convenient approaches to exploratory data analysis include (1) 2D or 3D representation of multielemental geochemical data involving computer-generated automated plots and contours to reveal the spatial trend and (2) frequency distribution plots to detect multimodality and possible outliers, if any. Understanding from exploratory data analysis of geochemical values is an essential prerequisite for the purpose of identification of population or subpopulations with reference to the geological controls that govern the ore forming processes. Statistical analyses are then carried out employing: (1) univariate data analysis, that provide estimates of population parameters of background ranges and anomaly thresholds for single elements and (2) multivariate data analysis, where one-to-one correlations, associations, and cross-associations among multiple elements are obtained. Detailed account of the procedures of statistical data analysis is available in various texts (Davis, 2002; Marshall and Merriam, 1987; Howarth, 1983) and readers may like to refer to these texts.

6.12 ANALYSIS OF EXPLORATION DATA AND IDENTIFYING GEOCHEMICAL ANOMALIES

The primary aim in all geochemical exploration surveys is the detection of compositional responses in naturally occurring media and then relating the responses to potential mineral bodies. Geochemical anomalies are expressions of geochemical dispersion patterns of elemental concentration that are in deviation with the normal background values. They are of two types, viz., (1) abnormalities that are related to potential mineralization and can be used as geochemical guides to ore and are called "significant anomalies" and (2) superficially similar abnormalities caused due to geological or geochemical processes that are unrelated to potential mineralization and are called "nonsignificant anomalies."

From the geology point of view, geochemical anomalies are distinguished by setting the threshold limits for a specific population of geochemical data. Data values falling inside the upper and lower limits of the normal variation are called "background values" and the ones beyond these limits are called "anomalies." Statistically, a geochemical anomaly can be described as the value that exceeds the background (mean) by ± 2 standard deviation at 95% confidence level or ± 3 standard deviation at 99% confidence level. Clustering of elemental concentration values beyond this threshold can be considered anomalous. Excellent reviews of identifying geochemical anomalies in exploration campaigns, using univariate and multivariate statistical analyses, are given by Howarth (1983) and McQueen (2008).

Statistical methods have been extensively used to translate geological and geochemical information to identify geochemical anomaly. Such techniques should be utilized mindfully on account of the specific pecuralities of geochemical information. In other words, geochemical data rarely signify one population but rather multiple populations which indicate operation of multiple geological processes that have led to spatially dependent geochemical dispersion of elemental concentration. In case of multiple populations, the data require population split prior to univariate or multivariate analysis. Statistical studies must make use of several methods to adequately understand the distribution behavior of geochemical data prior to identifying geochemical anomaly (Reimann et al., 2005). The spatial visualization of geochemical data, statistical treatment and analysis when integrated on a Geographic Information System (GIS) platform, provides an effective tool for process identification of geochemical dispersion patterns in the large data sets.

6.13 GEOCHEMICAL SURVEY INTERPRETATION

Interpretation of geochemical exploration surveys could extend from spatially correlating the anomalous values with known mineral occurrences to statistical data analysis with multivariate techniques. Prior to initiating statistical data treatment, following checks are required to ascertain the data acceptability. These include:

- Checking the reliability of techniques of sampling and analysis
- Review of the data to check for contamination issues
- Scrutinizing the data authenticity and check for missing data
- Review of quality control data.

On fulfilling the data acceptability checks, provisional thresholds and provisional anomaly for individual elements are determined using cumulative frequency distribution and correlation matrix with scatter plots, respectively. Interpretation is the final stage and conclusions should be drawn after a very careful consideration of every bit of information revealed. If the results are not understood, all the efforts spent in planning and execution of a geochemical exploration program would go in vain. More than one geochemical process can play and normally does play a part in formation of an anomaly. Attempt should be made to understand the relative effect of each of the factors contributing to the formation of an anomaly. Background value range should be carefully estimated. Recognition of regional and local threshold zones delineates regions where further detailed work is required. Certain types of rocks may enrich particular type of metals. Specific plants are usually enriched with particular elements. Such situations may lead to false anomalies. They may also result from contamination which may be caused by

- Fragments of ore scattered near mines and transportation paths
- Metal-rich mine water
- Mineralized waste rock used for construction purposes
- Industrial waste water
- Smelter fumes and slags
- Metal-rich manure, insecticides, etc.

It must be remembered that the presence of an anomaly, as established by geochemical exploration survey, is only indicative of possibilities of economic mineralization and not of an ore deposit as such. An uneconomic deposit near the surface and an economic deposit much below the surface may produce similar range of anomalous values. Favorable host rock and structures are essential prerequisites for the formation of ore deposits. Hence the necessity of an integration of the geochemical survey with geological exploration. Reference to adjoining areas already investigated is also useful. Anomalies are not necessarily vertical projections of the ore body, particularly when a survey has been carried out on transported soil cover. Thus, anomalies on the surface are only indicative of the presence of ore body somewhere in its vicinity or at some depth.

6.14 TYPICAL GEOCHEMICAL EXPLORATION PROGRAM

The first step in a geochemical exploration program is to plan an orientation survey. This aims at determining the type and size of the anomalies in the vicinity of a known mineral deposit in different media, namely, wall rocks, residual soil, stream sediment, water, plant, etc. In the course of an orientation survey, the best sampling medium is identified. Suitable methods of sample treatment and analysis including determination of the best grain size fraction for anomaly amplification and the choice of metal extraction techniques are also decided. The nature of the overburden, whether residual, glacial, alluvial, or wind borne origin is the first question that must be answered by an orientation survey. In residual soil, a thorough orientation survey starts with the collection of a series of vertical sections through the soil profile, reaching up to the bedrock. Comparable profiles from the background areas should also be sampled at the same time. The most practicable horizon for sampling would be the minimum depth at which an adequate anomaly contrast is obtained over the maximum width. In transported soil, the principal factors that are to be considered include the nature of sediment, pH, Eh, direction of ground water movement, glacial direction, seasonal changes in the environment, mobility of the elements sought for, among others.

A careful and critical orientation survey followed by detailed planning and organization of the relevant sampling campaign is crucial to the success of geochemical exploration program. Background values must also be determined by sampling barren areas in the neighborhood. The survey should be based on valid principles and should be carried out efficiently taking into account the following considerations:

1. Sampling pattern

Sampling pattern is determined by the size, shape, and topography of the target area under consideration. Rectilinear grids are preferred because of the ease in supervising the field survey and in plotting the data. If the strike direction is known, the traverse line interval should not exceed one-third of the minimum economic strike length. Interval between the samples should be fixed up by probable minimum width of the expected anomaly so that at least two samples fall within every important anomaly. If the anomalies are not homogenous, a corresponding chosen sampling interval would be required. If the strike direction of an elongated target is not known or if the anomalies are expected to be irregular or equidimensional, the most suitable pattern is a square grid. While collecting samples in an undulating terrain following a rectilinear grid, necessary slope corrections are to be made. Sampling pattern in a very steep terrain may be led out in a way that it conforms to the topography. For surveying a hydromorphic anomaly, the sampling pattern should be determined by the seepage or drainage pattern.

2. **Sampling procedure**

   Techniques of collecting samples depend on type of material being sampled.

   1. **Rock**: Elements occurring in rocks as primary dispersion may either be homogenously distributed or segregated. If it is uniformly distributed, one single grab sample may be collected. In cases where the elements are segregated, either channel or chip samples would be more representative.
   2. **Soil**: An orientation survey would determine the soil horizon to be considered. Care should be taken to collect samples from the same horizon. Soil samples at depth of 1 or 2 feet may be collected conveniently from small pits with the help of crowbar. Deeper samples may be collected by means of a soil auger. A quantity of 20–50 grams of sample, as a rule, will provide the analysts' need of 5 grams of dry prepared material (sieved or powdered). The optimum system for sample preparation should be determined as a part of orientation experiment. Generally, it is only necessary to sieve the sample. Rarely, the elements sought may occur preferentially as clastic grains (eg, such as Beryllium in Beryl). In such cases, it is necessary to retain and crush the particular coarse fractions. It may be necessary to dry the samples before sieving. Noncontaminating sieves should be used for the purpose.
   3. **Sediment**: Where heavy minerals content of sediment is to be determined, samples are collected from the bottom of the stream channel. About 5 grams of prepared sample is adequate for analytical work.
   4. **Vegetation**: Samples of vegetation should be taken far above the ground so as to remain free from contamination from rain spatter. Care should be taken to collect samples from the same organ of the same species of the plant for comparable data. Wood can be collected from the main trunk with the help of a carpenter auger. A quantity of 5 grams of dry sample would suffice the analytical requirement.
   5. **Water**: It is difficult to store water samples since traces of metals present are relatively unstable. Ions may be removed by adsorption to the walls of the container, metals may be coprecipitated with Fe–hydroxide precipitate that commonly form when water is stored. These effects may be minimized by acidifying the sample immediately after collection. Portable kit may be used for onsite spot analysis.

3. **Location of sampling sites**

   Location of sampling sites needs only be accurate adequately to enable any anomalous site to be revisited. Locations may be plotted on topographic maps. Samples should be numbered carefully so that there is no duplication and also no possibility of the sample numbers getting smeared.

4. **Elements to be determined**

Analysis of the major element is the most straightforward approach but analysis of associated elements and pathfinder elements also leads to the finding of mineral deposit. In a virgin area where nothing is known about the mineralization, analysis for expected ore metals should be carried out. However, since it is expensive to analyze for all the metals, orientation survey results should enable the selection of element to be determined.

References

Andrew, A.S., Carr, G.R., Giblin, A.M., Whitford, D.J., 1998. Isotope hydrogeochemistry in exploration for buried and blind mineralisation. Geological Society of Australia Special Publication, v. 20, pp. 222–225.

Averill, S.A., 2001. The application of heavy indicator minerals in mineral exploration with emphasis on base metal indicators in glaciated metamorphic and plutonic terrains. In: McClenaghan, M.B., Bobrowsky, P.T., Hall, G.E.M., Cook, S. (Eds.), Drift Exploration in Glaciated Terrain, Special Volume 1985. Geological Society of London, pp. 69–82.

Beus, A.A., Grigorian, S.V., 1975. Geochemical Exploration Methods for Mineral Deposits. Applied Publishing Ltd, Wilmette, p. 287 p.

Cannon, H.L., 1960. Botanical prospecting for ore deposits. Science v. 132, 591–598.

Cameron, E.M., Hamilton, S.M., Leybourne, M.I., Hall, G.E.M., McClenaghan, M.B., 2004. Finding deeply buried deposits using geochemistry. Geochem. Explor., Environ., Anal. 4, 7–32.

Clarke, F.W., 1942. The data of geochemistry. U.S. Geol. Survey Bull. 770, 10.

Davis, J.C., 2002. Statistics and Data Analysis in Geology, Third Edition John Wiley & Sons, New York, 656 p.

Fletcher, W.K. 1997. Stream sediment geochemistry in today's exploration world. In: Proceeding of Exploration 97: Fourth Decennial International Conference on Mineral exploration, Gubins, A.G. (Ed.), Toronto, pp. 249–260.

Fletcher, W.K., Hoffman, S.J. Mehrtens, M.B. Sinclair, A.J., Thomson, I., 1987. Exploration geochemistry: design and interpretation of soil surveys. SEG Review in Economic Geology, vol. 3, 180 pages.

Govett, G.J.S., 1983. Rock geochemistry in mineral exploration, Handbook of Exploration Geochemistry, v. 3. Elsevier, Amsterdam.

Govett, G.J.S., 1989. Bedrock geochemistry in mineral exploration. In Garland, G.D., Ed., Proceedings of "Exploration 87", spl. vol. no. 3. Ontario Geological Survey, Toronto, Canada.

Govett, G.J.S., Dunlop, A.C., Atherton, P.R., 1984. Electrical techniques in deeply weathered terrain in Australia. J. Geochem. Expln. 21, 311–331.

Gulson, B.L., 1986. Lead Isotopes in Mineral Exploration. Elsevier, Amsterdam, p. 246.

Hale, M., 2000. Genetic models of remote dispersion patterns. In: Hale, M. (Ed.), Geochemical Remote Sensing of the Subsurface'. Elsevier, Amsterdam, pp. 3–16.

Hawkes, H.E., Webb, J.S., 1962. Geochemistry in Mineral Exploration. Harper and Row, New York, p. 415.

Horsnail, R.F., 2001. Geochemical Prospecting. In: AccessScience@McGraw-Hill, <http://www.accessscience.com/>, <http://dx.doi.org/10.1036/1097-8542.285700/> modified March 29, 2001.

Lett, R.E., 2007. Drainage geochemical surveys—stream sediments, lake sediments, moss mats, heavy minerals. In: Geochem'07 Workshop, Toronto, 63 p.

Lett, R.E., 2009. Geochemical exploration from test tube to mass spectrometer. University of British Columbia, Mineral Deposit Research Unit. Society of Economic Geologists Student Chapter Invited Talk, March 29th, Talk Notes, p. 19.

Levinson, A.A., McCammon, R.B., Hitchon, B., 1980. Introduction to Exploration Geochemistry, 2nd Edition Applied Publishing, Calgary, p. 924.

Levitski, A., Filanovski, B., Bourenko, T., Tannenbaum, E., Bar-Am, G., 1996. "Dipole" CHIM: concept and application. J. Geochem. Expln. v. 57 (1), 101–114.

Luo, X.R., Toofa, Z., Hou, B.H., 2004. The geoelectrochemical extraction method (CHIM) in exploration for concealed ore deposits. Regolith 2004, pp. 230–233.

Luo, X.R., Wen, M.L., Shu, L.X., Hou, B.H., Keeling, J.L., 2010. Ionic conductivity anomaly in soil cover—Exploration of blind mineralization beneath regolith cover. Science China (Earth Science) v. 53 (n. 5), 649–656.

Marsal, D., Merriam, D.F., 1987. Statistics for Geoscientists. Pergamon Press, p. 176.

McClenaghan, M.B., Kjarsgaard, B.A., 2007. Indicator mineral and surficial geochemical exploration methods for kimberlite in glaciated terrain; Examples from Canada. In: Goodfellow, W.D., ed., Mineral Deposits of Canada: A Synthesis of Major Deposit-Types, District Metallogeny, the Evolution of Geological Provinces, and Exploration Methods: Geological Association of Canada, Mineral Deposits Division, Special Publication No. 5, pp. 983–1006.

McQueen, K.G., 2008. Identifying geochemical anomalies. CRC-LEME, Dept. Earth & Marine Sc., NU notes, 7 p.

Miller, J.K., 1984. Model for classic indicator train in till. In: Prospecting in Areas of Glaciated Terrain. Inst. Min. Metal., London, p. 69–77.

Reimann, C., Filzmoser, P., Garrett, R.G., 2005. Background and threshold: critical comparison of methods of determination. Sci. Total Environ. 346, 1–16.

Rose, A.W., Hawkes, H.E., Webb, J.S., 1979. Geochemistry in mineral exploration, 2nd Edition Academic Press, London, 657 p.

Smith, B.H., 1977. Some aspects of the use of geochemistry in the search for nickel sulphides in lateritic terrain in Western Australia. J. Geochem. Expln. v. 8 (1–2), 259–281.

Stajevic, B., 2004. Geochemical haloes of gold in Lece ore field, South Siberia, Russia. Ann. Geol. Peninsul. Balk. v. 63, 93–99.

Wang, X., Xie, X., Cheng, Z., Liu, D., 1999. Delineation of regional anomalies penetration through thick cover in concealed terrains—a case history from Olympic Dam.

Xianrong, L., Baohong, H., Meilanna, W., Nanshi, Z., Keeling, J., Fidler, R., et al., 2008. CHIM-geoelectrochemical method in search of concealed mineralisation in China and Australia. Chinese J. Geochem. v. 27 (2), 198–202.

CHAPTER

7

Geological Exploration

OUTLINE

Essentials of Mineral Exploration and Evaluation

DOI: http://dx.doi.org/10.1016/B978-0-12-805329-4.00014-4

7.1 INTRODUCTION

If there is no mineral exploration, there would be no mining, no processing, no industry, few of the cultural amenities and creature comforts. For the maintenance of any country's economy, it is inevitable to have adequate supply of mineral resources. Mineral exploration has its own characteristics that differ from other resources sector. They are: (1) mineral resources are largely hidden from view. Some part of the earth is well endowed with respect to mineral resources while other parts are poorly endowed. (2) Mineral deposits are "where they are in nature" and cannot be moved out to a desirable location by anybody. (3) After establishing a deposit, a mining camp is developed. Since many deposits are known to occur in cluster, the first developed mine and its infrastructure will facilitate for the development of additional finds, thereby reducing costs for the project. (4) Many a times, use of advanced technology is needed to define an economic mineral resource. Mineral exploration is a time-consuming "high reward, high-risk" event.

Many companies, Government, Private, and Local communities, are involved in mineral exploration worldwide. Although successful exploration yields good rewards/dividends to the participants and local communities, the risks involved are also high. Responsible mining companies can be a driver of economic and private sector development, as well as poverty alleviation at a national level. They can also offer technology transfer to both the local community and to the nation. The ultimate goal of mineral exploration is the extraction, beneficiation, and profitable and beneficial sale of mineral commodities.

7.2 MINERALS ACTIVITY PROJECT

In general, a minerals activity project can be divided into various stages involving the application of successively more discriminating (and more expensive) techniques, to successively smaller land areas, in order to identify, develop, and produce an economic mineral deposit. A "full sequence" of mineral project life cycle, from exploration to development and operation to closure would involve the following stages:

1. Regional appraisal
2. Reconnaissance of the selected region
3. Detailed surface investigation of the target area
4. Detailed three-dimensional physical sampling of the target area
5. Deposit development
6. Mine development and production

7. Mineral processing/mineral beneficiation
8. Mineral/metal extraction; smelting and refining
9. Marketing of mineral/metal produced
10. Mine closure, after the end of mine's life.

Target identification constitutes the first two stages whereas the next two stages are *Target investigation*. Hence, in short, mineral exploration entails the identification and investigation of targets to discover economic mineral resources. Once a mineral resource is discovered, all the associated infrastructural works to bring the deposit to the point of production is called *Development*. Actual mining of ore from the ground is referred as *Production*. *Mineral processing* involves comminution, separation of ore (economic) minerals from gangue (waste), and concentration of ore minerals. *Metal/mineral extraction (smelting and refining)* involves extraction of metals from the mineral concentrate and purifying them, if need be. *Marketing* involves shipping of the mineral product (can be concentrate or metal/refined ore) to market. The final stage is "*mine closure*" (after the end of mine's life) involving significant expenses to clean up and remedial measures of degradation due to mining and smelting sites. This stage will also involve employee retrenchment, social and community implications. The boundaries between these substages mentioned earlier are not precise but arbitrary. It is generally considered as a continuum of activities.

The various stages and activities of mineral deposit life cycle are given in Fig. 7.1. Only a few projects will see the full sequence since a project can be abandoned at

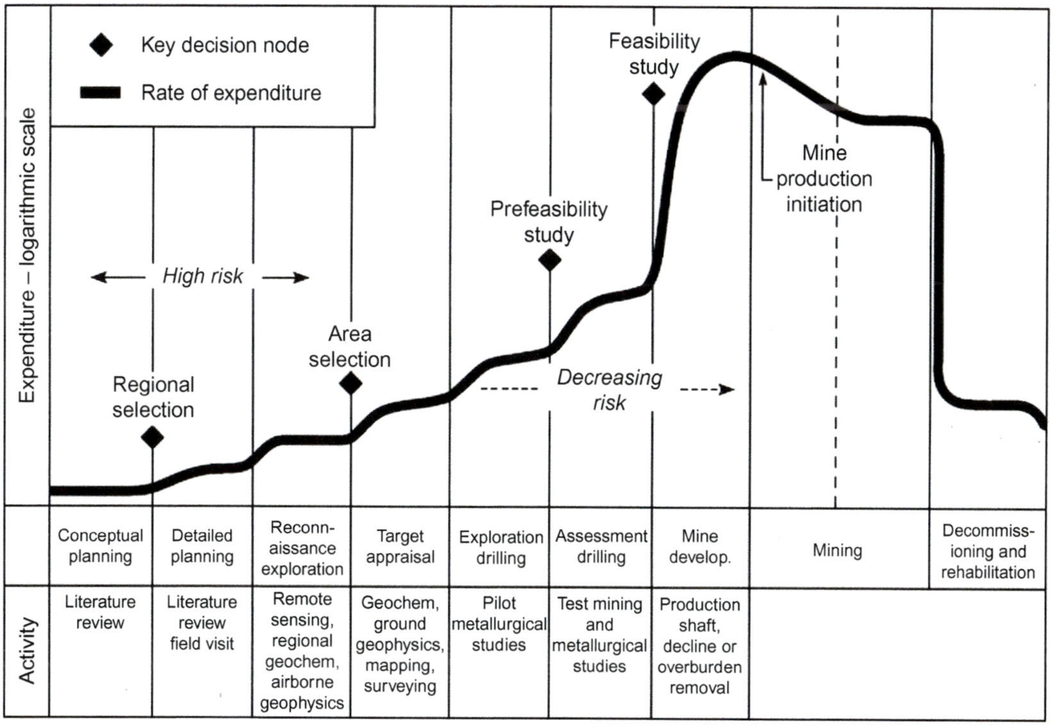

FIGURE 7.1 Various stages and activities of mineral deposit life cycle. *Source: From Moon et al. (2006).*

any stage, if the results are discouraging. It is possible to skip or abbreviate some of the initial stage activities, if the information is already available to the company either by its own earlier work or work done by other agencies. Only certain types of deposits in certain type of geological environments demand some techniques, owing to cost/physical characteristics.

In the above referred mineral exploration/supply chain, a Company, depending upon its strengths and weaknesses, can enter at any stage with many options. Grassroot exploration can be conducted by a Company and later involve in advanced exploration by joining with a partner to undertake further exploration. It can also opt to develop a known "undeveloped deposit." Alternatively, a company can invest in R&D work that can make activities at any stage of production more efficient. The scientific/technical inputs of exploration and mining must be fast, efficient, and effective in order to add value to the project (Woodall, 1992).

7.3 MINERAL EXPLORATION

Mineral exploration can be technically defined as "all the activities and evaluation necessary before an intelligent decision can be made establishing size, initial flow sheet, and annual output of new extractive operation." The purpose of mineral exploration is the discovery and acquisition of new mineral deposit amenable to economic extractive operations now or in future. The prime objective of mineral exploration is to find and acquire a maximum number of such economic mineral deposits at a minimum cost and within minimum time.

Mineral exploration acceleration is due to (1) increasing demand for metals that were not sought earlier, (2) growth of industrial output, (3) new ore types, and (4) greatly improved geological knowledge and exploration technology. Global demand for industrial commodities has doubled every 20 years. This growth is likely to continue. Global reserves and resources for some commodities are sufficient for several decades. In major producing districts, the reserve and resource grades are declining owing to depletion of high-grade surface deposits. Additional resources will be augmented by ongoing "greenfield exploration" in frontier countries and also by deeper discoveries from known mineral districts. These discoveries will become a major new source of future mineral supply (Woodall and Duncan, 1993). Despite some of the countries have the capacity to support production, it is probable that might be undermined by a few risk factors.

Mineral exploration represents the highest risk of all—risk of failure is great and the cost is high. Even the successes must be followed by capital-intensive long-term, high-risk operating investments. Investments to bring exploration discoveries into production may range in the tens of million dollars. Large mines require from hundreds of million dollars to billion dollar investment. The typical lead time span, from inception of exploration through the investment in mine/mill facilities that will return this investment out of production, is 12–15 or more years. To receive a profitable return on investment commensurate with the risk requires a minimum investment longevity of an additional 10 years, for a total of 20–30 years.

Since most of the surface or near-surface resources have been explored, the search for new mineral resources has to rely on more sophisticated prospecting/exploration techniques. Future ore bodies will probably be more costly to find, mine, and process than those in operation and production now, because most of them will either be of lower grade or found at greater depths. The chances of a mineral occurrence being developed as a mine are low (may be 1 in 1000). Lesser number of identified resources does not imply low mineral potential in an area. It might be due to many reasons like, incorrect geological theories, poor infrastructure and policy of the Government. A typical example is the "Diamond discoveries" in Northwest Territories in Canada. There were no known occurrence of "kimberlite pipe" (host of diamonds) although the geology was known to be favorable. But once the kimberlite pipes were discovered, the area attracted many exploration teams and led to significant resources.

Most exploration projects will not advance to mines. In areas with good potential for discovery, the competition is usually high. Invariably an interdisciplinary team of geologists, geophysicists, and geochemists searches for mineral deposits in prospective terrains. The key to successful exploration programme is to know, when and where to drill, when to "hold the properties" and when to "walkout."

7.4 EVOLUTION OF EXPLORATION TECHNOLOGY

The exploration has evolved in the past several decades into a highly sophisticated and expensive endeavor in which large corporate entities are involved with trained specialists in myriad fields. The evolution of exploration technology spanning over the past century can be broadly summarized as under:

1900–50	Prospect submittal, mapping, sampling in/near ore bodies; exploration done by trenching, pitting, shafts, drifts, and little drilling
1950–60	Submittals, sampling, geological mapping, drilling; ground and airborne magnetics, electromagnetics, early geochemistry
1960–70	Grassroots become popular, geophysical and geochemical surveys, induced polarization developed, scant drilling, photogeology
1970–80	Mostly grassroots, more regional geophysical and geochemical surveys, ore body modeling, less prospecting
1980–90	Plate tectonics, remote sensing, computer modeling
1990–2000	Increased use of high-tech office techniques, less field work, poor exploration success ratios
2000–	Increasing high-tech office work, extensive use of state-of-the-art 3D and GIS modeling techniques in synthesis and interpretation, increased discovery costs, very less field work

Searching only for outcropping minerals had given most of the present day mines in many areas, but this is approaching a point of diminishing returns in areas that had been thoroughly prospected. Exploration is now increasingly aimed at potential areas within which anticipated mineral deposits are covered by postore cover in the form of alluvium or volcanic sedimentary rocks.

7.5 DEVELOPMENT OF EXPLORATION TECHNOLOGY

In the past six decades, technological acceleration in other sciences, viz., computer science, telecommunication, genetic engineering, rocket science, etc., has more than doubled the sum of human knowledge in shorter time intervals, but the same is not the case in Mineral Exploration Technology. Any new technique in exploration technology needs to be tested for its worth in the field and if successful only, it can be adopted. Hence, it takes a lot of lead time (Fig. 7.2A). The slope of the curve depicting the development of exploration technology represents successive postwar development of magnetometrics, EM systems, that found many massive sulfide ore bodies in the 1950s and 1960s. Magnetic surveys helped to find many iron ore bodies and radiometric surveys found uranium ore bodies. Development in geochemical techniques caused later bumps in the curve. Geochemical techniques found many mines. In the late 60s and early 70s, many porphyry copper–molybdenum deposits were discovered due to better understanding of deposit geology. The improvement in understanding the origin of mineral deposits accounts for the upward slope of the exploration technology curve. Drilling technology and high precision, speedier analytical methods (XRF, mass spectrometer, ICPMS, ICPES, etc.) augmented geochemical exploration techniques (Lovell, 2000).

The exploration efficiency has slackened down in the recent years hence demanded an increased R&D efforts to develop techniques to locate new resources. In the recent decades, the exploration expenditure, both by the Government bodies and by the private companies, has gone down since the mining industry in general was forced to concentrate on survival during the very lean years.

With the easy to detect and easy to mine deposits nearing the end of their life span, one has to go deeper, hence, over the years, the exploration costs are increasing (drilling can be an expensive event). Similarly, the discovery/success rate tends to decrease, besides the overall decrease of grade of resource, as one does not have direct evidence of mineralization (Fig. 7.2B).

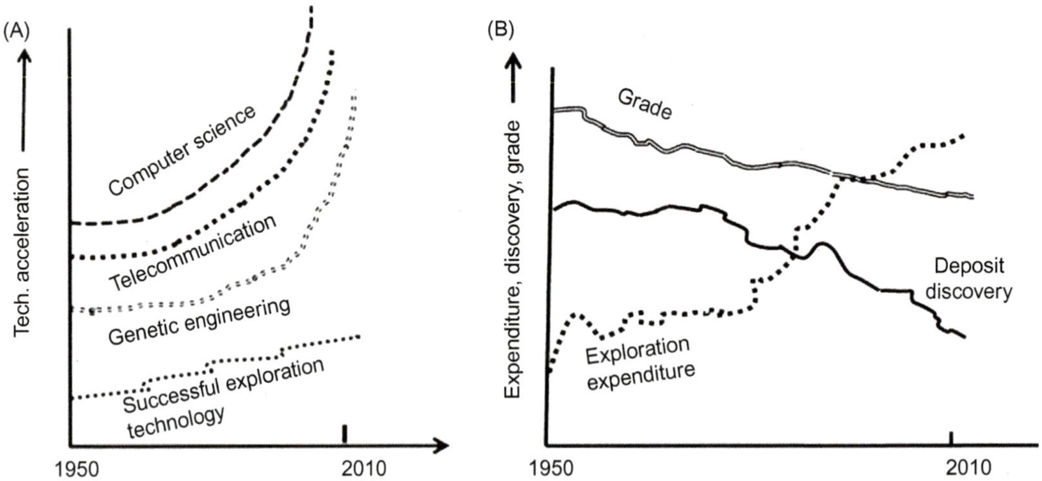

FIGURE 7.2 (A) Sketch map of development of exploration technology. (B) Sketch map of the trends of exploration expenditure versus deposit discovery and deposit grade, over the years. *Source: From Gandhi (2013).*

7.6 CHALLENGES FOR MINERAL EXPLORATION

Things in the mining business are not the same what they were, a couple of decades ago. Several things have happened, which have acted as disincentives. Dummette (2000) opines that "the mining industry practices its trade in a universe defined by prices, laws, and Governments. Prices, as in what the industry is paid for what it produces and the costs of goods and services. Laws have to be transparent and fair. Government that the industry prefers should be stable and supportive of the industry, catering not corruption." With the opening up of frontiers of many countries in the world for investment in mineral exploration, it has become a "borderless world." But at the same time, the companies need to be responsible corporate guests who respect and enhance the social, moral, and societal values of their host countries.

Some of the "emerging challenges" for the exploration industry are discussed here.

- Discoveries are harder to make; exploration costs have risen dramatically over the last three decades; unit discovery costs are rising. Drilling has become an expensive component (But if you do not drill, you would not discover any resource!).
- Rate of world class/large deposit discoveries (both size and grade) is declining with terrain maturity. (Existing mines are getting depleted fast and we need to replace them.)
- Recent undue emphasis on "brownfield" exploration—biggest discoveries tend to be in "greenfield" exploration.
- Growing importance of "native title" (land titles of mineral deposits in SW Pacific, Andean area, North Canada, India, etc.). The mining industry in a globalized business environment will be required to be responsible partners and guests.
- Startling imperatives of "sustainable development." Since the emergence of environmental awareness, society and mining have been in conflict and are at loggerhead.
- Mining sector is vulnerable to political risk because of the nature of high cost, long-term investments. "Expropriation, confiscation, and nationalization of assets, along with inadequate legal and regulatory systems that heighten contractual work (resource nationalism is the most evident in Latin America, Central Asia, and Africa, and it has become a notable trend in some of the world's richest territories also). Political risk is region- and asset specific and is not generic within a country/sector.
- Recent worldwide changes and developments in Government policies have become disincentive factors for many exploration companies (some countries have decreased/increased their taxes and royalty rates or introduced "nationalism-type" rules since 2011).
- Many exploration projects are often located in unstable locations that are prone to outbreaks of war and civil commotion or vulnerable to the risk of terrorist attacks. The risk to personnel of kidnap for ransom is high in many mining territories (Mali and Peru in 2012; Central African Republic, Colombia, Kyrgystan, South Africa in 2013). With the result, the importance of political risk insurance (PRI) as a risk transfer mechanism to minimize the impact of the country risk has increased.
- Inability to detect deep-seated mineral deposits and deposits beneath thick cover—a major impediment to success.

- Declining prices of most of mineral commodities.
- The "boom and bust" financial cycles restrict funding and capacity to advance exploration and technology development.
- Ever-increasing "production cost"; hence "low/diminishing returns" to company and its shareholders. Research by Rothschild has shown that for those major companies that have been around for 25–30 years, the average real return has been around 7–8% per annum.
- Decline of experienced exploration teams and strategies; brief time to assess new targets.
- Cost of collecting geoscientific data is the most expensive aspect of mineral exploration.
- A need to balance risk versus opportunity.
- *Massive restructuring of industry; mergers, joint ventures, partnerships, etc.* It is common nowadays that a mining company has merged with or acquired one, its competitor. The rationale for the merger is the creation of more valuable company because of low production cost and some synergies. The new company will be more profitable and more attractive to shareholders. In reality, study of many merged companies in the world indicates that one-third of the transactions created value, and another one-third were value neutral and the rest were value destroying. Mergers need not make better companies, in many cases, it is just "the opposite."

7.7 DESIGNING AN EXPLORATION APPROACH

It is necessary to have some knowledge or a hypothesis of how and why a mineral occurs in nature in a particular place in the earth's crust, in order to design an exploration approach. Gold, silver, and copper are particular type of elements with a metallic luster and have other unique physical properties that distinguish them from nonmetallic element. Metals not only occur in their native form but also occur as a part of a mineral.

The physical properties of minerals and their mode of occurrence facilitate in locating deposits of economic interest. The common mineral characteristics that are used are magnetic susceptibility, specific gravity, electrical conductivity, radioactivity, and velocity of seismic waves. Besides, some minerals of economic interest are found to occur in association with other rocks/minerals that may be easier to locate than the economic minerals themselves. Dispersal of chemical signatures of hidden mineral deposits over wider areas by physical/chemical processes such as weathering or erosion, thereby, making their presence known by sampling stream sediments, soil, glacial till, etc.

Exploration methods in increasing order of cost per square kilometer are:

In increasing order of Cost/km^2 ↓

- Remote sensing (satellite and imagery)
- Geological mapping
- Geophysical surveys
- Geochemical surveys (bulk sampling)
- Drilling

Number of factors like type of deposit (vein, disseminated, layered, etc.), location of the area, availability of infrastructure, the existence and quality of other available data about the area dictate the exploration approach to be adopted. An exploration approach designed for locating certain type of deposit (disseminated copper) would be generally not suitable in locating another type of deposit (high-grade gold).

Having developed an initial exploration concept, if the field work done in that area yields negative results, the same area attracts with renewed interest, a few years later, with the availability of new technology (may be satellite imageries) or a new concept of mineral formation (like diamond discovery in northern Canada). Hence, any unsuccessful exploration does not mean that mineral deposits do not exist. The probable causes could be, incorrect exploration method, insufficient effort or old technology incapable of locating deposits. Thus exploration in an area always follows many cycles by different agencies. In the process, the ownership can also change many hands.

7.8 THE EXPLORATION CYCLE

The exploration cycle includes geological concept formation and reconnaissance exploration, advanced exploration, feasibility studies, and possible development. The information gathered at the end of each stage of this process is evaluated before venturing into the next stage or revise the earlier concept or abandon it entirely. Different types and levels of information are obtained in each stage of the exploration cycle, environmental impacts, expenses and demand different skills.

In the early exploration stage, large areas are identified that may host certain type of ore deposit. The time taken can be long to locate areas that merit further work. Fewer companies are involved at this stage and the cost of exploration rises as drilling and detailed surveys are undertaken. Feasibility studies will be initiated if the results continue to be promising. At this juncture, junior companies may choose to spread the costs and risks of further development by joining hands with senior companies.

7.8.1 Reconnaissance and Preliminary Exploration: Geological Concept Formation

In order to find the type of minerals interested, the type of deposits in which such minerals are to be found and the areas where such resources are likely to be located require sound geological concepts. Geological knowledge gained through education and experience and the geological information of the areas where exploration is likely to be conducted are essential.

7.8.1.1 Typical Activities

Exploration starts with regional appraisal of large areas (in one or more countries), primarily for collating known geological information, reviewing of maps, surveys, and reports/records available from the National and Provincial geological surveys, Universities or Academic/Research Institutions, and accessing "company reports" containing prior proprietary knowledge. Selection of particular regions considered favorable for occurrence of the mineral or minerals being sought using air photo or satellite images, remote sensing technique, regional surveys (geological, geochemical, and geophysical), field reconnaissance, and preliminary engagement with local communities. Large-scale structural trends, potentially controlling mineralization, may also be identified through these studies. The main activities of the geologist are field mapping and sampling. There is nothing better than "boots on the ground" in assessing outcropping geology for its potential. However, many targets are obscured by intense weathering, soil cover, or more recent deposits and have to be detected through their remote geophysical or geochemical responses, as the mineralization cannot be seen at the surface. Grassroot exploration would help to identify promising areas which will be examined in greater detail in the subsequent stages.

The cost of early stage exploration programme depends on the extent of field exploration. Costs are relatively low, up to several tens of millions of dollars.

7.8.2 Advanced Exploration (Detailed Target Evaluation)

Since sufficient encouragement has been obtained from initial exploration, the activities warrant further expenditure at this stage. By now, geophysical and geochemical anomalies have been delineated, identified favorable geology and encouraging assay results of the samples collected. This stage demands greater density of sampling to be carried out than the grassroot level exploration.

7.8.2.1 Typical Activities

In this stage, more subsurface information is gathered, involving trenching, pitting, drilling, and delineation of the mineral resource. The detailed geological mapping, topographical survey, more soil sampling, ground geophysics, and some drilling may be undertaken. Preliminary ore amenability studies, environmental, social baseline studies are to be carried out. Constant interaction with the local communities is to be done. The outcome of these studies in advanced exploration stage leads to "scoping study" which is an "order of magnitude evaluation" of the deposit's commercial attractiveness. A "preliminary feasibility study" involving more detailed resource estimate, preliminary mine design, and preliminary cost estimates are to be done.

After examining the range of options, the prefeasibility produces a case for a plan that is commercially feasible. It addresses not only the economic outcome but also acceptable risk profile, while there are many areas of uncertainty, such as suitability of ground for foundations, costs have been estimated to within 20–30% of accuracy. If both "scoping and preliminary feasibility studies" suggest that a resource might be commercially feasible, the next stage (deposit development) will be taken up. The cost estimates for advanced exploration programme might be 10 times the cost spent in the initial stage (the expenditure could go up to several hundreds of millions of dollars).

7.8.2.2 Public Consultation

Exploration activities envisaged in advanced exploration stage will become more visible to the community around and probably may have potential impacts. At this stage, informal meetings, consultations, and public forums are to be carried out with the community, as and when necessary.

7.8.3 Feasibility Stage

This stage is taken up when the exploration results have been highly encouraging to justify detailed economic and engineering studies. These studies will facilitate investigation of possible ways of developing a property as a technically and economically feasible, environmentally sound, mining operation. If the "prefeasibility" leaves a few "fuzzy edges," the feasibility study will sharpen them. It is a focused process of determining exactly how the option chosen in prefeasibility study will come together and estimating costs to within 10–15% of actual. Between the discovery and delineation of a mineral resource, lying invisible beneath the ground, to the start-up and commissioning of the flow of ore through processing plant, comes the application of technical expertise and planning of the first order.

7.8.3.1 Typical Activities

The activities in this stage involve examination of possible mining and processing methods, estimates of mineable resources, continuation of environmental studies, and marketing and economic studies. In order to have additional confidence on the resource estimates, planning of detailed in-fill drilling is to be taken up. In essence, "feasibility studies" are detailed engineering investigations of the technical, economical, and environmental effects of a proposed mining operation. Besides, many ancillary studies are also to be undertaken which will form part of the feasibility report. Feasibility studies are considered as "major documents" of a project and can comprise of several volumes.

The expenditure in this stage of development will be in the order of US$5–10 million or more. Depending upon the location of the project and the availability of infrastructure, the companies may provide employment opportunities to the people of local community.

7.8.4 Deposit Development

Acquiring the necessary project capital funding and initiating construction activity are the next two important steps in mineral project development. Some amount of detailed planning and engineering work are often needed at this stage besides what is normally included in a feasibility study. Building a mine is a complex and demanding task fraught with technical obstacles, burdened with uncertainties, and costing sums in 1000s of millions, if not trillion. Mistakes can be very expensive. During all the earlier stages to this point in the sequence, no revenue has come from the campaign except for the expenses to delineate and develop an ore body. There are a few other industries where engineers and scientists, financial experts, and others combine to apply the sophisticated technological, mineralogical, analytical skills necessary.

7.8.4.1 Typical Activities

Typical activities in this stage include detailed, close-spaced in-fill drilling, mine planning, metallurgical testing, the likely environmental degradation of mine development, and continued community engagement. The land area, obviously, becomes much smaller in this stage (say about 1000 km^2). The expenditure may vary from project to project and may exceed US$1 billion. A company will begin to apply for permits and necessary statutory approvals to proceed to mining. The technical and economic assessment incorporating the latest information will form the basis for developing a mine. Every company prepares a "bankable feasibility study" report. This is a type of feasibility study that a company would take to a bank or other entity in its search for financing.

7.9 ENVIRONMENTAL IMPACTS OF MINERAL EXPLORATION AND DEVELOPMENT

Environmental impacts are bound to happen on natural environment and on local communities at every stage of mineral exploration and development. Fig. 7.3 depicts the important environmental impacts during different stages of exploration and the respective degree of severity. Preliminary exploration rarely spoils too much of Earth's surface. The disturbance is quite minor or of temporary nature. The impact is very minimal in grassroot exploration because the number of people involved is small and also most of the techniques are noninvasive. Geological mapping involves people taking traverses and collection of samples, etc. In some cases, certain amount of clearing the trees, road construction, etc. might be involved to access and collect samples.

Environmental and social impacts are quite significant during advanced exploration, but less, compared to those of mining. Drilling, trenching, access roads, vehicular traffic, etc. might affect air, water, flora, and fauna. By adopting suitable measures, some of these can be mitigated. Much more significant environmental and social impacts will be there in the deposit development/mining stage. Mine construction, actual mining, waste rock disposal, mineral processing facilities, tailing disposal, etc., although occupy a small acreage, there will be significant effect on the natural environment. The influx of outsiders (may be culturally different from local people) might affect local communities to a certain extent. Eggert (2010) opines that "many of the social impacts of a new mine can be minimized or controlled through deliberate planning but some community change is inevitable and permanent".

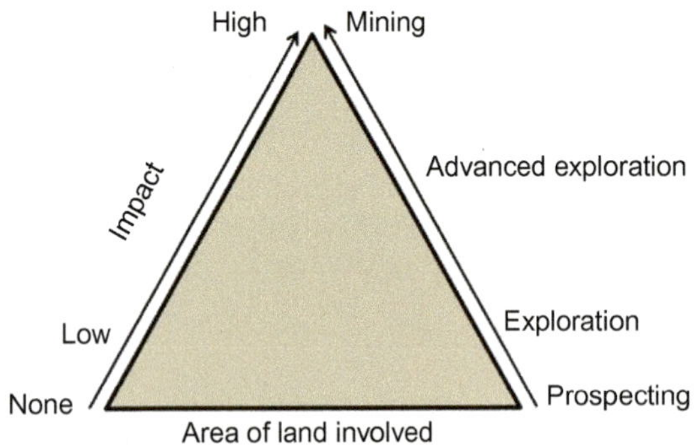

FIGURE 7.3 Sketch map showing environmental impacts during different stages of exploration and the respective degree of severity. *Source: From Gandhi (2012).*

7.10 MINE CLOSURE PLAN

Every mine faces closure after reaching its estimated life. A range of environmental, social, and economic factors are to be considered in closing a mine. Depending upon the commodity, location, legal framework of the specific region, processing methods involved, the mine closure challenges vary. Since no two mines are alike, the mine closure plan differs from mine to mine. It may involve rehabilitation or remediation of the town near the mine site and associated processing and tailing facilities. In many cases, closure costs constitute a major chunk of funds in developing a mine. But if the mine closure can be planned carefully, a company gets more time to remedy the issues. Postmining land use options need to be identified in advance so as to set out site-specific objectives.

Typically, mine closure involves (1) removal of buildings, plants, other infrastructures in the site, (2) reclamation of the decommissioned areas back to their premining land use, (3) restoration of areas where environmental damages caused by the mining activity, and (4) after care of the mine site till such period to show that the areas are physically, chemically, and biologically stable and safe and do not pose any threat to local community.

7.11 GREENFIELDS VERSUS BROWNFIELDS EXPLORATION

Greenfields exploration is normally conceptual, relying on the predictive power of the ore genesis models, to search for the mineral resource in an unexplored virgin ground. Brownfields exploration is conducted within geological terrains which are closer to known ore deposits. Brownfields is less risky, as the geology is better understood and the exploration methodology is well known, since large deposits have already been found. In such cases, the rewards are less. Greenfields, on the other hand, has a lower strike rate, since the geology is less understood at the conception of exploration programme but rewards are greater because it is easier to find a bigger resource with concerted exploration effort. Exploration done to expand a mineral resource that has already been found and developed into a mine is called "on-mine-site" exploration.

For greenfields exploration, geological analysis is a key component. A territory could have been explored and drilled earlier without the requisite success. But application of new exploration concept or new exploration technology, it is possible to find the area prospective for a mineral resource, not sought there before. Detailed mapping and application of advanced visualization of structural data are essential for brownfields exploration projects for success. Advances in 3D modeling enhance the probability of success in brownfields. Exploration targeting from geophysical data and geochemical data and their interpretation will help to design drilling programme and also search for buried resources. Rock outcrops, detailed drill core logs, evaluation of aerial photos, and geophysical data help an explorationist to decipher the tectonic structure of the area of interest. 3D modeling and visualization will give an idea of the resource and its projection.

7.12 RESOURCING THE FUTURE

Exploration is based on an understanding of the fundamentals of geological processes and the earth's history. In certain areas of the world, a higher density of deposits of particular type of metal or group of metals is a conspicuous feature. Such regions are called "metallogenic provinces" which represent enriched areas of crust where geological processes were suitable for the concentration of metals to form a group of ore deposits of a similar type. These provinces are said to be a manifestation of the regional control of geological processes by tectonic setting. They provide a valuable first step to exploration geologist in the search to further mineral resource (McKinstry, 1962; Lovell, 2000). Studies of how the earth is evolving help to predict which metallogenic provinces or basins have the best potential to host tier-one resources. As the easier and most obvious discoveries have already been made, attention is now turning to the deeper or hidden deposits that are both more difficult and expensive to find and develop. An area of interest is first identified by looking at prospective geology, known mineral occurrences, the maturity of previous exploration, and likely residual potential.

Generally speaking, the best place to start exploring is close to known mineral deposits in what is known as "brownfield" exploration. Brownfield exploration involves looking deeper or laterally for extensions or repetitions of mineralization at, or close to, existing mine operations. Greenfield exploration takes place outside the area of known deposits and requires more creativity, persistence, and some luck to find a significant discovery. In choosing where to explore, nature's distribution of resources needs to be weighed against the available technical ability to make a discovery. In addition, one needs to evaluate the relevant country risk, which involves considering the laws, regulations, and practices governing the minerals industry in that country.

7.13 PROJECT FUNDING

For major exploration companies, project funding is composed of "equity and debt" components. The lenders provide the "debt" portion after having conducted technical due diligence and satisfied that the project is technically and commercially viable. Besides they must be satisfied all necessary permits and environmental issues are properly addressed. Projected cash flow and the level of provision of corporate loan support will decide the amount of debt provided by lenders and the amount varies from project to project.

The sponsors of the project must meet certain cover ratios and other loan covenants prescribed by the lenders. Lenders require metal price hedging to reduce the risk associated with the project. They also ensure that firm marketing contracts covering the sale of mineral products, concentrates, or metals. The amount of equity can be determined once the loan amount and the parameters are established. A provision for cost overruns is also provided in the total project capital. Certain completion tests must be met once the project has been constructed and placed into operation, which ensure that the project has been constructed and is operating within certain design parameters. At this stage, most of the technical risks of the project are eliminated; the lender has recourse only to the project itself for loan security. Equity providers to the project and lenders are sensitive to issues like

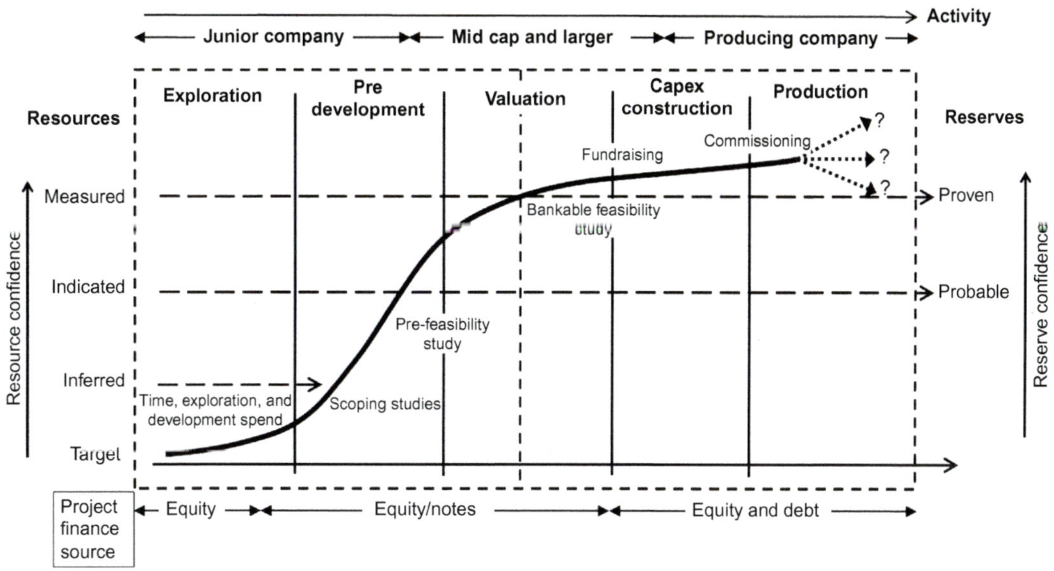

FIGURE 7.4 Funding options available as exploration project matures to production. *Source: From Russell (2012).*

project title, native land claim, environmental, permits, etc. The proposed project financing will suffer if there are impediments to a timely project development. Various funding options are available as an exploration project matures into production stage (Russell, 2012). The project finance sources for junior, mid cap, larger, and producing companies for various stages, from exploration through predevelopment, valuation, and capex construction to production stage are shown in Fig. 7.4.

7.13.1 Exploration Funding by Junior Exploration Companies

Given the inherent riskiness, many major exploration companies generally do not use debt to fund exploration programs. Instead, they mainly use their own equity to fund exploration. This rule applies to many major and junior companies. The major companies fund exploration out of their cash flows (sometimes up to 40%) from operations. However, every dollar spent on exploration is, one less dollar of profits and one less dollar available for dividends or debt repayments. Even though senior management recognizes the long-term necessity of exploration to sustain and grow their business, they often view these expenditures as discretionary in the short term—and so increase/decrease their level of funding inline with the ups and downs of the business cycle. Exploration expenditures by junior companies are more volatile. Junior exploration companies do not have an operating mine to fund their activities; instead, they have to rely on the support of their shareholder for cash—raised through issue shares (equity) in the company. The most high-profile example of this is the initial public offering (IPO) associated with when the company first lists on the stock market. The number of resource-related IPOs is dwindling in number over the years.

As a general rule, IPOs only raise enough money to fund the first 2–3 years of exploration. Consequently, to pay for follow-up work, the company has to go back to its shareholders for more funds, which involves issuing additional equity in the form of shares. To encourage investors to buy these shares, the company needs to demonstrate "good news" on their projects. This is usually in the form of positive exploration results and/or improvements in the business environment (such as strong commodity prices and lowering of business risk). Due to declines in commodity prices and concerns that the mining boom has come to an end, the market has savaged the share price of most junior resource companies. This has made it very difficult for companies to raise funds—as it requires selling the new shares at a heavy discount—thereby diluting the existing shareholder base. Given the difficulties in raising funds, most junior companies have responded by cutting back on expenditures. In extreme cases, in order to survive, they have gone into hibernation mode and are spending the minimal amount in the field. While most majors have resources to fall back on in hard times, their retreat on spending has left juniors in dire straits. Now the institutional investors have almost entirely stopped funding greenfields exploration, hence there is a dire need to tap into retail more than ever. But under the current securities regime, they have failed to do that in recent years and they need to win the retail investor back.

It is estimated that the average life expectancy of a junior exploration company is around 10–20 years and its longevity depends on ongoing funding from its shareholders. To achieve this, the company needs to generate a steady stream of good news on its projects. This, in turn, requires them to actively work their exploration leases. Given the above, it is clear that junior exploration companies do have the ability to fund exploration projects of +5 years duration. Shareholders will continue to invest if they believe that there is a (small but real) chance of a giant discovery. To make new discoveries and deliver new mines, the junior companies need good funding and access to good ideas and good tenements.

7.14 INGREDIENTS OF A SUCCESSFUL EXPLORATION CAMPAIGN

The important ingredients of a successful exploration campaign are:

1. Selection of right geological terrain
2. Optimum level of exploration expenditure,
3. Keeping pace with new technology.

7.14.1 Selection of Right Geological Terrain

The most critical issue in case of all mineral exploration programme is "Are we looking in the right terrain?" It is not only possible to find ore deposits but also to find them easily and cheaply, if the selection of most prospective area in a mineral find or geological terrain is made. By applying the theories of origin, the knowledge of known occurrences of ore and their mode of formation to known geological regions, it is easy to determine potential areas where particular ore type being sought might exist. New styles of deposits may be found that reveal chances to find look-alike deposit styles in rocks previously thought barren. The geophysical and geochemical exploration campaigns, detailed structural geology, basin modeling, petrology, etc., help to make predictions and draw parallels between known ore deposit and the unknown potential of finding "look-alike" deposits within the area selected.

Ore bodies occur in a typical geological setting with specific structural and tectonic environment. It will be difficult to predict mineral occurrences unless one develops a good understanding of the geological controls and their spatial distribution over wider region. Areas prospective for one mineral need not necessarily be prospective for other minerals. Many major mining companies will look for certain size/class of deposits because smaller deposits may not meet their economic criteria. Previous finds in a terrain ("nearology") will definitely help the area selection. Financial and taxation incentives and tariff systems of the host countries will also determine the area selection. Besides, the availability of good infrastructure is crucial since the ore/products have to be brought to market eventually.

7.14.1.1 Applying Conceptual Geometrical Models in Evaluating Mineral Prospects

During modeling process, data on geology, structure, alteration, etc. are underused many a times and assay grade takes precedence in interpreting the model. Such models based on grade alone have become unreallistic/unstuck at a later part of exploration, evaluation, and production. Geological structures observed in drill hole logs, conceptual geometrical model to the geology of the deposit can be judiciously incorporated in order to make a more accurate reflection of geology in the 3D models that could be used in realistic "mineral resource estimates." Large amount of data are usually available but often difficult to integrate and use effectively in brownfields exploration. Similarly the old ideas and exploration models hamper new interpretation and development of new targets. Such challenges are faced by the explorationists. 3D targeting is a new technique available that uses all available geological and other information to build 3D block model of the area of interest. Many canned commercially available 3D modeling software (like Geocad, Leapfrog, Geomodeller) are used to minimize time in making the necessary models in evaluating mineral properties.

7.14.1.2 Targeting and Target Generation

Modern exploration has grown intensely sophisticated with the advent of satellite, supercomputers, and modeling based on the platetectonic theory. Targeting constitutes the initial stage in any exploration programme. In the initial stage, the activities are focused into specific areas with the highest potential for discovery and develop plans to systematically explore them. Dedicated exploration targeting is often ignored. Integration of all available data helps efficient targeting to make the most well-informed predictions. It is to be ensured that targeting should not create too many "false leads" and should not be also simplistic by missing a "big one." Combination of available data and expert knowledge will help while assessing the exploration potential of vast areas. Characteristics of known mineral deposits serve as "predictive tools" in order to identify similar features in exploration datasets (such as geology, structure, geophysics, geochemistry). In predictive targeting, two approaches like "expert driven and data driven" are used.

Understanding the mineralizing system for a "mineral system" to occur is the basis for "expert-driven" approach to exploration targeting and is useful in greenfields exploration. An exploration model/rationale for future exploration is developed on sound geological reasoning, and it is necessary to compare with geological settings with known mineralization. In this approach, the concept of geological processes, in which mineralization and its movement from a source through a pathway, to the trap area where it is localized, is to be noted in detail. Based on explorationist's judgement, each element can be ranked and the features are combined in 2D or 3D GIS packages. The target areas are highlighted by the cumulative effect of all features.

In brownfields exploration, data-driven targeting model provides a robust/repeatable technique with vast amount of data. The empirical evidence of each element is ranked with little knowledge of actual mechanism of mineralization in "data-driven" approach. Such approach is called "weights of measure," where a spatial feature (rock type, structures like fault, fold, etc.) is assigned a statistical weighting, based on its relationship with identified mineralization. Features related to mineralization can be identified allowing exploration to focus on prospective regions independent of past interpretations. A robust geological framework and mineralization models will be provided by targeting processes that can form the base for exploration programme.

7.14.1.3 Target Selection by Airborne Surveying

To ascertain the geophysical aspects of an area having a potential target, many exploration companies use airborne surveying. Areas conducive to mineralization which may have accessibility problems from ground surveys (in remote areas) aerial surveying is done either with helicopters or fixed wing aircraft. The geophysical surveying methods like aeromagnetic, electromagnetic, gravimetric, radiometric, and "light detection and ranging (LIDAR)" have their own advantages. Certain methods may be better applicable than others depending upon the geological conditions and survey objective. Besides company's budget, the selection of survey method is such that the geological and geophysical data can be well understood.

The various advantages of airborne techniques over ground-based surveys are: (1) cost effective to cover large areas rapidly and efficiently, (2) helps development to take place on shorter time scales thereby reducing costs and improving efficiency, (3) can cover many areas inaccessible to ground-based exploration, (4) data coverage over remote terrain is uniform without any ground hindrances, (5) minimizes environmental impact on the ground, (6) helps to narrow down the area to be covered by detailed exploration, (7) can provide regional scale geology and structure for better interpretation and to select the target areas.

For planning an exploration programme, valuable pool of data are obtained by combining all the acquired data integrated into a powerful GIS. By conducting drilling campaign in areas of high potential, an effective exploration programme can be carried out.

7.15 MINERAL EXPLORATION AND DEVELOPMENT—GEOGRAPHIC LOCATION

A good geologic potential and the investment climate are the two factors that determine the geographic location where mineral exploration and development occur. Geologic knowledge gathered from previous exploration/mining and nonmining activities like road construction, etc. attract the investors. A message of mineralized core in a drill from one exploration agency often leads to a lot of claims staking or purchase of exploration rights in the area by other agencies. Public geological survey organizations play a critical role in attracting exploration in an unexplored area by providing geoscientific research and information. Many companies imitate by taking advantages of research carried out and information gathered by others for their use. Providing basic geoscientific information by Government can act as a catalyst for mineral exploration in unexplored areas. The information types the Government may provide are geological maps, regional geophysical surveys, regional geochemical survey data, etc. In most of the survey organizations, such data are available in digital format which could be downloaded from "internet."

Different exploration agencies view the same data differently. In fact the exploration programmes/drilling campaign by several companies in the same area led to the discovery of many deposits. Attractiveness of the same piece of land changes over time. A preliminary investigation could have been discarded by one company, but relinquished the area owing to unfavorable economic criteria or due to problems in mineral processing. As time advanced, technological capabilities, economic conditions, and exploration techniques change, chances of detecting subsurface mineralization get improved.

Private investment in mineral exploration and development depends upon the "investment climate" of a country. Political stability, attractive fiscal and tax regimes, attractive royalty policies, legal and permitting systems are the important factors. A nation's specific policies for mining sector, transparent and nondiscriminatory rules, and investor-friendly environment attract the investors. Besides, investors prefer a simplified single window policy for granting licenses, permits, and approvals.

7.16 EXPECTED REVENUES, COSTS, AND RISKS

The expected revenues and costs adjusted for time and risk are determined by the location and level of investment in mineral exploration and development. An investment opportunity is said to be "more attractive" when the expected revenues are higher or expected costs are lower. The longer an investor has to wait to get the revenues (the riskier the investment), the investment is said to be less attractive. As per Eggert (2010), the main four factors that influence expected revenues, costs, and risks in mineral sector are:

1. Geological Factors: Whether a mineral resource exists in a region? If so, what is the quality and quantity of the anticipated resource?
2. Technical Factors: With the existing or likely future technologies, Can known resources be extracted and processed economically?
3. Environmental, Social, and Political Factors: Whether a resource be extracted in ways which are consistent with a Nation's preference and policies for environmental protection. Whether it can be extracted in ways consistent with preferences and policies of local communities/people?
4. Economic Factors: Whether a mineral resource can be extracted at a profit? The economic risk is an "overarching" type, since it incorporates and reflects the three earlier categories of risks cited earlier.

How the investing companies manage risks and returns is shown in Fig. 7.5.

The projects which attract mining investments and the risk-return equation are shown in Fig. 7.6.

In their recent survey of business risks facing global mining and metals industry in 2014–2015, Ernst & Young summarized the top 10 "Business Risks Radar" as given below:

1. Productivity and improving
2. Capital dilemmas
3. Social license to operate
4. Resource nationalism capital projects
5. Price and currency volatility
6. Infrastructure access
7. Sharing the benefits
8. Balancing talent requirements

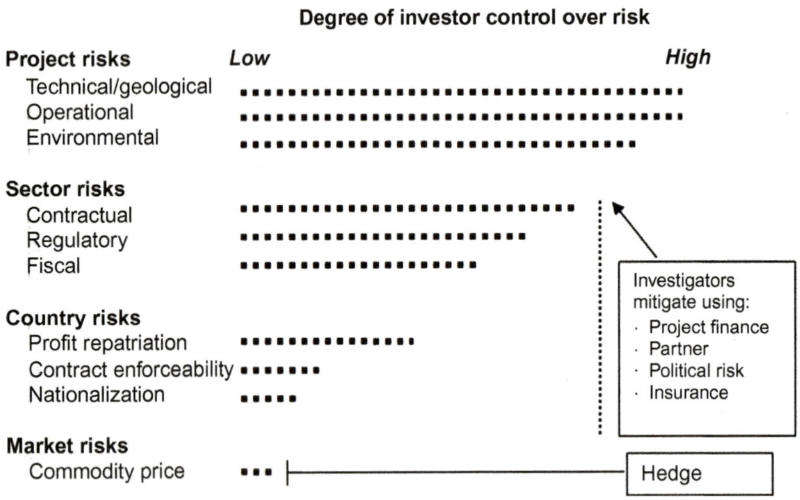

FIGURE 7.5 How the investing companies manage risks and returns.

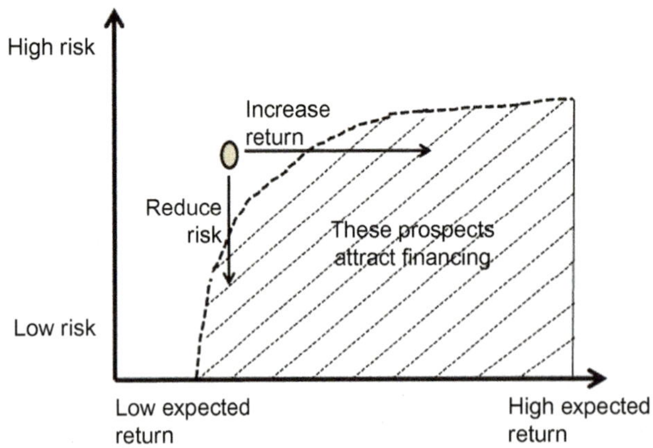

FIGURE 7.6 Projects which attract mining investments and the risk-return equation.

9. Access to water and energy.

They have also listed the following under "the Radar Business Risks":

10. Cyberattacks and information security
11. Threat of substitutes
12. Pipeline shrinkage
13. Fraud and corruption
14. Competing demands for land use
15. Climatic change concerns
16. New technologies.

The Mega Trends: The top 10 business risks are 1–2 year priorities of mining and metal sector, as influenced by the global "mega trends" that are impacting business, society, culture, and the economy. These include (1) digital transformation, (2) changes in the way we work, (3) the global market place, (4) the urban world and its demands on infrastructure, and (5) health reimagined to meet growing needs. The mega trends by their nature are medium to long term in relevance (EYGM Ltd., 2014).

7.17 EXPLORATION EXPENDITURE

7.17.1 Sources of Exploration Financing

There are various sources of financing the high-risk exploration activity, viz., (1) the personal funds of the individual prospectors or the funds from friends/families; (2 Governments can fund for exploration surveys either wholly on their own or as partnership programmes; (3) multinational mining companies spend funds from their internally generated cash flow, and (4) private individual investor or institutional fund. In many cases, both major and junior companies are working together jointly sharing their funds and expertise.

The key factors for a gradual increase in exploration expenditures, as per Ericsson (2012), are:

- The location of new found deposits is likely to be in inhospitable and exposed to more extreme weather conditions (deserts and arctic conditions). High altitude and submarine areas are likely to be investigated. Thus new locations of deposits will be far away from metal market centers which will indirectly add transportation cost of the products mined.
- Since surface or near-surface deposits have been located, search has to be gradually concentrated on targets of greater depths with greater expenditure budget.
- General depletion of high-grade ores leaving behind exploration of low-grade ores and the associated metal extraction problems.
- The feeling of local community (wherein the resource is found), being neglected, from not getting the benefits that they are entitled to, besides other political problems.

The exploration expenditure over a period of time will be increasing to cater to the resource depletion and growing due to difficulties of geological, technical, financial, and political/geopolitical character. Mineral exploration activity is a labor-intensive activity and relies on many consultants, contractors, and other service providers. Exploration has large indirect and induced employment effect. It is said that "three jobs" are indirectly generated for every job directly attributable to expenditure. Although it is difficult to estimate, the employment multipliers are greater due to high labor-intensive activity.

7.17.2 Optimum Level of Exploration Expenditure

Mineral industry companies are spending a lot of money on exploration. The exploration funding is entirely driven by the price cycle of the commodity (gold, base metals, etc.). The size of exploration investment is determined normally by the quality of the opportunity and the project risk, as well as assessment of long-term prospect of the commodity. The total worldwide exploration budgets in 2012, and the forecast, by Region and by Commodity, up to 2020 are given in Figs. 7.7 and 7.8 (Schodde, 2014a,b). The forecast of exploration budget shows a downslide from the peak level of spend in 2012. By commodity, gold is projected to fall the most, whereas spend on uranium is expected to increase a bit.

The percentage of total exploration investment in 2014 in the world, by region, is shown in Fig. 7.9 (SNL-MEG, 2015). The industry is reasonably confident and optimistic that the present trend might show a change in the years to come.

The worldwide mineral exploration investment in 2012 *(in million US$) in different places* is given in the table (SNL-MEG, 2013).

Latin America *(Mexico, Peru, Chile, Argentina, Brazil, Columbia)*							5160
Africa *(West & Sub-Sahara region)*							3455
Canada	3440	Australia	2580	USA	1720	China + Mongolia	1505
Pacific Isles	1290	Europe	860	Russia	645	Rest of the World	645

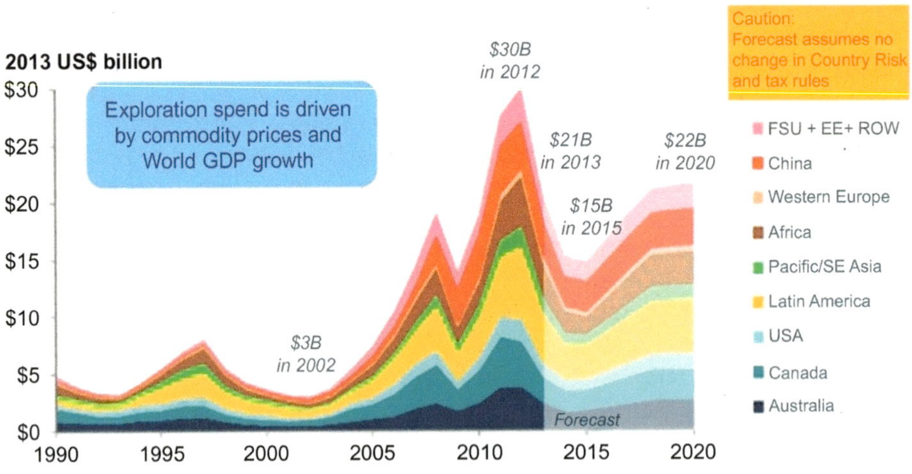

FIGURE 7.7 World exploration expenditure and forecast by region *(forecast based on multiple regression analysis of data $R^2 = 0.95$). Source: From Schodde (2014b).*

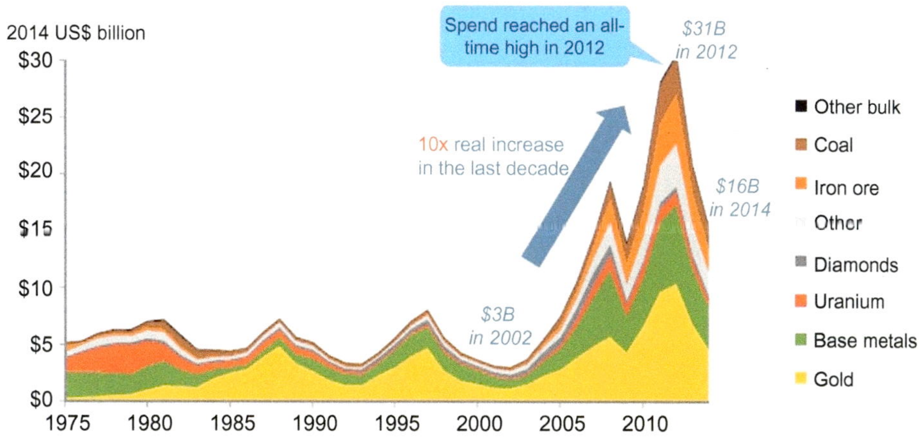

FIGURE 7.8 World exploration expenditure by commodity. *Source: From Schodde (2014c).*

Compared to these countries, the overall exploration investment in developing countries, in the corresponding fiscal year, was very less/abysmal. With the ever-increasing price and the lure of gold, Latin America continues to be the most favorite place for investment by exploration companies. "*After all, the honey bees go where the honey is!*" As in the previous years, globally, gold remained the top exploration target, followed by base metals, PGM, uranium, diamonds, etc. Gold and base metals constitute almost 85% of the global exploration target.

The commodity price, availability of funds, and long-term growth in metal demand are the key drivers for exploration spend. There are a few secondary drivers like new geological concepts and search techniques (like Cu-porphyry model), new processing technologies that make low-grade ores economic (CIP Au recovery, SXEW of Cu), and bioleaching for base metals. Historical review of exploration investment in major countries like the United States, Canada, and Australia indicated that to discover a world class mineral deposit with an anticipated gross metal value in excess of US$600 million, the cost for discovery may be in the range of US$100–150 million (1990 prices). One has to spend more (~1.25–2 times) to improve higher probability of success. Traditionally, mining companies allocate between 10% and 40% of their posttax cash flow to exploration investment, in order to stay in business (Woodall, 1993).

When there is a cyclical behavior in exploration expenditure in the world over, China's exploration expenditure is increasing strongly. With the present pace, China may probably end up as the largest exploration country of all. Chinese exploration expenditure figures do not show up in the common figures published over global exploration. The total exploration expenditure for all minerals in China is said to be 25–26 billion RMB (~4 billion US$) in 2010. Excluding the figures for coal and iron, the total Chinese expenditure is reduced to 14 billion RMB (a little more than 2 billion US$).

7.17.3 New Technology Adoption

Adoption of right technique or a combination of techniques is warranted to conduct exploration in a cost-effective manner. The impact of technology on mineral exploration is

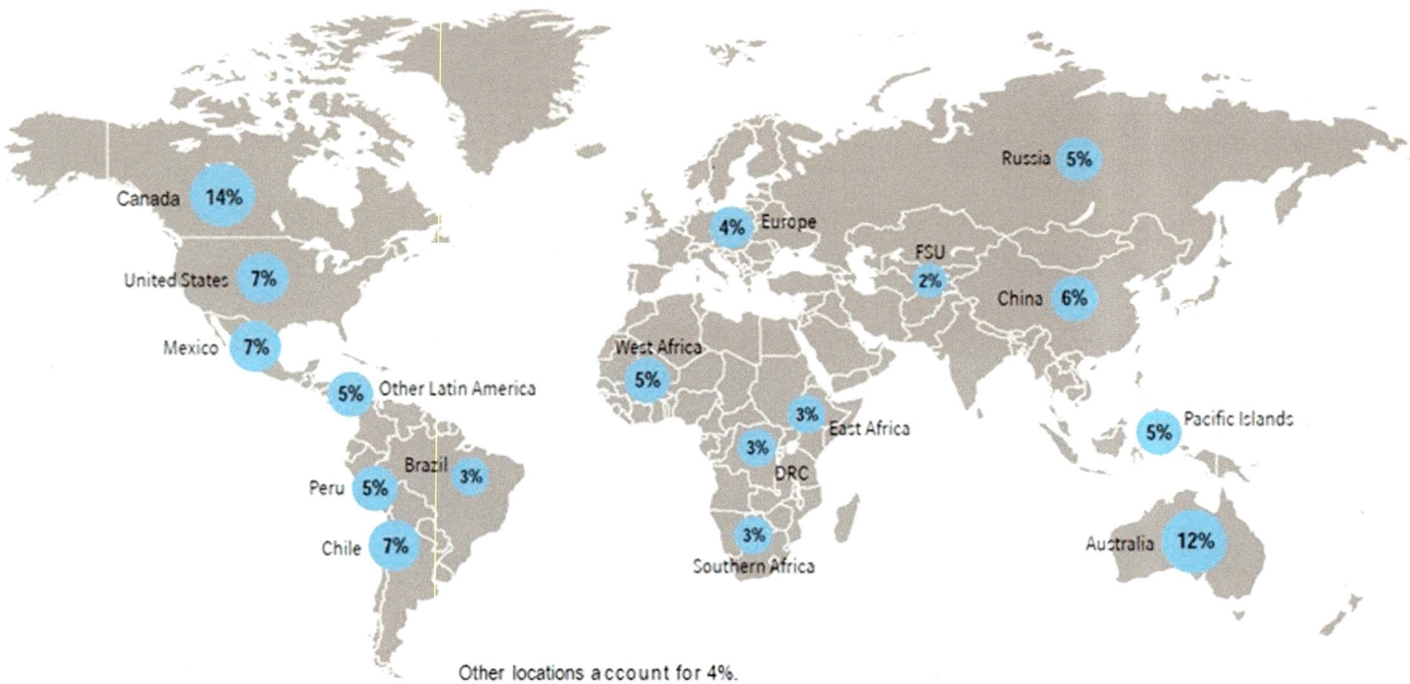

FIGURE 7.9 The percentage of total exploration investment in 2014 by region. *Source: Courtesy from SNL-MEG (2015).*

dramatic in the 21st century, compared to the decades of discoveries in the mid-20th century. We now process "terabytes of data" in minutes and days. A variety of GIS software systems integrate layers of a wide range of data types and formats. High-resolution (100–400 m line spacing) airborne surveys for electromagnetics, magnetics, and gravity mapping guide exploration targeting. GPS navigation and digital revolution have brought huge improvements in airborne geophysical technology. Multiarray and multichannel geophysical setups detect electrical conductors at 1000 m depth and map fault structures and rock alteration to 2 km depth. Challenges, however, impede the rate of discovery in the 21st century. Detailed geological integration does not keep pace with geophysical surveys. Drill targeting and discovery at depth and in remote areas is largely unsuccessful or difficult. New remote discoveries are the result of a combination of persistent teamwork of experienced specialists and timing. Many companies are in the race to develop better methods to integrate multiple datasets to obtain more comprehensive and confident interpretation of geological structure and targets.

Unmanned aerial vehicles (UAVs) are being used now, for a number of applications, viz., aerial photography, geophysical surveying, mapping mission, etc. The low altitudes of UAV can result in higher accuracy than conventional aerial photogrammetry techniques, ease of operation, and overall visibility.

7.18 DISCOVERY DEPENDS UPON VARIOUS FACTORS

Discovery is the crucial moment of exploration when the first unequivocal high-grade, and potentially economic, mineralization is encountered. Some discoveries are obvious from the first outcrop or drill hole, while others are only made after dozens or even hundreds of drill holes and often after the project has passed through many owners, each finding subtle and elusive signals without discovering the prize. Often, the significance of a discovery is not immediately understood, and it may be some time, after follow-up drilling, before the true potential of a resource is recognized and appreciated. Only a small number of discoveries ever lead to large-scale mines that make a material difference to the producer, customers, host government, and community. The long sweep of history suggests that we have not run out of mineral deposits. The number of discoveries and the exploration spend (red line) in the world are shown in Fig. 7.10 and the number of discoveries by commodity is shown in Fig. 7.11.

Out of worldwide deposits, about 30% were *"accidental or chance discoveries."* Some of the accidentally discovered mineral deposits in the world were: Rio Tinto (Cu, Au), Sar Chesmeh (Cu), Cornwall (Sn, Cu), Chiquicamata (Cu), El Teniente (Cu, Au), Yanacocha (Hg, Cu, Au), Kiruna (Fe), Rammelsberg (Ag, Pb, Zn), Mexican (Ag), Sudbury (Ni), Cobalt (Ag), Kalgoorli (Au), Broken hill (Pb, Zn, Ag), Erstberg (Cu), Rand belt (Au), Lake Superior (Fe), Bisbee (Cu, Ag), Bingham (Cu, Au, Pb), Porgera (Au), Red Dog (Pb, Zn, Ag), Rampura-Agucha (Pb, Zn), Sukinda (Cr), etc. About 32% were due to complex, purposeful exploration by Private Exploration Companies/Corporations; about 20% were discovered due to unsophisticated prospecting and about 18% were discovered due to Government enterprises or agencies. Literally, thousands of properties must be examined in regional surveys carried out, before a success is registered. Generally accepted yardstick

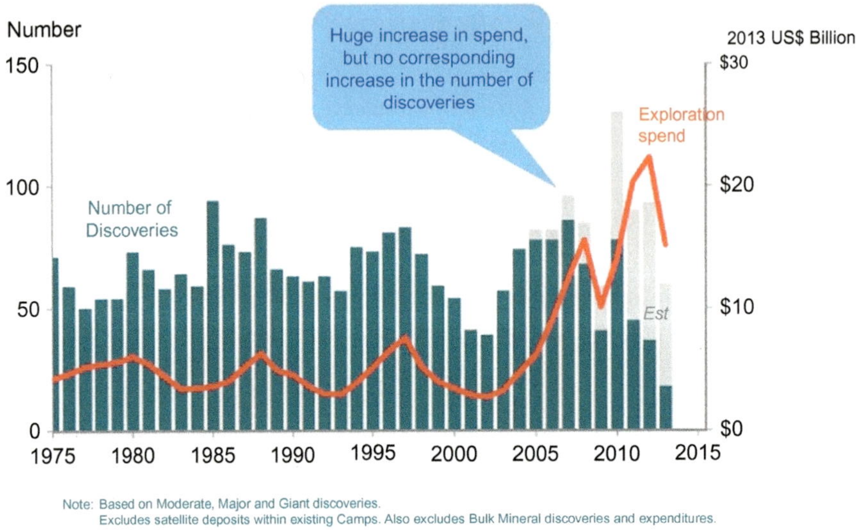

FIGURE 7.10 Exploration spend (red line) and number of discoveries in the world. (For interpretation of the references to color in this figure legend see the color plate.) *Source: From Schodde (2014b).*

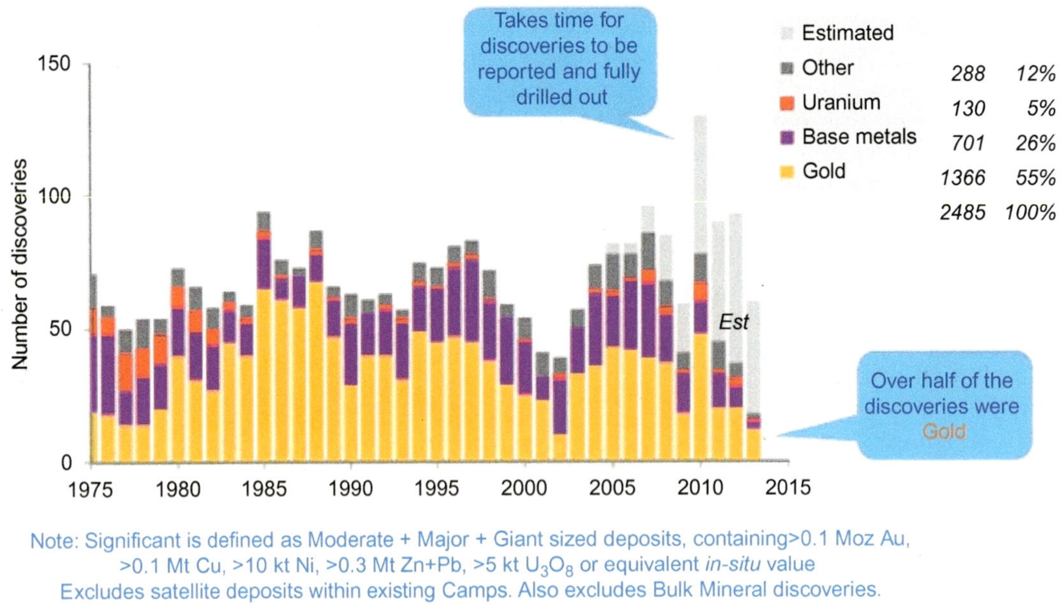

FIGURE 7.11 Number of discoveries, by commodity, in the world. *Source: From Schodde (2014b).*

is that 1 in 1000 discoveries turns into an ore body. The typical time frame from the inception of exploration through the investment in mine/mill facilities that will return this investment out of production is 12–15 years. To receive a profitable return on investment commensurate with risk requires minimum investment longevity of additional 10 years, hence a total of 20–30years.

Many fine, productive mines were discovered in the 1950s and the peak of discoveries was between 1960 and 1975 with the advent of new geophysical survey tools and geochemical techniques (Blain, 2000). Since then, the rate and number of discoveries have fallen off drastically later. When the genetic models of mineral deposits became accepted and widely known, great discoveries were made through the years: during the 1950s, roll-front uranium deposit discoveries; in the 1960s, porphyry copper–molybdenum deposits; in the 1970s, "sedex" zinc–lead deposits, unconformity uranium deposits; in the 1980s, "carlin type of gold deposits"; and in the 1990s, IOC gold deposits. The application of such models by the exploration teams facilitated to locate many potential deposits yielding a burst of discoveries. Many of these discoveries were brought to production much later owing to various reasons. Besides, the development of new technology enabled mining of many low-grade ore deposits.

The exploration costs have risen dramatically over the past few decades. The most expensive component is exploratory drilling. With the terrain maturity, the rate of discovery of world class deposits has declined. Concomitantly, the overall grade of the ore deposit has shown a declining trend over the years (Fig. 7.2B). Most discoveries have been made not only in the right geological terrains but in countries that traditionally have a sound set of mining tenancy and investment laws, these include the United States, Canada, Australia, South Africa, some countries of the SW Pacific, and some countries in South America, particularly Chile, Argentina. Mining industry has traditionally chosen to put its money, where it is safe. The worldwide significant mineral discoveries in the period 2000–2013 and the current hot spots of exploration activity are shown in Fig. 7.12.

Discovery of economic resource depends upon various factors. In exploration stage, mineral activity is a very risky business since there are many unsuccessful projects before finding one successful project which may end up in an economic deposit. So, the cost of discovery includes the total costs involved in both "successful and unsuccessful" ones. Generally, a few good successes must be profitable enough to compensate the other failures (USBM, 1977). In mineral exploration, the "key performance indices" are drilling density, the amount spent per discovery, and the \$/kg of metal found. These factors will show the "measure of efficiency."

In the past decade, owing to higher input costs for drilling, salaries, and administration, etc., the unit cost of discovery has doubled or tripled. The impact of increasing depth of cover reflects on the discovery performance. Especially for brownfield areas, there has been a gradual move to deeper discoveries (Fig. 7.13).

The overall conversion rate of all the discoveries made since 1950 into mines is only 45% and in terms of contained metal, it is 57%. It has been found that gold prospects have higher rate of conversion. The size and depth of cover have no bearing on the delay of the project except that "brownfield" discoveries were quicker to develop. Besides, the country risk was also a critical factor; in low-risk countries, the projects were developed 40% quicker.

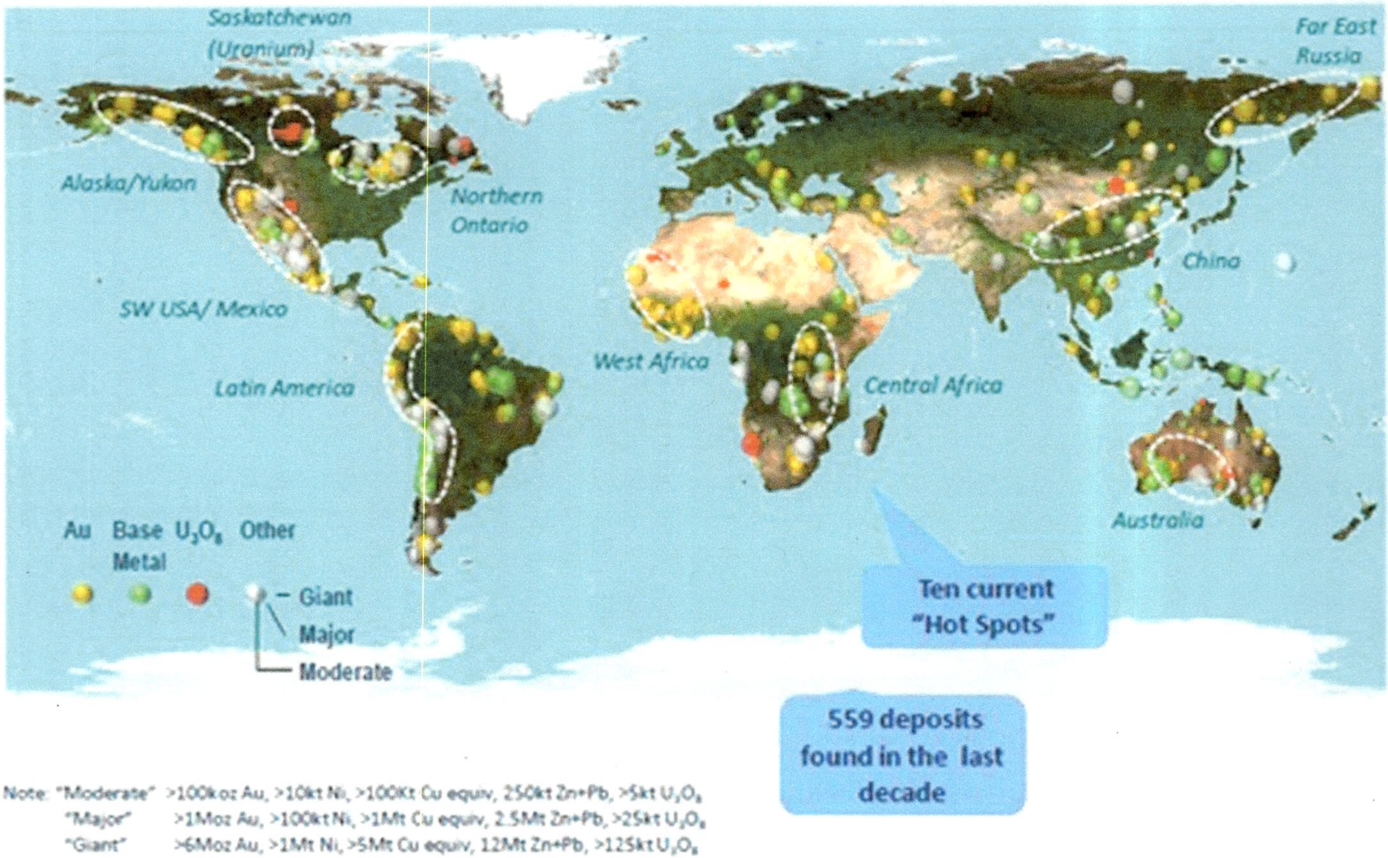

FIGURE 7.12 Significant mineral discoveries, worldwide and current hot spots. *Source: From Schodde (2014a).*

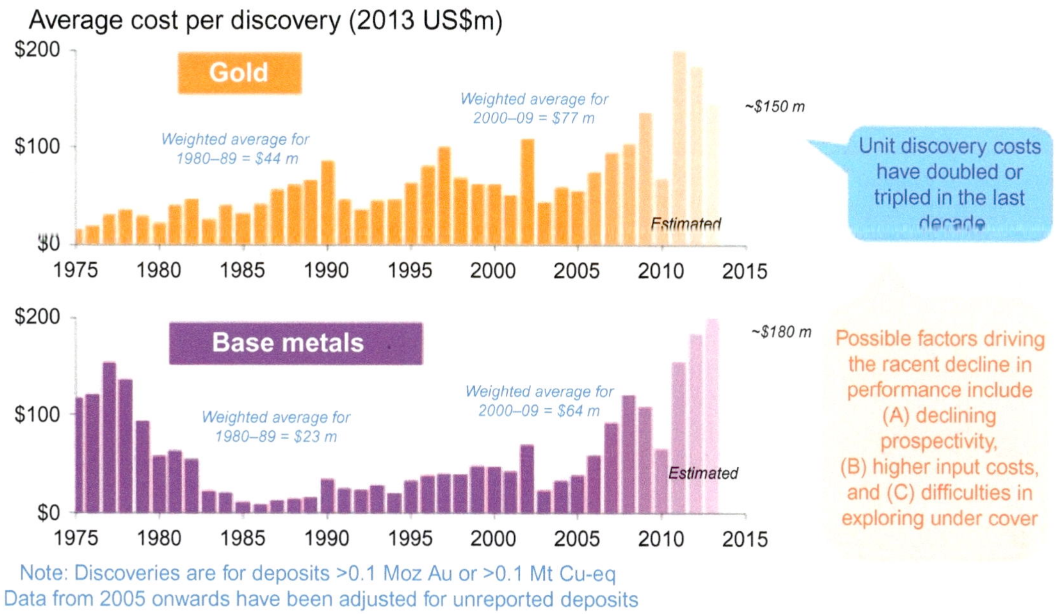

FIGURE 7.13 Rising discovery costs (unit cost/moderate-sized gold and base metal discovery in the world). *Source: From Schodde (2014c).*

The rates of success calculation and interpretation, and the cost per success (including cost of failures) for mineral exploration projects are complicated by many factors. The very basis of division of projects as "success or failure" is artificial and misleading. Exploration success is defined as a discovery of significant concentration of mineralization, currently developable. But many a times, an economic resource is discovered which may not be currently economical but commercially developable in the course of time, as a result of advance in technology, infrastructure or higher prices for the minerals/metals (Woodall, 1984a–c). Such discoveries are termed as "technical successes" or "on-the-shelf" deposits. Even if there is no concentration of mineralization, the detailed information gathered by exploration will be important additional information, which in the days to come might be very valuable when new technology/new theories of ore formation are considered. It is normally difficult to assess trends of exploration productivity. Exploration productivity can be determined by dividing the expected financial return by exploration costs, after these have been adjusted to take account of inflation.

Blain (2000) analyzed the mineral exploration success rate with the available data. With the advent of modern geophysical exploration surveys and geochemical exploration techniques in the 1950s, large number of significant mineral discoveries was made, but many of them remain undeveloped. The discovered deposits developed since 1980 (giant Andean copper-porphyry deposits of Chile/Peru) are low-cost producers. Since then, the rate and number of discoveries have fallen off drastically later. The decline of discoveries in nongold sector (from mid-1970s onwards) is largely due to the difficulty explorationists have in finding a viable deposit in an unfavorable economic climate.

The discoveries were a series of waves; offset over time, in different commodities, viz., copper, gold, base metals, and nickel. Most of the discoveries in the 1980s and 1990s have been of gold deposits, but the rate of discovery drastically fell later. After World War II, the price of gold remained low with little fluctuations for about three decades. When the fixed link between US$ and gold was removed by Nixon administration, the price levels soared to record high (about US$800/oz), an inconceivable figure at the beginning of the 1970s, it then fell back during the late 1990s. The rise in gold price after 1972 led to an increase in prospecting and the discovery of many large deposits. The trend continued until 1993 and gold production increased and reached the peak in 2001. Unfortunately, the demand for gold is not determined by industries but by unpredictable and variable factors like fashion (jewelry), sentiment, and investors. Many companies had to adopt a cautious approach in their exploration strategy, not concentrating on a commodity of volatile and uncertain price levels but putting more emphasis on exploration of other commodities (like base metals), besides gold.

The success of the discoveries should be measured in terms of outputs (ie, number of economic discoveries) rather than inputs (such as expenditure). Not all discoveries turn into mines, the conversion rates vary between 60% and 75%, in different commodities, depending upon various factors. Taking into consideration of losses on mining and the demand of the metal in question, the discovery of more resources is a must. Given the long delays to convert a discovery into a mine and considering the future size of the market, we need to find resource, at least twice (2 ×) as much as we mine, in order to sustain.

Mackenzie and Woodall (1988) studied *"base metal exploration"* only and compared the period 1955–78 for Australia with 1946–77 for Canada. Australian exploration was found to have been uneconomic but Canadian exploration, financially very favorable. Although exploration expenditures in Canada were double those in Australia the resulting number of economic discoveries were eight times greater. Finding an economic deposit in Australia costed four times as much and took four times longer to discover and assess as one in Canada. By contrast, the deposits found in Australia were generally three times larger than those found in Canada. Comparable exploration expertise was used in two countries. Australia is either endowed with fewer and larger deposits, or the different surface blankets (glacial vs weathered lateritic) render exploration, particularly geophysical, more difficult in Australia, especially when it comes to searching for small- and medium-sized deposits. The latter is the more probable reason for the difference in productivity, which suggests that improved exploration methods in the future may lead to the discovery of many more small- and medium-sized deposits in Australia. Mackenzie and Dogget (in Woodall 1992) have shown that the *average* cost of finding and proving up *economic* metallic deposits in Australia over the period 1955–86 was about $A51M or $A34M when discounted at the start of exploration. The average reward discounted in the same manner is $A35M. This is little better than a breakeven situation.

Schodde (2014a, 2014b, 2014c) after having examined all worldwide discovery of deposits, their sizes and the costs involved between the periods 1980 and 2011, came up with the costs per discovery, by the sizes, are given in Table 7.1.

Exploration is increasingly getting tougher. Despite a 10-fold increase in the amount of money spent on exploration over the last decade, the pace of new discoveries remained relatively tame. This means that more amounts was spent per discovery owing to doubling

TABLE 7.1 Discovery Rates and Costs, by Size

Period	No. of Discoveries Worldwide	Cost per Discovery, US$
MODERATE DISCOVERIES		
1980–89	520	$85 m
1990–99	559	$79 m
2000–11	412	$189 m
MAJOR DISCOVERIES (INCLUDING LARGER SIZE RANGE)		
1980–89	260	$171 m
1990–99	313	$133 m
2000–11	221	$329 m
GIANT DISCOVERIES		
1980–89	55	$742 m
1990–99	87	$475 m
2000–11	44	$1564 m

of input costs such as drilling, salaries for geologists, engineers and for general administration. Besides, with the ebullient market, much amount was wasted on projects with negligible odds of success. Given the volatile outlook for price and exploration, exploration agencies need to develop new strategies to survive and position themselves for growth in the longer term.

7.19 MINERAL EXPLORATION UNDER DEEP COVER

New mineral deposits must be found to replace deposits currently being mined and to meet the anticipated increases in global consumption rates of many minerals/metals. Since many of the identifiable surface or near-surface mineral deposits have been found thus far, it is necessary to focus the efforts in locating mineral resources from depth or under deep soil cover. The depth of cover of various base metal deposits discovered between 1900 and 2014 is shown in Fig. 7.14. The increase in the depth of cover, past 1945, was basically due to the innovations in geophysical exploration methods.

There has been a growing trend in the discovery of buried deposits (a few 100 m below surface). The challenge of exploration under cover will be in the order of about 1–1.5 km deep. The traditional exploration methods adopted to find new deposits will be of little help for locating mineral resources under deep cover which normally masks the clues/guides to mineralization. Deeper ore targets although have the ability to show up the detectable primary footprint, it dilute the signals. So, in order to detect the secondary or tertiary footprints of mineral systems against a regional level, it is necessary to understand the processes controlling the formation of cover and its mineralogical, geophysical, and geochemical character.

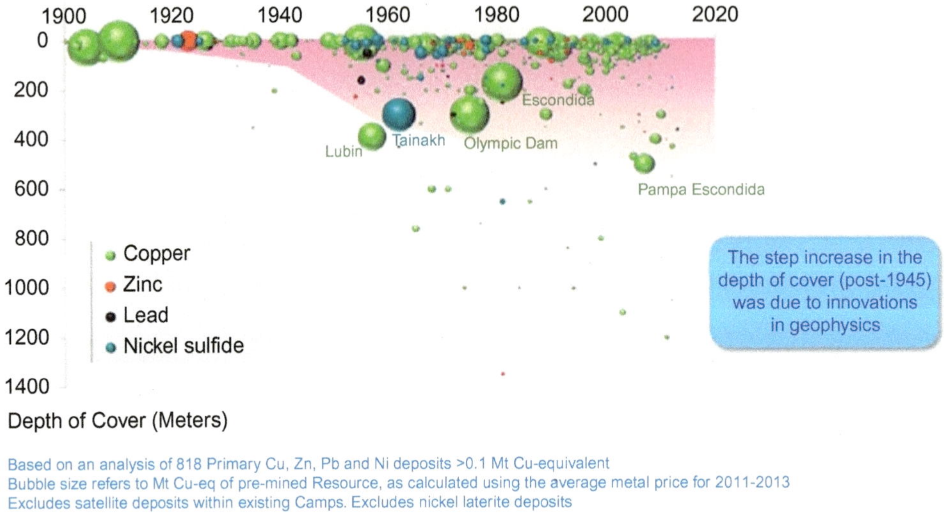

FIGURE 7.14 The depth of cover of various base metal deposits discovered between 1900 and 2014. *Source: From Schodde (2014c).*

With many incentives world over, more areas of the world have been opened up for exploration and development including unexplored areas which are in remote areas which were unavailable for exploration due to various reasons (political, economical). Besides these, the areas which were previously felt too deep to explore also demand a "reinvestigation." Thus the "search space" for greenfield exploration has widened and may be good for locating new targets. In areas where the bedrock exposure is minimal, mineral exploration efforts have to rely heavily on predictive quality of geological, geophysical, and geochemical data. The exploration efforts in areas under cover can demand new approaches of locating prospecting areas, viz., well-defined target models, and emphasis on geological and geophysical modeling in a 3D environment.

The exploration style has to have much reliance on subsurface imaging using innovative techniques to subsurface geology in order to detect deep-seated ore bodies rather than the customary outcrop-based exploration style. Better tools and techniques are required to accurately describe the geology of the Earth's crust in four dimensions, viz.

- New and improved targeting models to guide good selection of potential targets
- New insights that are not geological, geophysical characteristics of mineral systems and mineral deposit
- More powerful geophysical exploration tools to directly detect possible mineralization of a wide variety of types of depths
- Cost-effective drilling technologies to test-drill the zones of interest/deep mineralization.

The targets for exploration under cover have to be of a size much larger so that they can be detected from distance and also have a value that permits their exploitation a profit.

7.20 INTERPRETATION AND 3D MODELING

With the advent of technology comes an avalanche of data. We can do things faster and more efficiently than we ever thought possible. It facilitates to make decisions instantaneously. The volume of data generated in many exploration campaigns is tremendous and most of it remains unused, sitting in server or on the cloud gathering dust. The technological challenge in mining industry is not how to create more data but how to manage and leverage whatever data available. The companies that fail to adjust to the rapidly changing environment could see their operation, efficiency, productivity, and profitability affected. Those companies which embrace change to look for innovative ways to address these challenges may uncover distinct competitive advantages.

Big data utilization and extraction is the "next boom" in the mining industry. The key is identifying problems, then using the data help solve them. One has to sort through massive amounts of data to find the correlation of different datasets and disregard the incorrect data and deviations in order to find patterns that could lead to improved decision making. It is up to the exploration teams to prioritize what they want to see and do what the data available. Sometimes it is difficult to make decision when you do not know where to start.

Generation of 3D subsurface models which are capable of visualizing, analyzing, and interpreting of many complex tectonic setting is possible now by the synthesis of QA/QC-validated geological data. Data from different exploration methods such as geophysical surveys (seismic, gravimetric, geoelectric, well-logs, etc.), drilling or remote sensing can be integrated into a single data model (McGaughey, 2007). It supports the geologist's task to achieve a comprehensive evaluation of the underground by reducing uncertainties.

Of late, many innovations have been made in graphic design and 3D modeling in exploration and mining owing to the modern day advances in computational power. Based on a thorough understanding of the deposit geology, structural analysis, etc., a 3D model is built for validation. Plenty of canned softwares (like Leapfrog, FracSIS, Gemcom, etc.) are now available for construction and validation of 3D models. Innovation rarely comes without pitfalls; hence, care must be exercised to construct interpretation models. Prolific use of advanced 3D capabilities and more time spent generating interpretations facilitate an explorationist to focus on their projects, by reducing risks before taking a final decision to initiate drilling campaign. By application of modern technology to the data of the old projects, many recent discoveries (HudBay's Lalor Lake Rio Tinto's Bingham Canyon mine) have been made. These were possible because of fresh perspectives that came from deeper exploration techniques and 3D visualization capabilities.

First-order distribution of numerical datasets (grade, magnetic susceptibility, geotechnical parameters, etc.) can be modeled rapidly using "Leapfrog," which often highlights unforeseen geological trends which were poorly represented in traditional geological models. Multiple 2D–3D datasets can be integrated using "FracSIS," thereby interrogation can be possible of drill hole versus geological mapping data, geophysical data, satellite imagery, and existing 3D geological models. 3D graphics and geological modeling capabilities have been expanded in "Gemcom." Although no single software program package provides a complete solution for geological modeling, the integrated use of all the three softwares cited earlier can offer significantly advanced geological models in lesser time frame and many a times highlight unforeseen exploration potential especially in "brownfield targets."

Geosoft has developed its robust exploration information management solution (EIMS), a web-based technology around its core DAP server and search technology, to integrate plethora of exploration data (Fig. 7.15). It helps to work faster and smarter by organizing and sharing data and information across teams as they can manage and make sense of large amount of digital data for exploration decision making.

Effective data use and management: In the past decade, the greenfield discoveries have fallen every year despite steep rise in exploration expenditure. It is necessary to harness the power of "big data use" and "cloud computing," probably to narrow down the gap. In several cases, plenty of data are available with the exploration agencies but got nothing remarkable out of it, mainly because of time constraints and inability to find a noteworthy thing (discover) from that data. The key factors in the current challenging mineral exploration market are to make best use of available data through better modeling and interpretation, linked with new technology (Fresh Perspective).

7.21 ECONOMIC CONCEPTS FOR EXPLORATION STRATEGY

The process of discovery of an economic resource is characterized by a high element of risk, large time span, and sustained cash outflow, which is a total loss, in case of failure to make an economic discovery. A typical exploration process evolves from a generative stage that results in:

- Identification of targets
- Testing and evaluation of the targets culminating into discovery of mineral occurrences
- Definition and delineation of the deposit
- Development stage.

The full process involves inherent risk at each stage that requires testing of a large number of ventures of mineral occurrences to discover and delineate an economic deposit. The risk is related to the factors of uncertainty associated with mineral endowment with respect to their location, size, shape, quality, depth, and the techniques of exploration used. The risk, therefore, could be minimized within given limitations by evolving an exploration strategy which judiciously combines sound geological concepts with economic and commercial judgement at each stage of exploration.

Brant (1968), Mackenzie (1973, 1981), Snow and Mackenzie (1981), Mackenzie and Belodeau (1984), Eggert (1987), Mackenzie and Woodall (1987), Tilton et al. (1988), and Silver (2000) have developed several criteria for assessing risk on investment in mining through mineral exploration using probability considerations to examine actual or likely returns. The "Expected Net Present Value (EV)" and "Expected Rate of Return (ROR)" concepts of Bilodeau and Mackenzie (1979) provide an effective means to estimate the possible risk in investment in mineral exploration.

The Expected Net Present Value can be expressed as

$$\mathrm{EV} = R - E$$

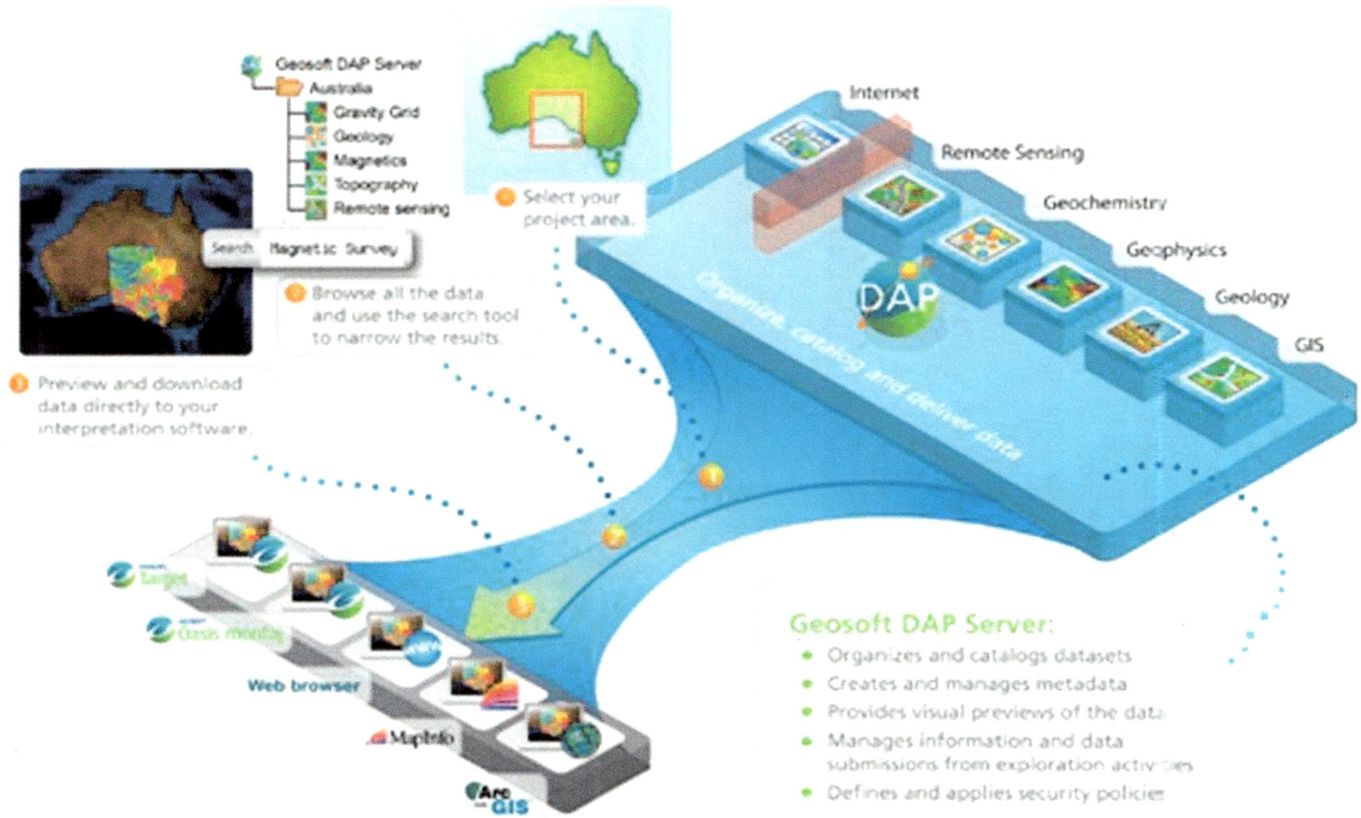

FIGURE 7.15 Exploration information management solution. *Source: Courtesy from Geosoft.*

where

EV is average expected net present value per economic discovery
R is average return associated with an economic deposit after discounting cost of development and production
E is average exploration cost required to find and delineate an economic deposit.

The relationship between the time distributions of an average cash flow to the expected value of an economic deposit is illustrated in Fig. 7.16.

Any investment in exploration which would yield satisfactory return on capital should satisfy the following conditions: (1) EV > 0 or $R > E$ and (2i) risk criterion is minimum.

The relationship of positive EV in the long run when successes and failures of large number of discoveries are considered would depend on minimizing cost of exploration (E) which in turn is related to discovery risk or the number of exploration ventures required to discover and to delineate an economic deposit. The exploration cost is mathematically defined as

$$E = C/Ps$$

where C is the cost of discovering a mineral occurrence and Ps is the probability of discovery of an economic deposit.

Thus, the lower the probability of a discovery, the higher is the risk involved in the process. The link between cost and return is, therefore, the probability of failure $(1 - Ps)$, where

$Ps = P1 \times P2 \times P3$, in which

$P1$ is the probability of occurrence of an economic deposit in a prospect
$P2$ is the probability of actual discovery of that deposit
$P3$ is the probability that the deposit has sufficient economic value to compensate the cost of exploration.

These probabilities are interdependent on each other. If any one of these values is zero, the net result is a failure ($Ps = 0$). These probabilities are substantially affected by man-made constraints such as selection of proper geological environment, exploration techniques used, identification of minimum acceptable economic targets, monetary and time cost of exploration together with the quality, skill, and motivation of scientists and technologists involved in the search.

7.22 RESEARCH AND TRAINING

Successful exploration relies on efficiently managing exploration campaign based on effective application and interpretation of geology, geophysics, and geochemistry, to generate and test exploration targets. In order to discriminate and select best terrains at the regional scale, it is imminent to have a good baseline exploration data (geological, geophysical, geochemical) and drilling data. In addition, access to such data (updated high quality) should be possible, either online or web-based data portals in order to have real-time collaboration. Access to new collaborative tools, viz., net meeting, web conferencing,

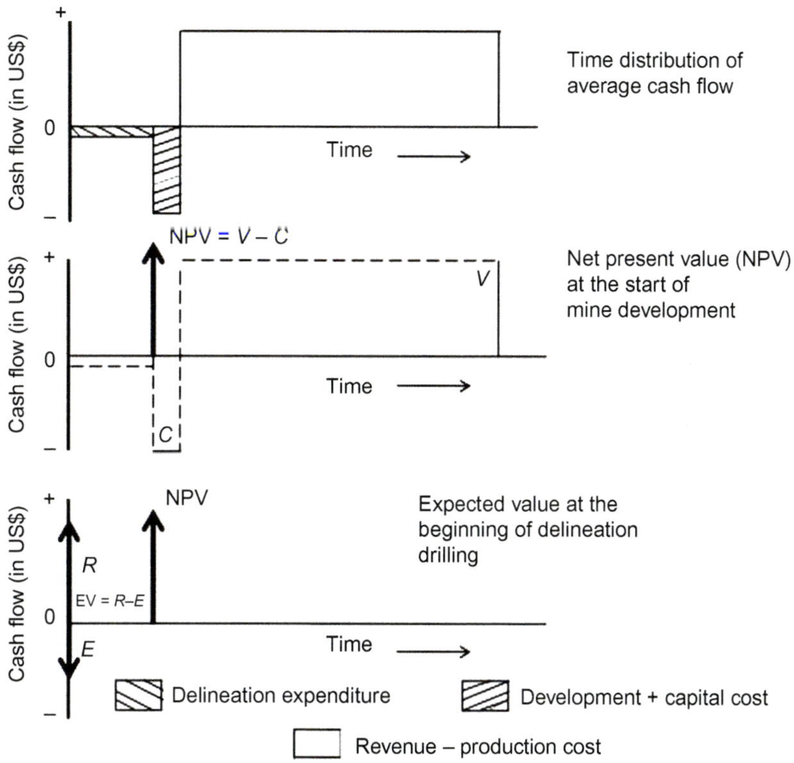

FIGURE 7.16 Relationship of time distribution of average cash flow to net present value and expected value of an economic mineral deposit. *Source: From Kala et al. (1994).*

data streaming, data and metadata harvesting and to innovative thinking, new ideas and interpretation through effective collaboration and knowledge sharing initiatives are imminent. It is needed to access–integrate–manipulate data more effectively by live 3D modeling, advanced processing, and visualization. One has to have free access to supercomputers, more web-enabled access to public processing and research software or the ability to use proprietary software and also seamless access to large amounts of data with fast display speeds. Exploration teams should have an access to researchers and specialists. Normally, the researcher's timelines are not always aligned with exploration requirements. Establishment of portals such as "innocentive" will bring industry and key researchers, consultants together more effectively. The research teams need to focus on "mineral discovery" not "mineral genesis."

A combined R&D effort of Universities, Industry and Government agencies in geoscientific field and exploration techniques will help to achieve the desired result of better success rate in future exploration efforts. Of late, one to one research collaboration is becoming more popular between Researchers and Exploration companies with good funding. It is desirable to have teams from Industry, Government, and Academic, to develop team-oriented 3D interpretative approach focusing on the 3D maps of the mineral endowed belts.

Exploration is based on an understanding of the fundamentals of geological processes and the earth's history. There is a need for exploration geologists to go back to basics too. It is seen that there is a decline of experienced exploration teams, because there is a declining trend in students opting hard rock geology in the universities and the focus has increasingly shifted to environmental studies, sustainable energy, etc. (which need to be explored too). This trend has significantly affected the exploration industry.

There is more to the process than rushing to initiate a drilling campaign by many mining companies, after selecting anomalies outlined from remote sensing data. Knowing when and where to concentrate drilling campaign, where to retain the properties, and when to relinquish them are the key elements to a successful exploration programme. Good geologists carry out diligent field work, geological, structural mapping, measurement of rock properties, and other evidences before advancing a theory or model. Basic structural mapping by experienced people using the newest technologies has proved to be a cost-effective exploration method. Field exploration geologists, geophysicists, and geochemists should work in tandem in an exploration campaign and the combined effort is critical to identify potential mineral targets. It is necessary to motivate more cross-training of disciplines in geology, so that everyone is clear on what to get from the data and what kind of data to use on each situation.

Nowadays, the proliferation of software packages makes life way easier. It is not just software training that is needed but one should know what to do with the data and how to use the software. University cooperative programme can work with students concentrate regular university terms with working in the industry, it helps to expose them with real problems using real tools. It may take a while to evolve, but a beginning has to be initiated, so that ultimately will facilitate to produce exploration teams for the industry. With the arrival of computers, satellites and platetectonic theory, modern geophysical and proven geochemical techniques, etc., mineral exploration has grown intensely sophisticated. In spite of space age technology, mineral exploration still remains some vestiges of the time-honored, empirical "pick and hammer" artistry, that flourished in the past, that helped to discover giant deposits like Butte, Bisbee, etc., about two centuries ago. The combination of time-proven technology of exploration with unproven but promising discovery techniques illustrates that the foundation of minerals exploration is cumulative as well as deeply rooted in tradition.

It is concluded with a famous conundrum "Whilst success in mineral exploration is clearly achieved through new high quality mineral discoveries and also comprehend what, represents best practices in industry research collaboration in exploring geosciences, we still remain poor at closely linking the two together and in tracking the impact and influence of research collaboration upon discovery" (anon.).

7.23 SCARCITY OF EXPLORATION GEOSCIENTISTS

Exploration geoscientists are becoming a scarce resource, since in the recent past, not many come forward to opt the curricula in geosciences in the collegiate level, as the avenues of employment opportunities in other fields like computers/biotechnology/telecommunication, etc. have increased manifold. In order to meet the current and planned exploration projects in pipelines, there is shortage of geoscientists/exploration team. Many exploration companies face difficulties in retaining experienced geological staff as there is an acute shortage and the demand is high. The retirement of many experienced people adds further scarcity. As the exploration budgets increase, more projects come up and need more geoscientists to man the projects (Taranik, 2009).

Employment in Geosciences has been cyclic, in a period of 3–5 years. The uphill periods and the peaks are always exciting. The employment opportunities of geoscientists vary in different parts of the world. Depending upon the skills acquired both in the academics and field, the geoscientists are hired by exploration companies and put them with their teams to enrich experience and other skills. Generally, multinational exploration companies expect the following traits for employing field geologists in the entry level:

- Good, balanced all-round geological education (undergraduate degree, OK; Graduate level, useful)
- Good computer and numeracy skills
- Good mapping and field skills, capability to identify minerals/rocks in the field
- Willingness to travel and work in remote areas
- Language skills, ability or willing to try, ability to get on with people
- Enthusiasm, cultural sensitivity, and a Driving Licence.

As the years pass by, the companies also expect them to keep themselves up-to-date and stay current in their field by associating themselves in "national/professional geological bodies" (like Geological Societies/Economic Geological Society, etc.). The field geologists are expected to develop interest in Economics, Finance Law, etc. Besides enhancing the core skills and try to get experience, they should develop courage, communication/negotiation skills, and interest in social, environmental and current world affairs.

Whenever the unpleasant event of retrenchment happens, that time should be constructively used for company move and acquiring higher/extra qualification. Finally, a geoscientist should be an "optimistic realist" and look for the uphill periods/peaks to come.

References

Bilodeau, M.L., Mackenzie, B.W., 1979. Delineation drilling investment decisions. In: Weiss, A. (Ed.), Computer Methods for the 80s in the Mineral Industry. AIME, New York.

Blain, E.F., 2000. Fifty-year trends in minerals discovery—commodity and ore type targets. Exploration Mining Geol. 9, 1–11.

Brant, A., 1968. The pre-evaluation of the possible profitability of exploration prospects. Min. Deposita 3 (1), 1–17.

Dummette, H.T., 2000. How ore bodies are found. Mining Engg, 31–36, July 2000.

Eggert, R.G., 1987. Metallic Mineral Exploration: An Economic Analysis. Resources for the Future, Washington, DC.

Eggert, R.G., 2010. Mineral exploration and development. Notes of Int. Conf. on Mining "Staking a Claim for Cambodia", Pnom Penh, Cambodia, May 2010, 16 p.

Ericsson, M., 2012. Mineral exploration expenditures. POLINARES Working Paper No.25, March 2012, 6 p.

EYGM Ltd, 2014. Business Risks facing Mining and Metals 2014–2015. E&Y Global Ltd., survey 2014, 56 p.

Gandhi, S.M., 2012. Sketch map showing environmental impacts during different stages of exploration and the respective degree of severity. Mineral Exploration Course Notes 32.

Gandhi, S.M., 2013. Challenges and Emerging trends in Mineral Exploration. Keynote address in Int. Conf. on Future Challenges in Earth Sciences for Energy and Mineral Sector, ESEMR 2013, ISM Dhanbad, November 2013.

Kala, P.P., Bahadur, S., Gandhi, S.M., Sarkar, B.C., 1994. On some economic concepts for planning an exploration strategy with special reference to lead–zinc deposits in India. J. Minerals Metals Fuels, 163–171, July 1994.

Lovell, J.D., 2000. Mineral resource development- our new challenges. Mining Engg, 37–48, July 2000.

Mackenzie, B.W., 1973. "Corporate Exploration Strategies" Application of Computer Methods in Mineral Industry SAIMM, Johannesburg.

Mackenzie, B.W., 1981. Looking for the improbable needle in a haystock: The Economics of base Metal Exploration in Canada. CIM Bull. 828, 115–125.

Mackenzie, B., Woodall, R., 1987. Mineral Exploration Economics: The Search for Base Metals in Australia and Canada. Centre for Resources Studies Working Paper No. 40, p. 112.

Mackenzie, B.W., Woodall, R., 1988. Economic productivity of base metal exploration in Australia and Canada. In: Tilton, J.E., Eggert, R.G., Landsberg, H.H. (Eds.), World Mineral Exploration. Resources for the Future, Washington, pp. 21–104.

Mackenzie, B.W., Belodeau, M.L., 1984. In: Glenside, S. (Ed.), Economics of Mineral Exploration in Australia—Guidelines for Corporate Planning and Government Policy. Australian Mineral Foundation, Australia, p. 171.

McGaughey, J., 2007. Geological Models, Rock Properties, and the 3-D Inversion of Geophysical Data. In: Proceedings of Exploration'07: Fifth Decennial International Conference on Mineral Exploration, B. Milkereit (Ed.), 2007, pp. 473–483.

McKinstry, H.E., 1962. Mining Geology, second ed. Asia Publishing House, New Delhi, p. 680.

Moon, C.J., Whateley, M.E.G., Evans, A.M., 2006. Introduction to Mineral Exploration, second ed. Blackwell Publishing, Oxford, p. 481.

Russell, D., 2012. Exploration funding—trends across the mining and metals industry. In: Global Mining Finance Spring Conference, Ernest & Young, London, 42 p.

Schodde, R.C., 2014a. Challenges and opportunities for under-cover exploration in Australia. Presentation to the UNCOVER Summit 2014, Adelaide, March 2014.

Schodde, R.C., 2014b. Key issues affecting the time delay between discovery and development—is it getting harder and longer. Presentation to the PDAC, Toronto, March 2014.

Schodde, R.C., 2014c. The global shift to undercover exploration—how fast and how effective? In: Soc. Econ. Geol. Conference 2014, Keystone, CO, USA.

Silver, B.D., 2000. The power of perseverance in mineral economics & finance. Mining Eng. 7–8, June 2000.

SNL-MEG, 2013. World Exploration Trends. SNL-Metals Economic Group survey for 2012.

SNL- MEG, 2015. World Exploration Trends. SNL-Metals Economics Group survey for 2014. For earlier years, summary of the entire survey is available online at www.metalseconomics.com.

Snow, G.G., Mackenzie, B.W., 1981. The Environment of Exploration: Economic, Organizational and Social constraints. Econ. Geol.871–896, 75th Anniv. vol.

Taranik, D., 2009. Remote sensing for mineral exploration. In: Irish Assn. for Econ. Geol. and Joly Geol. Soc., Trinity College, Dublin, March 2009, 75 p.

Tilton, J.E., Eggert, R.G., Landsberg, H.H. (Eds.), 1988. World Mineral Exploration: Trends and Economic Issues. Resources for the Future, Washington, DC.

USBM, 1977. Management of fuel and non-fuel minerals in Federal land. Status of Mineral Industries, Ch. 2, pp. 37–75.

Woodall, R., 1984a. Success in Mineral Exploration, a Matter of Confidence. GeoScience Canada 2, 41–46.

Woodall, R., 1984b. Success in Mineral Exploration, Confidence in Prosperity. Geoscience Canada 2, 83–90.

Woodall, R., 1984c. Success in Mineral Exploration, Confidence in Science and Ore Deposit Models. Geoscience Canada 2, 127–132.

Woodall, R., 1992. Challenge of mineral exploration in the 1990s. Min. Eng. 679–684, July, 1992.

Woodall, R., 1993. The third dimension: A geoscience challenge for the 21st Century. Keynote address to The Aust. Inst. Min. Metal. Centenary Conference, Adelaide, pp. 39–40.

Fresh Perspective—technology trends, solutions for mineral exploration, 44 p. http://www.geosoft.com/.

CHAPTER

8

Drilling

OUTLINE

Essentials of Mineral Exploration and Evaluation
DOI: http://dx.doi.org/10.1016/B978-0-12-805329-4.00015-6

8.1 INTRODUCTION

Drilling of hole in the earth's crust is the most effective means of prospecting for exploration of mineral deposit of all types and also for other geological activities. In a mineral exploration process, drilling finally facilitates to define the subsurface geometry and the third dimension of the prospect. Exploration drilling is directed mainly toward determining the presence or absence of the ore target and gaining a preliminary idea as to possible grade and size. Exploration drilling, to a larger extent, indirectly tests geological information, viz., rock types, stratigraphy, structure, wall rock alteration, geochemical zoning, and other ore guides and control. Drill holes are also used to take samples of the ore and to provide all the information that is required for an estimation of tonnage and grade. In the early phases of exploration, emphasis is usually upon rapid progress and low costs with less than perfect assay and geologic sample data. Once the initial high risk has been diminished by favorable preliminary results, consideration can be given to more accurate drill samples, etc. The basis for this prospecting is the proper recovery of core from earth crust obtained in cylindrical form by annular cutting action of bit and can be regarded as true and unaltered specimen of rock with hard mineral deposit.

Drilling is equally important for Engineering Geology, Geochemical, and Geotechnical investigations. Investigation by drilling is a preliminary prerequisite and precise method for geological assessment of the ore deposit. It is usually resorted to only when preliminary geological studies like mapping, geophysical surveys, etc. indicate probability of deposits in a particular area. The accuracy of data generated by drilling is the most important factor in predicting the reserves correctly.

Following the surface geological work of mapping and geological, geophysical, and geochemical inputs wherein promising mineral occurrences are anticipated, a drilling program is initiated to gain knowledge on the continuity of ore and to establish the quality of mineral deposit. Drilling becomes necessary for detailed exploration. Based on the above inputs, a fair geological map depicting lithology, structure, and altitude, the program of drilling in an area is finalized by the geologist with quantum of drilling, number of holes, location, and the tentative depths of the borehole. Drilling provides most of the requisite information for the final evaluation of the prospect. Chemical analyses of drill core samples help the explorationist to provide basis for calculating grade of the ore resource. Detailed logs of drill core samples provide important geological and structural details to define geometry and calculate volume of ore.

8.2 CATEGORIES OF DRILLING RIG

Based on the choice of the power source for drilling, the rigs can broadly be classified into various categories, viz., mechanical, electric, hydraulic, pneumatic, and steam. The drilling rigs can either be "small/portable" or "huge," depending upon the purpose and the depth to be drilled (Atlas Copco, 2014).

8.3 DRILLING METHODS

Diamond, rotary, percussion, auger, and sonic are commonly used drilling methods in mineral exploration, and the choice normally depends upon the depth of the drill hole, location of drill site, geometry of ore target, and the kind of sample needed. The contributions by Hartman (1968), Chugh (1971), Heinz (1994), Hartley (1994), ADITC (1997), Devereux (1999), Dunster (2004), GSI (2010), and DMP (2012) provide the most comprehensive details of various drilling methods, applications, and other related aspects.

8.3.1 Diamond Core Drilling

Probably ancient Egyptians, in 3000 BC, invented the earliest core drills. Diamond core drill is used for myriad applications. It remains the single most versatile approach, capable of drilling at any angle with every range in size of machines from machines coring less than 3 cm in diameter to large machines capable of drilling more than 1.5–2 km deep (Fig. 8.1). Diamond drill machines are usually equipped to set and remove casing, pump the various drill muds and cement when bad conditions are encountered. The drill core is generally regarded as the most useful type of drill sample for analysis, visual inspection, and testing. In this method a coring bit connected in the core barrel is used for drilling. The core gets collected in the barrel which is recovered after completion of the run. The core collected gives information about the rock types and information about the various layers of rock. Surface set and impregnated diamond bits are used for core drilling operation. Much detailed account on this is given elsewhere in this chapter.

8.3.2 Air-Rotary Drilling

Being a quick and economical method of producing a sample, air-rotary drilling is used frequently in mineral exploration. Compressed air is forced down the drill pipe and when it returns to the surface, it carries rock cuttings/chips produced by drilling bit. Samples generated by this method may sometimes get contaminated by getting into contact with other rocks as the rock chip cuttings are blown up. This method is ideal in soft rocks up to 25 m depth. The structural features and their relationships (like fracturing, veining, bedding, etc.) are not preserved in the sample. Drill cuttings can be collected in a cyclone and are often split at the drill site for assay, geological, and mineralogical examinations. The principal disadvantages of this type of drilling are that most equipment can drill vertical

FIGURE 8.1 Heavy-duty drilling rig capable of both diamond and reverse circulation drillings. *Source: Photo by S.M. Gandhi.*

holes only; it is difficult to set casing and the individual samples are considered too small to provide needed information in some kinds of exploration drilling.

8.3.3 Mud Rotary Drilling

For drilling through soft rocks (sand, clay layers, etc.), mud rotary drilling technique is used. In this method, as the name implies, a mixture of clay and water (mud) is pumped into the drill hole, through drill pipe fitted with the drill bit. The return mud from the hole carries back to the surface, the rock chips/cuttings which are collected for sampling and further studies. A mud logger is used to identify the chips. This type of drilling can be done to greater depths from ships/offshore platforms for oil and gas exploration also.

8.3.4 Auger Drilling

Auger drilling is cheap and fast and is usually done during the reconnaissance stage of mineral exploration besides for taking samples for geotechnical studies, soil surveys, environmental studies, construction purposes, etc. It is a popular method when an overburden penetration is necessary to obtain a small bedrock sample to check geology and/or

FIGURE 8.2 Large diameter mobile auger drilling machine. *Source: Photo by Jacob.*

geophysical or geochemical anomalies before proceeding with major exploration drilling. It is done with a helical screw which is driven into the ground with rotation. The earth cuttings are lifted up to borehole by the blades of the screw to the surface. The rate of penetration in Auger drilling depends upon the type of formation being drilled. In order to hydrate dry holes, to improve penetration, and to help lift the cuttings, water is used. These drills can drill up to a depth of about 20 m, depending on the material being drilled and size of the rig and stem. Continuous flight augers with hollow stems are often used for sample recovery in geotechnical operations. Short flight augers are used for large diameter holes.

Augers are available in various sizes. Hand auger drilling is normally done for getting soil samples up to 2–3 m depth. Tractor-/tire-mounted augers are available and the usual hole diameter will be 75–80 mm and can go up to 50 m depth (Fig. 8.2). Small truck-mounted auger drills are often used for mineral exploration projects, whereas the large augers are used for Engineering geology, Hydrology survey, Construction drilling (eg, sinking foundation piles), Seismic exploration, etc. The only limitation in this type of sampling is that undisturbed samples cannot be obtained. Auger drilling is effective in soft and medium hard formation and also in weakly cemented pebbles (Barrett, 1987).

8.3.5 Percussion Rotary Air-Blast Drilling

Rotary air-blast (RAB) drill units consist of either "Down-the-hole hammer drills or Top hammer drills" and are the most frequently used in mineral exploration. They are of two types—handheld and large truck mounted. The latter drills are capable of drilling large diameter holes to several hundreds of meters. This method employs a blade bit, tricone rotary bit, roller bit, or down hole hammer (DHH) to penetrate the rock. The drill bit is hollow solid steel and is fitted with tungsten carbide (TC) tips of about 20 mm, protruding from steel matrix as buttons at the end of the drill pipe. Compressed air is pumped down the hole while drilling. The upward flow under pressure between the drill pipe and the wall of the hole brings up the cuttings to the surface which are sampled. Sometimes some debris may clog/obstruct the way between the drill pipe and the surrounding formation preventing the drill cuttings to come to surface. To circumvent this problem, reamer tubes are used that prevents any external material to fall inside. RAB drilling proceeds rapidly (sometimes 200 m in an 8-h shift) and the cost can be very low once the correct bit and equipment have been selected. Unlike diamond core drilling, this method does not precisely provide the location of samples despite RAB is rapid and cheap. RAB method is widely used for blast hole drilling in mines and for water bore drilling and is usually noisy in operation owing to compressed air equipment.

8.3.6 Air-Core Drilling

In order to drill into unconsolidated ground, air-core drilling technique is employed. The drill bits are made up of hardened steel or TC which sits inside the hollow outer rod barrel. The bit has three blades around the bit head. The hollow dill rods are fitted with inner tube. Small compressors are used for air-core drilling to drill through weathered, loose formations (regolith). These drills are unusable for drilling fresh rock. The drill cuttings are removed by injection of compressed air into the hole via annular gap between the inner tube and the drill rod. The drill cuttings are blown back to surface through the inner tube and pass through the sample separating system. Samples are collected through a cyclone in bags and buckets. This method is preferred to RAB drilling for mineral exploration because it provides cleaner samples. Since the cuttings are removed inside drill rods, contamination is less when compared to Auger and RAB drilling methods. A good support vehicle is needed for air-core drilling rigs, to carry diesel, water, and other supplies.

8.3.7 Dry Drilling

Dry drilling using TC-tipped cutting tools at low "rpm" is useful where the ground is loose and has easily friable earth material (like gossans), and also in areas where the use of any drilling fluid results in loss of core. It is not possible to do dry drilling in hard rock formation as the heat generated will be too high to be sustained by the string. Dry drilling can provide uncontaminated core but is suitable for shallow holes only.

8.3.8 Rotasonic (Sonic) Drilling

This is particularly useful for areas with extremely thick glacial sediments (hundreds of meters) and used when more detailed stratigraphic information is required. This technique uses high-frequency resonant vibration and rotation to drill through glacial sediments and recover a continuous core. Sonic drilling retrieves core but without the contamination caused by drilling fluids. It uses a variable frequency drill head to transmit vibration energy through the drill pipe and core barrel to allow continuous core sampling to take place. It can penetrate overburden, fine sand, boulders, and hard rock, is capable of collecting samples up to 254 mm in diameter, and can drill up to 200 m vertically or in inclined holes. The big advantage of sonic drilling is that uncontaminated, undisturbed sample scan can be collected because no air, water, or other drilling medium is used. This technique can realize 100% core recovery, even in glacial till, clay, sands and gravels, as well as hard rock. It offers rapid penetration, reduced on-site costs, and minimal environmental impact.

8.3.9 All Hydraulic Drills

In conventional drills, the major cost component is labor and bit cost. The drilling time available is very low on account of time lost for hoisting and lowering of drill string. The labor cost component alone is 50% of the total cost. In all hydraulic drills the rotation, penetration, hoisting and lowering, screwing and unscrewing of drilling rods, and even lifting of drilling accessories from rod rack to feed frame and vice versa is done automatically with the help of hydraulic power. Nowadays computer-controlled drills are available in which all the drilling parameters are fed into a computer which is connected to the control panel and the entire drilling operation is performed by a single person. These machines have safety features like safe operations as control panel is away from feed frame, no fatigue to the workers, no manual operational error, etc. In all hydraulic drills, good-quality impregnated bits are used which give productivity of 30–40 m per shift of 8 h, in different drilling environments using a crew strength of 2 persons and 1 person in case of computer-controlled drills.

8.4 SELECTION OF DRILL

The suitable drilling technique is selected depending upon the depth and size of the borehole, mineral to be investigated, sampling requirements, and nature of formation. Conventional core drilling is used for shallow-to-medium depths in difficult clayey and fractured formation. Wire-line drilling is used for deep holes with consolidated formation. When representative soil samples are required, Split Spoon Sampler device is generally used, preferably with self-tripping assembly. Selection of drilling method is also determined by the availability of proper equipment and accessories. Selection of drilling fluid depends on nature of formation. Selection of methods of drilling is also affected by the availability of equipment and relevant standardized series accessories.

The selection of drill depends on following factors:

Borehole depth: The proposed depth and size of the borehole are the guiding factor for the selection of the drill. The capacity of a diamond core drill is based on its capacity to lift the drill string.

$$\text{The capacity of drill(HP)} = \frac{\text{W(weight)}}{100} \times \text{V(velocity)} \times \text{n},$$

where

W = Weight of string with extra 20% for frictional losses as per DCDMA standards.
V = Hoisting speed (m/min).
n = Efficiency.

The power consumed by the auxiliaries is around 20% and so the required capacity of the drill is to be increased by 20%.

Character of rock: The character of rock determines the speed and feed requirement. The drillability depends mainly upon formation hardness and toughness. Hard formations require hydraulic feed rigs.

Terrain condition: In difficult and hilly terrain lightweight rig and pumps are preferred. The Geologist-in-Charge determines the borehole location, drilling procedures (coring/noncoring), total depth, and sample intervals and documents the requirements in the Drilling plan and Sampling plan prior to the start of field activities. The Drilling plan and Sampling plan may be prepared for each borehole. The Geologist-in-Charge will observe all handling of the core by the drill crew, will prepare a lithology log, and will collect core samples at specific intervals for purposes of chemical analyses and physical testing.

8.5 SELECTION OF DRILLING FLUID

Drilling fluid is any fluid that is circulated in the borehole to help in carrying out a cost-effective and efficient drilling operation resulting in stable and gauged borehole to targeted depth with minimum possible damage to prospective formations. A major component in drilling operation success is drilling fluid performance. The successful completion of borehole and its cost depend, to a great extent, on the properties of the drilling fluid. The functions of drilling fluid, which are critical to the drilling process, are: (1) aid formation stability and productivity; (2) clean the bottom of the hole; (3) lift formation cuttings to the surface; (4) suspend cuttings while circulation is stopped; (5) cool the bit; (6) control subsurface pressure, and (7) lubricate the drill string and assist in its corrosion control.

Selection of drilling fluid mainly depends on the type of formation and the borehole depth. The various kinds of drilling fluid normally used are water, bentonite mud, cutting oil, and polymers (both water-based and mud-based).

Water: It is freely available fluid and is primarily used in core drilling operation. It has very good cooling properties and acts as moderate lubricant and vibration dampener.

Bentonite Mud: This increases the viscosity and gives better cleaning of the hole. It acts as very good lubricant and also reduces the vibrations in the string considerably. It forms cake on the borehole walls and thus protects the caving in the borehole. Bentonite mud solution has gelling properties and thus it keeps the cutting in suspension even when the circulation is stopped.

Polymer: The polymer has similar properties as that of bentonite mud except gel strength. It has excellent flushing capacity. The only drawback is that circulation is to be maintained all the time else the cuttings will quickly settle at the bottom and jam the string.

8.6 SELECTION OF PUMP

In diamond drilling operations the fluid requirement is basically of low flow rate and high pressure. These requirements are best met by reciprocating pumps. These pumps are capable of delivering constant discharge at various heads. The replaceable liners provide extra flexibility for different discharges and pressures. The reciprocating pumps used in drilling can be of duplex/triplex, single acting, or double acting, depending upon the head and discharge required. For shallow boreholes and in hilly areas where water is used as drilling fluid, smaller pumps are used. Duplex double-acting pumps are used where mud is used as fluid. For deep boreholes, that is, drilling for Coal/Coal Bed Methane, etc., RD-900, FM-45, RD-395, Vol 6x6, etc., are used. The return water velocity should be in the range of 30–50 m/min for good flushing of the borehole. It depends on the annular space between the string and the wall which changes at different depths due to telescopic holes, rheological properties, viscosity and density of cutting, etc. The return velocity has to be such that it does not erode the borehole walls, wash out the core, and at the same time lift the cuttings.

8.7 EXPLORATION DRILLING METHODS

Diamond core drilling and reverse circulation (RC) drilling are the principal types of drilling among many other types.

8.7.1 Diamond Core Drilling

There are three parts involved in the Diamond core drilling system: (1) a rotary rill which provides power, (2) a drill rod which will contain core as and when drilling is in progress, and (3) a circular diamond bit that has an open central area to clear entry into the ground. The cylindrical section of the earth removed after drilling, for inspection/examination, is called "core" which differentiates other drilling methods wherein one gets rock chips/cuttings. The core sizes differ from 25 to 65 mm in diameter. The larger the core one desires, it will involve bigger drill bits, large drill rods, higher power to penetrate, and higher costs. The larger the core one deserves, which will involve corresponding bigger drill bits, the larger drill rods and more power and of course higher costs. The drill hole diameter and core diameter of different sizes of drills are given in Table 8.1.

The drill rod, which has a "diamond bit" at the end, penetrates into the ground by the rotary action provided by drill. As the drill penetrates into the ground, a solid column of rock (core) slides slowly into the drill rod/pipe. After completing drilling one drill rod length, another one is added to drill further. The drilled cores are retrieved from the ground and stored in sectional boxes with proper indexing to correlate the drill section from which they were obtained. On the other hand, in a "wire-line" drilling, there is a core tube inside the drill pipe that has a latching mechanism attached to a cable. The latter assists to winch the core tube containing the core to the surface at the end of each 10-ft run.

Invariably all drill rods are 10 ft long and after drilling 10 ft, new section of pipe is added up, to drill further depth into the ground. The diamond bit slowly cuts the earth as gentle pressure is given with water as the lubricant to prevent overheating. The number of drill rods used will give the drilled depth. If the drilling bits get overheated, they might

TABLE 8.1 Drill Hole and Core Sizes

	Hole Dia. (mm)	Core Dia. (mm)
AMERICAN SIZES **Conventional**		
AX	48	30.1
BX	59.9	42.0
NX	75.7	54.7
Wire-Line		
AQ	48	27
BQ	60	36.5
NQ	75.8	47.6
HQ	96.1	63.5
METRIC SIZES **Conventional**		
36T	36.3	31.7
56T2	56.3	41.7
66T2	66.3	51.7
Wire-Line		
56ST	56.3	35.3

get stuck inside the hole. If fractured rocks/loose rocks are encountered while drilling, there will be heavy water loss and might lead to overheating, resulting in the drill bit get stuck in the hole. In such cases, "drilling mud" (or saw dust or other materials) is sent into the drill hole to plug the fractures, thereby preventing escape of fluids.

There are drill rigs which serve as "multipurpose rigs"; sometimes these rigs start drilling in RC and later change to diamond coring when the target depth is reached, thereby avoiding the necessity of another rig. The drill rods of RC drilling are very strong compared to light diamond drill rods. In diamond drilling there is a small gap (3–4 mm) between the rod and the wall of the hole through which the coolant/lubricant travels to keep the cutting edge cool. Diamond core drilling is possible up to depths greater than 2500 m. The drill penetration in diamond core drilling is much slower than other methods owing to hardness of rocks encountered and also the time involved in retrieving the core from depth. On an average, 30–40 m of core, per shift, can be recovered.

8.7.1.1 Diamond Drill Bit

Owing to their extreme hardness, diamonds are used to make tools that are needed for hard jobs in mineral exploration. There are two types of diamonds (1) gem quality (to make jewelry) and (2) industrial diamonds (to make drill bits for diamond drilling). One set of drill bits designed to penetrate a particular formation may not be suitable for other type of formation. Diamond bit's performance and life is influenced by so many factors. Diamonds embedded into a drill bit provide the cutting edge, the drill bit needs to drill. Diamonds may either be attached to the surface or be integrated into the bit by a sintering process. With the progress of drilling the bit and the diamonds normally wear away and the new diamonds are exposed for cutting in a sintered bit. By this way, the life of the bit is prolonged and also eliminates the need to change the bits during drilling or when the hole is very deep.

Diamond drill bit is usually a hallow tube with diamonds set at one end. Occasionally, the diamonds are set to rise up into the tube which is meant for larger holes. A stabilizer and an anchor are provided in the middle of the bit to hold in place. Diamond bits when they are used for hard materials, lubricants such as oil or water are needed in order to reduce the friction, lowering the risk of breaking the bit or the material it is drilling. Drilling bits of different types are available for specific applications (Fig. 8.3). The operator must be careful and cautious to use the appropriate speed and pressure while drilling

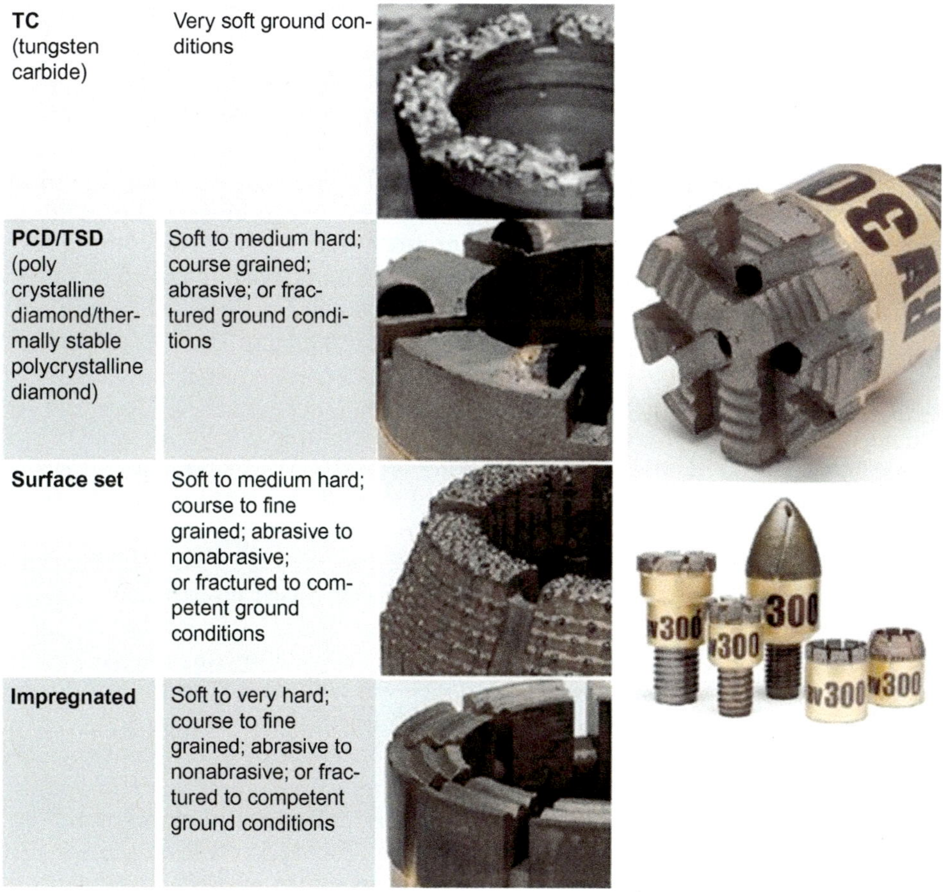

FIGURE 8.3 (Left) Types of drill bits; (right) directional and wedging bits. *Source: Sandvik Mining & Construction (2012).*

operation. Many manufacturers of bits even issue charts and recommended diamond bit types along with speed and pressure recommendations, etc. The standards for manufacturing core drilling tools globally are set by Diamond Core Drill Manufacturers' Association (DCDMA) but now the standards are controlled through the ISO 9001 quality process.

8.7.2 RC Drilling

In terms of both equipment and sampling, RC drilling is different from diamond core drilling. The major difference being the rate of penetration, cost per foot, and the sample type. RC drilling equipment is relatively larger with a high-capacity air compressor in order to handle larger down hole equipment. Pneumatic reciprocating piston or DHH drives the drill bit, which has round protruding "TC" buttons to cut the rock, and the drill rods rotate at speeds of 30–50 rpm. Another type of bit is called "hammer bit," which rapidly pounds and pulverizes the dense and hard rock formation normally above the water table. When the bit encounters the water table, the shattering becomes less effective. Bits consisting of "three revolving cone shaped grinders" (Tricone bit) are very effective in softer formations and in wet drilling conditions but slower in hard rock conditions.

A PVC or metal piping is installed at the surface to prevent collapse of unconsolidated material into the hole before commencing drilling. The stability of the surface formations dictates the depth. Air is pumped down the drill rods between inner and outer tubes. The returning air through inner tube brings up cuttings to the surface. The compressor forces air down the outer space of a double-wall pipe (Fig. 8.4). A cyclone chamber slows down the rock chips that travel at high velocity and then make the chips spiral downward to the cyclone bottom. As the drill advances, continuous sampling of rock chips can be done. The RC drilling is usually either 6″ or 8″ in diameter.

Drill cutting samples are collected at regular intervals (5′). Owing to large diameter hole, huge volume of sample material constitute each sample which are usually split into a reasonable volume to handle (one-eighth of total) and sent to laboratory. A dry sample splitter (Jones splitter) is used in dry drilling conditions (Fig. 8.5A). On the contrary, when the drill reached the depth of water table, a rotary wet splitter is used (Fig. 8.5B), which spins around and splits the sample using a series of fins like the fins of a turbine engine.

Samples of small rock chips are collected in plastic boxes (chip trays). Geologist-in-Charge observes and carefully logs the samples. It is obvious that structural details, etc., of the formations drilled may not be available since the samples obtained are rock chips, but systematic microscopic examination of the samples will reveal valuable information. Besides, the other tests like fluorescence and effervescence can easily be carried out. Compared to RAB or air-core rigs, RC rigs are larger, slower in operation, expensive, and can achieve better penetration in hard rocks, sometimes up to the depths of about 500 m. RC drilling is one of the preferred methods for preliminary mineral exploration work, since it is less expensive than diamond core drilling. Owing to their size, these rigs need good roads to mobilize them to drill sites. RC drills generally are ably supported by auxiliary vehicle(s) for diesel, water, maintenance supplies, spare compressor, booster compressor, etc.

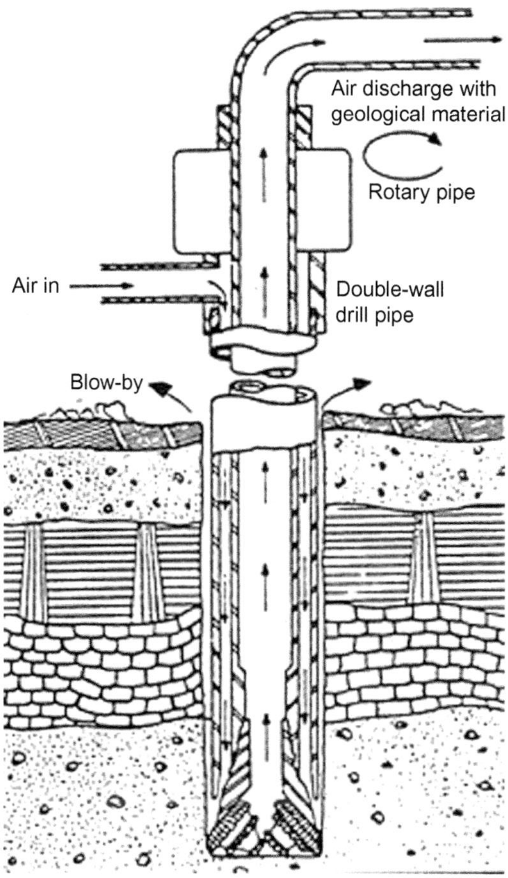

FIGURE 8.4 Sketch of RC double-wall drill pipe (with tricone bit) showing air path.

8.7.3 Wire-Line Core Drilling

Normally, core is extracted inside the barrel by advancing the drill by its rotary action. When the depth of drilling increases (say beyond 300 m), withdrawing the entire heavy drill pipe is not only time-consuming but cumbersome; hence, a method has been developed to retrieve the core inside the barrel. Suppose a solid granite area is drilled, the core would normally break at the drill bit when drilling is stopped, then a grabbing device is lowered by a wire to pull up the core. With the objective of cutting the downtime spent in hoisting and lowering of the drill string, the wire-line system of drilling was developed. It is a special type of core drilling most commonly used for mineral exploration. In wire-line drilling in medium hard and soft formations, as are usually encountered in coal areas (Sandstone, Shale coal), it is seen that in deeper holes, more time is spent in hoisting and lowering of drill string, for taking out core sample after every 3-m run in comparison to the time spent in cutting the 3-m depth. In order to remove the core from wire-line core barrel, an overshot assembly is lowered on the end of the wire-line. It gets attached to the

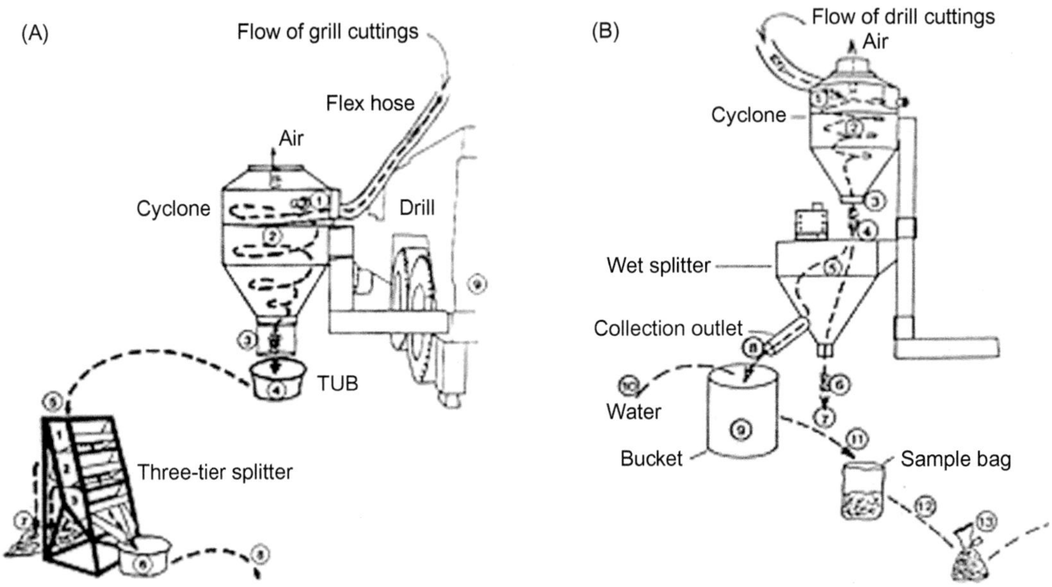

FIGURE 8.5 Typical dry splitter sampling (A) and wet splitter sampling (B) arrangements.

back of the core barrel inner tube and the wire-line is pulled back to retrieve the core sample. Only core sizes of NX and BX sizes are possible with the wire-line system.

Advantages of wire-line core drilling:

- More time is available for drilling, hence speedier progress can be made.
- Hoisting the drill string needs to be done only when the bit is to be changed.
- Wire-line drilling is possible up to a depth of 1000 m.
- Best suited in coal formations drilling where depths involved are in the range of 200 m.
- Ordinary drilling equipment can be adopted to this system with necessary modifications.
- Wire-line rods are as per specifications as laid in "Q" series of DCDMA.

8.8 THE COILED TUBING DRILL RIG

Kanck (2014) has described that a prototype of world's first coiled tubing drill rig (CT rig) for mineral exploration (Fig. 8.6) has been developed by Deep Exploration Technologies Cooperative Research Centre (DET CRC), which will facilitate a spurt in mineral exploration campaigns. Coiled tubing drills have been used for years, in the oil and gas industry, primarily for "work over drilling," where existing holes, lines, or wells are redrilled or cleaned out to improve productivity and delivery of oil or gas." These holes can be thousands of meters deep; however, the coil tube operates inside existing casing or pipes, not unlike a drain cleaner. The typical application in mineral exploration is in the host material known as "soft rock." Unlike traditional rigs, which employ drill rods,

FIGURE 8.6 The coiled tubing drilling rig. *Source: Courtesy from DET CRC.*

the CT rig uses a "continuous reel of tubing" that enhances safety and maximizes the length of time the drill bit is able to operate "at the bottom of the drill hole." Compared to conventional rigs, the CT rig is much smaller in size and lighter and uses a motor at the base of the shaft rather than at the surface, thereby reducing fuel costs and environmental impact.

The challenges faced by hard rock explorationists include developing economic methods that can drill hard rock of 150–200 MPa with a single pass, and at a high speed and penetration rate. The CT rig technology, the DET CRC, is developing aims to work within this challenging space, using a cost model that is a fraction of existing hard rock exploration, and a very small fraction of oils and gas budgets. Coil tubing needs to become a long-life composite and new drilling tools that are significantly more efficient and effective must be developed. Cutting materials and hardware that are purposely designed and developed for the new operating parameters, plus control systems and down hole and surface tools that provide cost-effective, decision-ready data, are also needed. It is anticipated that the use of the DET CRC hard rock CT rig will facilitate to define mineralization haloes faster and expedite the prospecting and exploration process, hopefully improving the successful strike rate. CT rig, concomitant with real time sensors, would rapidly speed up the rate of testing targets. By using this rig, one gets an advantage of instant feedback on the nature of ore, likely grade, etc. This technique will help to expedite and focus on the mineral exploration campaigns by the exploration companies, whose dictum is "time equals money."

8.9 SAMPLES FROM DRILLING CAMPAIGN

In a drilling campaign three kinds of samples are recovered:

1. Core samples from diamond drills or from the casing drills
2. Dry cuttings from air-flushed diamond, rotary, auger, or percussion drilling
3. Wet cuttings or sludge from churn drills, diamond drill or wet rotary, or percussion drilling

8.9.1 Core

The core consists of one or more cylindrical pieces of rock (Fig. 8.7). In solid ground, it may be a single piece for the length of core barrel. The normal core barrel length (10′/3.04 m) limits the core intercept and if the core and sludge assays are to be combined, the two footage intervals should match including all the sludge collected in a tank. The core more often consists of sections from a few centimeter or more in length or even some small fragments. The driller places the core in a box having longitudinal compartments of proper size. Each run is divided from the next by a small block of wood marked with meterage. The depth of the hole is ascertained by keeping track of number of drill rods (usually standard length) and correcting for the distance between the collar of the drill hole and the top end of the last rod. The core barrel should preferably be emptied directly into its permanent storage tray or box, to avoid the loss of material during handling from one place to the other, rehandling, etc.

8.9.2 Dry Drill Cuttings

In drilling programs where dry samples are collected, cyclones are used to draw cuttings away from drill collar and settle the sample into bottom of the apparatus where it can be dropped out into storage containers. The sample is to be run through a splitter and the samples are collected in plastic bags and sealed tightly. Care is to be exercised in the handling and transport of dry cuttings because the fine and heavy material has a tendency to settle to the bottoms of containers.

FIGURE 8.7 Core pieces recovered from diamond drilling.

8.9.3 Wet Drill Cuttings

The sludge from diamond, churn, rotary, and percussion drilling is caught in various devices, ranging from simple overflow buckets hung under the drill collars to elaborate multiple sludge tanks where all cuttings carefully settled out of drilling water. After siphoning off water for recirculation, the sludge is air-dried or dried on stream tables. Extreme temperature and direct contact with flame during drying are deleterious, as, for example, partial roasting of sulfides. One should be careful against gravity settling of heavy minerals since they often are the very object of interest and of high unit volume and great error can result from small involuntary enrichment or impoverishment of samples.

8.10 CORE RECOVERY

Core recovery is defined as the percentage of core length recovered with reference to drilled length of run.

Core recovery (%) = (Length of core/Length of run) × 100.

Core recovery of 100% is difficult to attain when drilling soft rocks/fragmented formation. Standard core recovery in medium to hard formation is between 80% and 100%. The core recoveries in soft/friable and loose soil/sand can be as low as 10–20% or even lower.

Core recovery is classified as per the following table.

Very good	90–100%
Good	75–90%
Fair	50–75%
Poor	25–50%
Very poor	<25%

However, it may be emphasized that core recovery depends on the nature of strata, condition of drill accessories, and type of drilling. It is largely independent of human intervention. During all phases of core drilling, recovery should be watched constantly, as too often core recovery is poorer in mineralized zones. If the core recovery is poor, the sludge samples then become of utmost importance. In exploring veins, bedded ore, or other confined mineral zones, it is possible to case the hole immediately prior to making the ore intersection or to ream and case, if the ore zone is penetrated unexpectedly and to deflect the hole for several additional ore intersections, taking every precaution for good core and sludge recovery through the ore zone. The drilling crew takes absolute care for enhanced core recovery but it can be achieved only when better accessories and technology are used. Apart from nature of formation, there can be various reasons for low core recovery.

The main factors that affect core recovery are:

- type of formation encountered for loose formation, there are chances of poor recovery,
- set of equipment selected,
- excessive washing of core in barrel-bit assembly,

- improper selection of bit speed and feed,
- excessive vibration of string,
- improper drilling fluid, etc.

The diameter of core is determined by inner diameter (ID) of drill bit under use. The bigger core size is better for analysis. Typical core sizes in diamond core drilling are 44.45 mm (BWT size) to 80.95 mm (HWT size) under DCDMA series. If Swedish standards are used, the core size may be 47 mm (for 66 mm T-6) to 93 mm(for 1160 mm T-6). The bigger the size of the core, the higher is the cost of drilling.

8.11 CORE STORAGE

Drilled core is normally kept in flat wooden/aluminum trays partitioned lengthwise into compartments, little wider than the diameter of the core (Fig. 8.8, left and right). Since the core is heavy, it is preferable to have a box of about 1–1.5 m, so that it could be handled easily. In some places, galvanized iron or aluminum boxes of different sizes are used. The core samples should be protected yet to be available for continual reference as preliminary project data begin to take shape. The boxes are kept one over the other in a rack so that the desired box could be taken out for reexamining, etc. Each box is properly labeled with all details, including the drill hole number, meterage details, etc. The core recovered after spending substantial cost deserves proper care. In many exploration projects, boxes containing economic mineral intersection with about 5 m of footwall and hanging wall portion of the rocks are preserved for future reference. Sometimes it is possible to donate the core to permanent reference "core libraries" maintained by National or Provincial Geological Survey departments, which could be accessed by interested parties.

8.12 CORE SPLITTING

In order to have a continuous sample of core for geological record and at the same time chemical assay of it, core splitting is done usually with an impact core splitter or a rock

FIGURE 8.8 (Left) Diamond drill cores in wooden boxes; (right) aluminum core box.

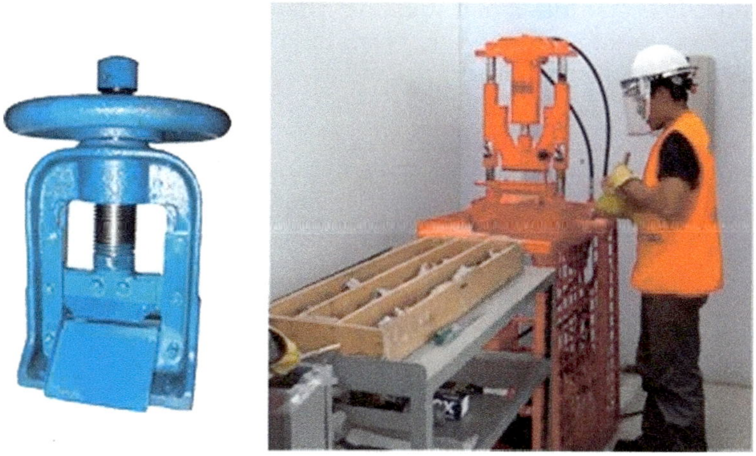

FIGURE 8.9 (Left) Mechanical core splitter; (right) hydraulic core splitter.

saw. Depending upon the nature of core and its constituents, there is always a problem of getting exact representative split of core. Each section of core is split longitudinally. The core is split, with half for geological logging, one-fourth for chemical assay and the remaining one-fourth for metallurgical tests. A mechanical core splitter consists of a clamp to hold the piece of core and a blade to which force can be applied with a hammer (Fig. 8.9, left). In the recent past, the force to the blade is applied using hydraulic and pneumatic methods (Fig. 8.9, right). Besides, there are machines, wherein a longitudinal piece of core is split by diamond saw. The assaying practice of core varies from place to place:

- Some prefer to split the entire core and half of it sent for assay whether the material is ore or not.
- Core is not assayed except for intervals in which the sludge assays show values.
- Only, when it shows mineralization, the core is sent for assay.

When the character of the ore and its distribution are well known, the geologist can decide one of the above methods.

8.13 CORE LOGGING

The information concerning drill holes and the samples collected from them, usually, is kept in two different logs, viz., engineering and the other a geologic log. Details pertaining to the location of the hole, under conditions it was drilled, sample recovery data, drilling problems encountered if any, push in drilling, and other features are given in engineering log. A geologic log is a run-by-run description of all the pertinent geologic observations made of the drill hole samples, whether they be core, cuttings, or sludge.

Usually the drill core is wetted first, not only to remove the dirt/dust on it but also to bring out clearly the texture of rock type, mineralization, etc., before logging by a qualified geologist. If need be, the split core is also examined since certain features show up better on the broken surface. As drill core is obtained after so much of effort, time, and money, the logging of the core is to be done very carefully. The geologist's log is a systematic and quantitative description of the core on a standardized log sheet/form. The form has columns for each of the types of information which will be recorded. The information typically shown includes the run length, core length, core recovery percentage, lithology (changes in rock types), alteration facies, mineralization (ore-forming minerals), rock forming minerals, rock quality data (RQD), visual estimation of grade of ore minerals, structural details, specific gravity, etc. The visual estimation of ore grade gives an idea to compare the chemical assay received from laboratory. A variable power binocular microscope with built-in light source is invaluable in all phases of drill sample logging (Blackbourne, 1990). A copy of the standard geological log sheet is provided as an "Appendix."

From drill cores, strike and dip of the planar features are not discernible but the angle of planar features like foliation, bedding, vein, fault trace, etc., with respect to drill hole axis will give a clue about the possible geometry of the features. While geological logging, testing for the presence of noteworthy/special minerals is also to be done using various techniques; testing for fluorescence (scheelite); testing for effervescence (carbonates and calcareous alteration); and mineral stain testing (feldspars and carbonates). Many a times, exploration companies photograph the core for their record. In order to illustrate surface geometry of the orebody, drill sections are constructed using the data from the detailed geological log.

8.14 HIGH-TECH CORE SCANNING AND INTERPRETATION

The most significant cost factor in an exploration project is drilling and recovering drill cores. As indicated in the previous sections 8.9–8.13, various studies or investigations can be done on drill cores. Geological sampling by splitting of core involves damage of the core, besides corrosion pressure relief, etc. Some of the petrographic, tectonic queries that might arise at a later day of the project cannot be answered, if the core is damaged for sampling and assay purposes. In the recent years, high-tech core logging systems which enable hyperspectral scanning, digital imaging, processing and analysis of drill core, rock chips, other geological samples, etc., have been developed by the companies, "DMT CoreScan" and "CoreScan." With these innovative systems and high-quality images, creation of "digital core library," qualitative and quantitative structural analyses are now possible. The "core scan" system is composed of (1) an imaging device (drill core scanner), UV fluorescence in some cases; (2) analyzing software, and (3) data bank (Fig. 8.10). The system can generate high-resolution, full circumference images of drill core samples and can provide high-resolution optical images of the full core box in its true color. The drilled core is scanned immediately following drilling just before cutting. Mineralogical maps are generated and provided directly to site geologists. The instrument is designed to use in rough environments, even in exploration camps.

The system "CoreScan's HCI-3" is known to integrate reflectance spectroscopy, core photography, 3D laser profiling to map mineral phases, automated quantitative mineralogical, geochemical and textural information both for exploration and process studies of a

FIGURE 8.10 High-tech drill core scanner. *Source: Courtesy from DMT CoreScan.*

deposit. The greatest advantage is that the digital records of all these information can be transmitted electronically to anybody that could be linked to various other system software packages available for further analysis, both qualitatively and quantitatively. Some systems analyze cores at a later point of time by producing high-resolution digital images of the drill cores. With additional input of structural details (taking into consideration of core loss, etc.), they can even be oriented to "North." The systems also permit retroactive structural and sedimentological interpretation of the strata that were drilled through.

8.15 DEDUCTIONS FROM DRILL CORE SAMPLES

8.15.1 Grade

The drill hole assay is comparable to channel sample, if the core recovery is complete. Despite its smaller volume, the drill core will have more uniform dimension. Depending upon the uniformity of the orebody, a series of drill holes can afford to get an accurate sampling of an orebody. However, in precious metal deposits (narrow gold vein type), drill holes may not be satisfactory to determine the average grade of the ore, unless they are more numerous and closely spaced. Assays of core give some indication as to whether or not, a vein is mineable. The mineralogical nature of the vein matter in the ore may give an idea, whether the vein is worth developing or not. Peters (1967) gives a detailed account on the costs and the desired value of information, from drill core samples.

8.15.2 Stratigraphic Thickness (Width)

In order to obtain true stratigraphic thickness, it is necessary that a drill hole should penetrate bedding at right angle. Practically, situation gets complicated due to structural features like faults, folds, etc., and borehole deviation. The true stratigraphic thickness could be calculated by using the true dip and hole deviation. As per Dunster (2004), the "stratigraphic efficiency" is the ratio of equivalent true stratigraphic thickness to meters drilled, expressed as percentage. There are several 'freeware' software packages available which can calculate these.

The distance of orebody intersection measured along the core is not a true width of the orebody that needs to be corrected, for the amount by which the intersection between borehole and vein varies from a right angle (Fig. 8.11). Inclination and bearing of the drill hole is compared to the dip of the vein. The dip is determined by drawing a cross section of the vein through its outcrop or underground exposure and the point of intersection. On irregular veins, this may involve some error and all data as to the shape of the vein shown in various plans and sections must receive consideration, in arriving at the most probable attitude. In wider orebodies, the correction should be not only for the angle of intersection but also for the curvature of the hole, as "the length of an arc is greater than the length of chord." The probability that a single hole or a limited number of holes will afford a correct measure of average width of an orebody depends upon the habits of ore, whether the width of ore is habitually regular or not. The possibility could be determined on a statistical basis in the same manner as reliability of grade (McKinstry, 1962).

8.15.3 Structure

Besides the indicators of ore, diamond drill holes give valuable structural information. They not only determine the location of rock contacts but also furnish evidence as to the attitude of contacts and other structural planes. If a particular bed, vein, cleavage plane, fault, or any other planar feature can be identified in three holes, its dip and strike can be calculated by the three-point method. This consists in taking the point which is

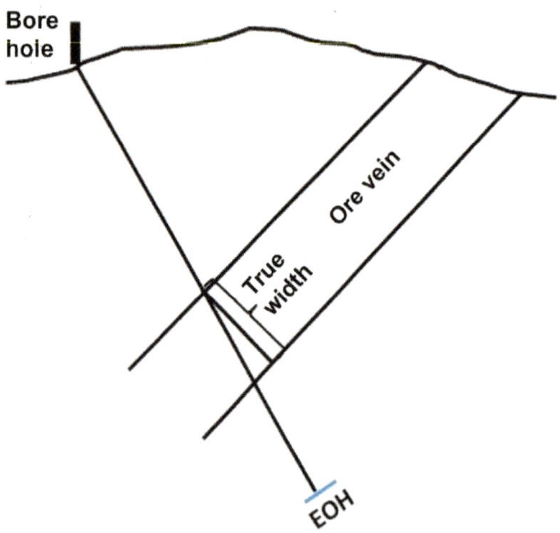

FIGURE 8.11 Relation between true width of a vein and width as measured along drill hole.

intermediate in elevation (as compared with others) and computing the position of a point of equal altitude along the line between the other two. A line joining the two points of equal altitude will give strike and from this the dip could be calculated.

If the data are available from only one hole, the angle which the bed makes with the axis of the core can be useful, since the core has been rotated from its natural position by an unknown amount, it does not in itself furnish full evidence unless core-bedding angle happens to be 90 degree. In general case, the attitude of the bed is indeterminate, but if the strike is known from other data, there are only two possible angles of dip and the geologist may feel confident in selecting one of them as the more probable, if the geologist knows about the local structure.

In many areas, the planar or linear features that have uniform orientation over considerable distances can be used to determine what portion of the core occupied, when it was in the ground. In many areas, regional cleavage is likely to be nearly constant, even though the bedding changes in attitude abruptly from place to place. Linear structures such as parallelism of amphibole prisms may vary very little throughout a district. When either of these structures is placed in its correct geographical position, the attitude of bedding may be read directly. If planar features are parallel to core axis, two possible orientations are possible; if they are normal to core axis, the numbers of options are infinite. If the linear structure is parallel to core axis, the number of possible orientations are infinite; if they are perpendicular to core axis, two possible orientations are possible.

8.16 PORTABLE XRF ANALYZER

With the advent of technological advancement in instrumental analysis, portable X-ray analyzers are of various designs and are available in the market, which can be used for nondestructive, quick on-site detection of minerals, metals, and contaminants, saving time and cost. The instant results of core/rock samples enable the exploration geologist to determine the next course of action for the entire process including ore grade, exploration, and environmental sustainability.

In order to test the chemical composition of materials, portable X-ray fluorescence system (XRF) is used. It is a nondestructive elemental analytical technique for quantification of elements in diverse materials (Fig. 8.12). The principle behind the system is that "illumination with X-rays causes each element to fluoresce at a characteristic wavelength. The pattern and intensity of wavelengths are analyzed by an XRF analyzer to determine the identity and quantity of elements present in the sample in question." Handheld XRF system is a battery-operated, lightweight system and requires miniature X-ray source and detectors and has become an indispensable tool for an exploration geologist. This system nullifies laborious method of sample preparation and is designed for high-throughput testing. The quick results obtained by portable analyzers offer economic and time saving in exploration projects, especially in remote setting. The quick results assist an exploration geologist not only to manage where to concentrate but also what to follow up in the field. Portable analysis systems help for immediate decision-making, faster drill hole target generation, and for managing the time frame for the project development, and ultimately lead to quicker "finds."

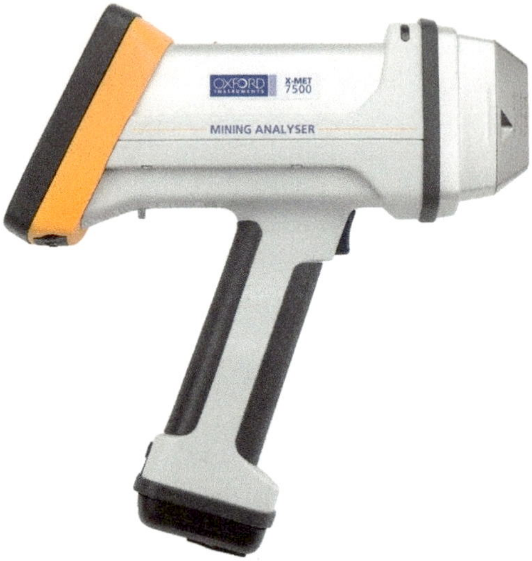

FIGURE 8.12 Handheld portable XRF analyzer. *Source: Courtesy from Oxford Instruments.*

Another type of X-ray analyzer is "Scanmobile" (by Finland based, Mine-on-line services), a vehicle-mounted X-ray analyzer, designed to analyze, full uncut drill cores before logging, drill cuttings, and till samples. This can visit customer's sites and carry out analyses. If further analysis is needed, the samples are selected based on the Scanmobile analyses.

8.17 DEVIATION OF DRILL HOLES

It is always essential to know the orientation (bearing and inclination) of drill hole throughout its length, to draw correct and meaningful inferences from drill hole data. Owing to various reasons, the orientation of a drill hole at the collar of the drill hole will not be same in its course and rarely drill holes are straight. Deviation is usually negligible in short holes but in longer holes (>100 m); it is likely to be quite appreciable and for lengths of 800–1000 m, it may amount to as much as 25 degree or even more. Careful drilling minimizes the tendency to crooked drills, but even with the best technique, some curvature is unavoidable (Killeen et al., 1995; Marjoribanks, 1997). Drill rods become increasingly flexible with the "torque" of the rotating bit works against cutting face. The direction of deviation is influenced by the nature and structure of the rock. Drill holes at a small angle to bedding may curve toward parallelism with the bedding plane, but when a drill hole intersects bedding or cleavage at a large angle, it tends to assume a direction at right angles to the laminated structure. Besides, the horizontal holes tend to curve upward, perhaps because of sag in the drill rod string, behind the bit. It is hard to straighten a hole that has started to curve, but often local experience makes it possible to anticipate deviation when laying out the drilling program and thus hit the target through a curved trajectory. Deviation also depends on the drilling technique. Fig. 8.13 shows the plot of borehole deviation survey readings, by cross section and by azimuth.

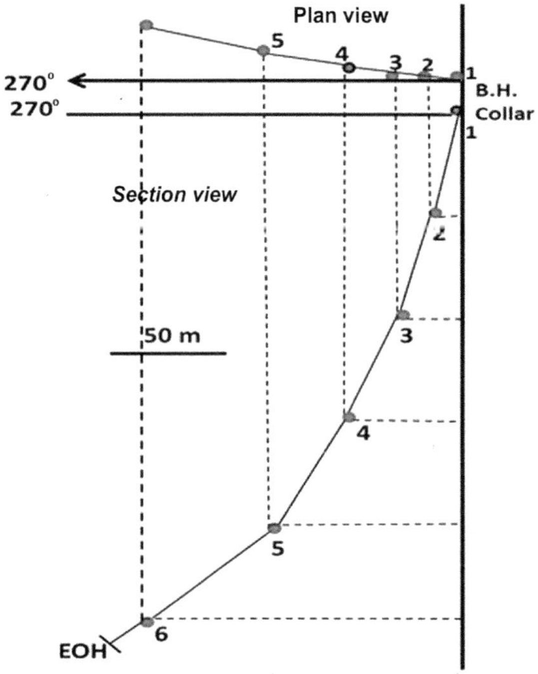

FIGURE 8.13 Plot of borehole deviation survey readings, in cross section and in azimuth.

Borehole Survey Readings

Reading	Depth (m)	Inclination (in degree)	Azimuth (in degree)
1	0	80	270
2	50	78	272
3	110	76	274
4	155	73	278
5	200	70	284
6	260	66	289

Although the usual intention is to drill holes as straight as possible, it is entirely feasible to cause a deflection intentionally, usually by lowering a metal wedge into the hole. Thus a change in direction may intersect the vein in question, at a better angle or after a vein has been cut, a branch hole will give a second penetration and an additional sample.

8.18 DIRECTIONAL CORE DRILLING

Many drilling companies have come up with techniques for directional core drilling (DCD). A steerable wire-line operation system "DeviDrill core barrel" has been developed

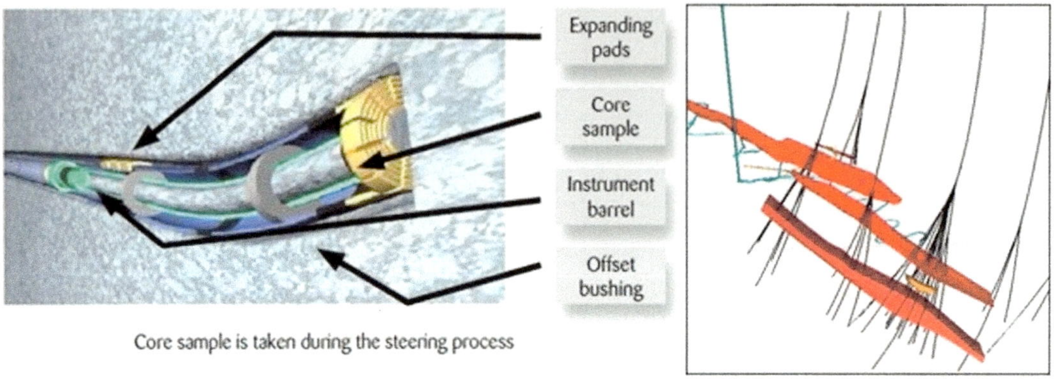

FIGURE 8.14 (Left) DeviDrill's directional core barrel; (right) drilling several branch holes.

by "Devico," a Netherlands company. It is "a drive shaft running through a bushing offset from the centerline of the tool. Expanding pads operated by different pressures keep the "DeviDrill" in a fixed orientation while drilling in a curve. The inner assembly carries an inner tube collecting the core, a "mule shoe" system, and an instrument barrel with the survey tool recording inclination and tool orientation" (Fig. 8.14, left). By using this directional core barrel, boreholes can be steered to a desired direction and inclination (Devico, 2010).

Specified targets can be pin-pointed with DCD boreholes with accuracies of about 10 m in 1000-m long boreholes. Reduced drill length, accuracy, continuous coring, etc., are significant benefits of DCD. Since one borehole can be used to hit several targets, DCD reduces the length of drilling with a considerable savings of cost (60–80%) and time. This system can help to steer a borehole, controlling natural deviation and permitting borehole within given limits of hitting targets with high accuracy. Concomitant with the steering process, core samples can be collected. When a drill hole is finished, the facility of sidetracking is possible by steering toward a "second target" by cutting straight in a "curved section" of the first drill hole. By adopting this technique, there is no necessity of employing any wedges or cement plugs. By sidetracking and making many branch holes (Fig. 8.14, right), the overall length of drilling program may be reduced thereby saving time and money remarkably (Devico, 2010).

8.19 SURVEYING BOREHOLES

The amount of curvature can be determined by surveying the hole after drilling. Several methods are being used for this. To determine inclination, an age-old method of using a glass partly filled with hydrofluoric acid, which is normally enclosed in a bronze case (which, for survey purposes, replaces the core barrel of the drill) and lowered into the hole. The acid etches on the glass in the position in which the liquid stands, thereby giving the inclination of the test tube, hence the borehole. The azimuth of the hole, indicating the

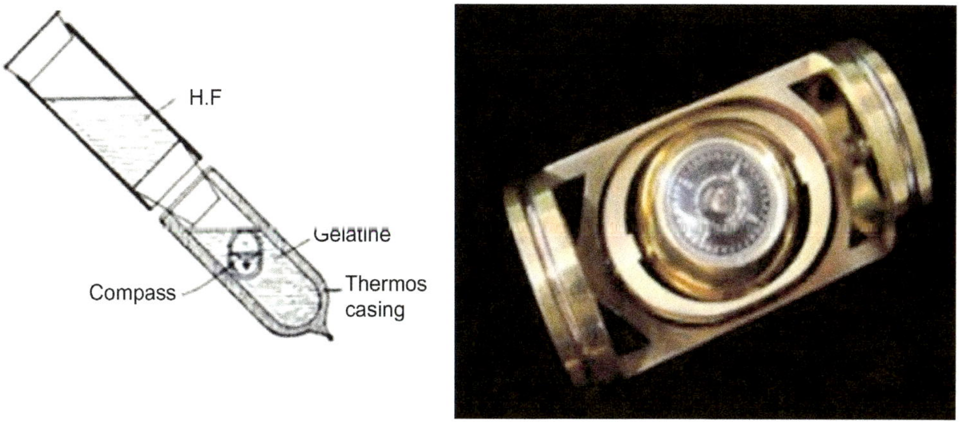

FIGURE 8.15 (Left) Old method of using HF etch tube and compass for surveying borehole; (right) borehole survey equipment "Tropari." *Source: Courtesy from Pajari Instruments.*

horizontal component of deviation, is measured with some type of "compass" (Fig. 8.15, left).

Tropari, an instrument which is a single-shot, micro-mechanical borehole surveying equipment operated by timing device, was introduced by Pajari Instruments (Canada) in the late 1960s (Fig. 8.15, right). The attitude (direction and inclination) of the borehole at the survey depth can be obtained using this instrument. The timer mechanism can be set, depending upon the depth to be surveyed and the time taken to lower the drill string into the holes. One can set the time in minutes, and keep the instrument in an aluminum core barrel and lower the drill rods, to a desired depth. After the time elapsed, the locking mechanism will lock both the compass dial and the inclinometer pin. In this way, one can determine both the inclination and azimuth of the drill holes. The Tropari provides magnetic azimuth and inclination readings to an accuracy of $\pm^1/_2$ degree, ease of use, easy to transport, cost-efficient surveying, and longevity of instrument if properly maintained.

Various optical and photographic methods of borehole surveying have been developed and television console has become a useful tool in mineral exploration. Optical Penkopes that can be lowered in a borehole and the direct observations are made by the operator at surface. Multishot Borehole survey Camera, of late, is being used widely to survey the deviation, both in azimuth and in inclination. Borehole photography has been used for a direct record of the hole walls, etc.

A nonmagnetic, electronic multishot assembly (Deviflex) for surveying inside drill holes has been developed by "Devico." It is a fairly simple one, employing the wire-line system, with no influence by the magnetic disturbances (Fig. 8.16). Three accelerometers and four strain gauges constitute two independent measuring systems, to calculate the changes in azimuth and inclination. It also records and stores gravity vector, temperature, battery capacity, etc. All these results are viewable on the PDA screen in the field. It is also reported that an advanced multishot borehole survey instrument, "Devishot" (developed by Devico) is available that records azimuth, inclination, gravity vector, magnetic field vector, magnetic dip, etc.

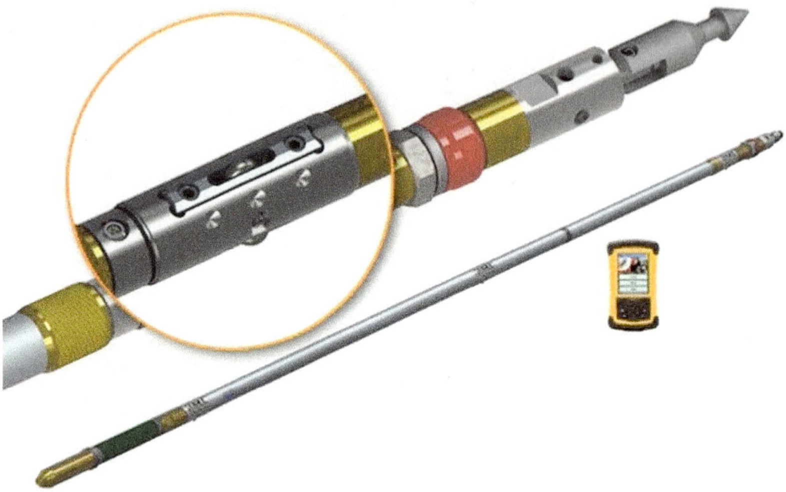

FIGURE 8.16 Multishot borehole survey assembly. *Source: Courtesy from Devico.*

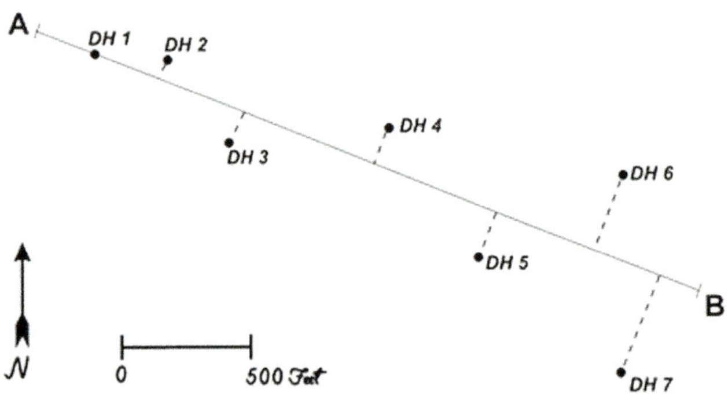

FIGURE 8.17 Map showing the projection of several drill hole collar locations onto drill section line.

8.20 DRILL SECTIONS

Analogous to cross sections, drill sections are constructed which show drill holes in a vertical profile in order to interpret drill data. By creating topographic profile, drill sections are also constructed in the same manner as that of geological cross sections. The collar locations are plotted along topographic profile. The drill holes are plotted: a vertical hole will plot as a vertical line on the drill section and an angular hole showing appropriate inclination. The scale of the drill section will determine the length of line representing the drill holes.

Whenever the boreholes are located away from the drill section line, it is necessary to project them onto the plane of drill section. The projection is done along a line perpendicular to drill section line (Fig. 8.17). "If the inclined drill hole does not plunge directly into the vertical plane of the drill section, its inclination on the drill section will appear as an "apparent dip." The apparent dip angle is a function of the true dip and the angle between the drill section line and the drill hole surface trace in a map view."

If a drill hole intersects a tabular-shaped ore zone at 90 degree angle, then the thickness in the drill core will represent the "true thickness." If the drill hole intersects a mineral zone at an angle less than 90 degree, then the observed thickness will be "apparent thickness." It is always necessary to calculate the true thickness of the mineralized zone in order to calculate the "volume of ore." Once the dip of mineral zone and the inclination of drill hole are known, the true thickness can be calculated, using simple trigonometry.

Every drill hole in the mineralized zone typically shows intervals of significant ore grade values which are often highlighted/colored in drill sections. The overall geometry of the ore zone is interpreted by the concerned geologist, by extrapolating the zones of interest between drill holes. This will involve connecting the upper and lower contacts of ore zones from one drill hole to the next. It is possible that different geologists interpret the geology in different ways (Fig. 8.18). For better interpretation, it may be necessary to construct additional drill sections showing different aspects of drill data, viz., specific alteration zones noticed, mineralization types, etc.

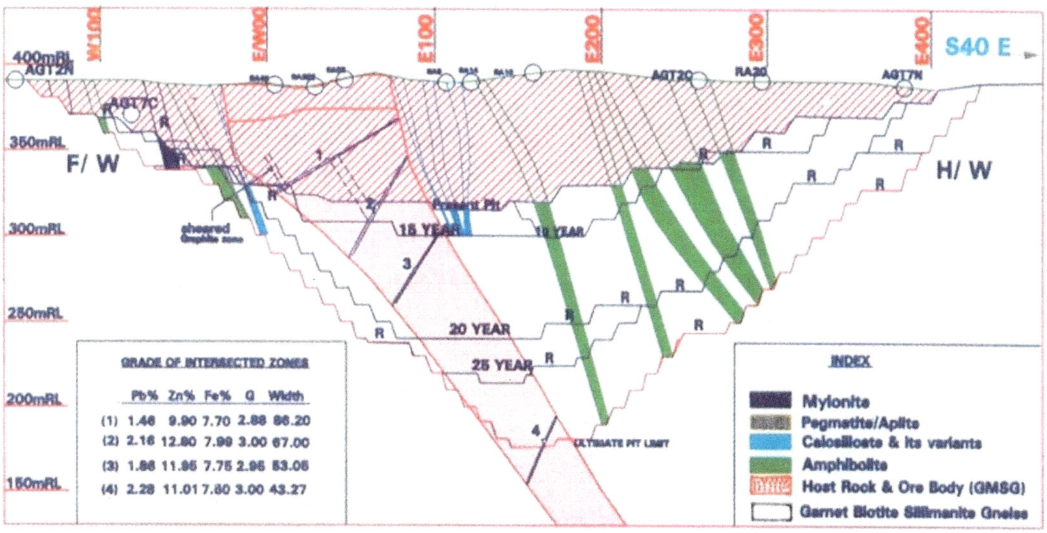

FIGURE 8.18 Drill section with ore intercepts and geology, Rampura-Agucha mine. *Source: From Gandhi (2003).*

8.21 PLANNING A DRILL CAMPAIGN

While planning a drilling campaign, the choice of the method, whether Diamond drilling or Churn drilling, is based on the purpose whether to test the ore grade and width of an orebody; to secure geological information; or to find new ore. Huges (1965) opines that "the principal factors which further influence the choice are (1) the nature of information required, (2) the shape and attitude of the target, and (3) the physical nature of the ground." The respective advantages of Diamond and Churn drills are given in Table 8.2 (McKinstry, 1962).

Churn drill is economical and cuts a larger hole of broader diameter, hence yields a large, representative sample. The best results are obtained in rocks of medium hard to soft that do not slump or cave-in to the hole. A drill method (Diamond or Churn drilling) should be selected to insure obtaining samples that will provide a rough "yes" and

TABLE 8.2 Comparison of Diamond and Churn Drills

Diamind Drill	Churn Drill
1. Can drill in any direction-downward Horizontal, inclined, and (from underground) upward	Can drill only vertical downward holes Usually drills from surface but can drill From underground if large station is prepared
2. Core sample gives valuable geological information: texture of rock, distribution of mineral grains, attitude of bedding, cleavage, veinlets, etc., to axis of core	No core sample. Cuttings (sludge) examined in laboratory give considerable information regarding nature of rock and mineralization
3. Sample is small, though shape and diameter are uniform	Sample more accurate insofar as size is larger. Shape and diameter more subject to variation
In precious metal deposits, gives accurate sample if core is good	Large particles difficult to raise though Performance improved by thick mud and suction bailer
4. Slower than churn drilling under average conditions	Faster, at least up to 300–400 m
a. Successful in hard rock	Slow and expensive in really hard rock
b. In fractured, blocky ground gives Incomplete core and slow progress	Successful in fractured blocky ground
c. Seriuos "bit-wear" in poorly consolidated conglomerate, or soft rock with hard veinlets and nodules	Progress satisfactory in conglomerate (if not too hard) and chert-bearing limestone
d. Gives unsatisfactory core in unconsolidated material	Gives good samples in unconsolidated material (sand, gravel, and clay)
5. Hole usually serves no purpose other than testing	Hole being large may serve for ventilation, drainage, or (in open pits) for blasting

'no" answer at the lowest cost, in the shortest time possible. If the answer is "yes" or 'perhaps," a more sophisticated drilling method and hole pattern will be necessary. High-risk exploration campaigns are set up in a series of steps, the decision to proceed from one step to the next being made only after a thorough evaluation of all results obtained to insure that risks have been minimized. The advantages of By analysis of sufficiently large number of developed orebodies are possible to establish drill hole pattern that can gain most efficiently the data needed for estimating the grade and size of a given type of ore occurrence. The most modern exploration work is blind to some extent–vertically or laterally beneath cover. The explorationist usually has rather-definite ideas of how and where the geological trends project with some support from geophysical and geochemical data.

In the beginning of the exploration campaign itself, it is to be spelt out clearly, the results desired before commencing exploration work. The exploration company might agree that their first stage drilling should consist of "x" number of vertical holes, each about 60–100 m deep spaced in a grid pattern across the mineralized belt. It might also be specified that two adjacent holes should each contain the correct target host rock lithology with mineralization in a continuous intercept of at least 5 m, assaying "x" metal content. If the drill results are better or worse than the stipulated condition, the exploration group might decide to advance to continue or to drop the project. Once the initial high risk has been reduced by discovery mineralization, the results can be compared with similar projects elsewhere so that efficient exploration can be initiated.

The highest degree of judgment is necessary in laying out preliminary drill hole pattern in the initial phase of exploration. A more reliable approach is to have a clear concept of the controls of and characteristics of the expected target and to plot all known geology in plan/cross section, to insure the proper positioning of drill holes. After careful analysis of the geology of the prospect area and geometry of the target, the drill grid should be such that two drill holes will penetrate the minimum-sized ore zone. Drilling in a flat lying tabular ore deposits, a rectangular drill grid might be used. It is found that combinations of diamond and churn drilling are sometimes advantageous.

8.22 DRILLING FOR SAMPLING PURPOSES

Drilling, the method best suited to the local conditions, can furnish a satisfactory means of sampling and blocking out ore if an orebody can be drilled at a sufficient number of places at a reasonable cost. This normally depends upon the depth of the ore and the spacing necessary to give an accurate indication of grade of the orebody. In flat, shallow deposits, short holes are sufficient and if the grade is relatively uniform (like in iron ores, porphyry copper, residual bauxite deposits, etc.), the drill holes need not be closely spaced. Sometimes, it may warrant to drill a few scout holes to deeper levels to test the persistence of ore. In steeply inclined deposits, inclined drill holes are to be drilled at an angle more or less perpendicular to the orebody to determine the width and the grade of the orebody. If the grade is uniform, the drill holes need not be many.

8.23 ANGLE OF INTERSECTION

In any drilling campaign, it is always economical if one can drill several holes from one station, as it involves sizeable amount of funds, to shift the rig and setup again in a new place. Particularly in underground workings, establishing drill stations is expensive involving good amount of excavation. The number of holes that can be drilled in "fan pattern" is controlled by the length of the hole necessary to reach more remote points of the orebody and the angle at which the hole will intersect the plane of the ore vein. Too acute an angle is to be avoided since the accuracy of the hole as a measure of width decreases rapidly with the angle of intersection. Thirty degree is usually the minimum permissible, and 45 degree is safer.

8.24 DRILLING FOR NEW ORE

There are two methods of planning a drilling campaign for finding new orebodies: (1) to test a given block of ground completely by drilling holes on a preplanned pattern, and (2) to "feel your way" by letting the results of each hole determine the position of the next. The former has definite advantages. Since the number and depth of the drill holes have been preplanned, it will be economical, several machines of different capacities, could be deployed, serviced, and supervised. The method of "feel your way" is much more flexible, since it gives scope to judgment and to the application of geological knowledge which is acquired in the course of work. Many a times, a compromise between the two methods is better than one alone. A sketch showing an idealized drill grid campaign is given in Fig. 8.19.

If the ore is found at an early stage in the campaign, it is better to test its extent rather than probing in an unknown area. If the initial holes indicate sufficient ore to warrant

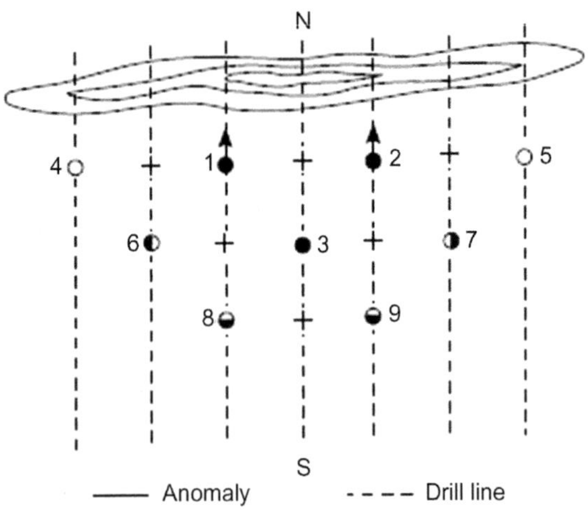

FIGURE 8.19 Sketch showing an idealized drill grid campaign. *Source: From Annels (1991).*

underground development, additional holes in that area may be superfluous. If the ore is expected to occur in a large area, it is best to explore the entire area by wide-spaced holes in reconnaissance fashion. If the ore is found, the adjoining holes are drilled to test its extent. On the contrary, if the results are discouraging, further drilling can be terminated.

Knowledge of the habits of the ore in any district will help to decide the widest permissible spacing to avoid missing an orebody. If the smallest ore shoot that would be of interest is about 30 m wide, measured in the plane of a vein, then the holes spaced to cut the vein at intervals somewhat less than 30 m would penetrate any orebody that existed. If the ore shoots are long horizontally and short vertically, the holes may be spaced farther apart horizontally than vertically. On the other hand, if the ore shoots have steep pitches, the spacing should be wide vertically and close horizontally. If there is a definite target to aim (an extension of known ore shoot), first hole should be aimed at this target; additional drill holes could be located to test the most likely place first and a few holes to exhaust the possibilities. If the pitch of the ore shoot is not known, a couple of shallow holes may be drilled to determine the direction before undertaking deeper drilling.

8.25 WHEN TO STOP DRILLING

In any drilling campaign, the hardest decision will be when to stop drilling. Some of the situations which may warrant this decision are as follows:

- no mineralization was encountered;
- mineral resource intersection was not up to desired grade and width;
- limited continuity of mineralization of economic resource grade, size might be small for the investigating agency. Type of prospect not interested by one party might be of interest to some other party;
- delineation and establishment of an orebody of economic grade and size;
- exhausion of budgeted funds.

Reexamining a prospect (recycling) that a previous party has discarded for various reasons is a tempting proposition, but the buyer must be satisfied that they can be more meaningful than the earlier exploration company. This success might result from different exploration models or improved exploration techniques. There were instances of good successes.

APPENDIX

Drilling Log Sheet

ESEMGEE EXPLORATION	**Location**: Vardhine Prospect, Rajasthan, India		**Latitude**: N 200	**Drill Hole Deviation Measurements** (multi shot camera)					
DIAMOND DRILL CORE LOG	**B.H. No.** VP-08		**Departure**: E 240	**Depth**	**Angle**	**Azimuth**	**Depth**	**Angle**	**Azimuth**
Logged by: Dr.S.M. Gandhi	**Commenced on**: 3 March 2012		**Collar RL**: 290 MSL	Collar	70°	270	240 m	63°	276
Sampled by: Dr.S.M. Gandhi	**Completed on**: 16 June 2012		**Inclination**: 70°	50 m	68°	272	290 m	60°	279
Analysed by: High Tech Labs. Jaipur.	**Date**: 22 April 2012	**Sheet No.** 17	**Closed depth**: 356 m	110 m	67°	274	340 m	58°	282
				170 m	65°	276			

GEOLOGICAL LOG SHEET OF THE DRILL CORE

Drilling Detail						GEOLOGIC INFORMATION				VISUAL ESTIMATION		Sampling Details (m)				Assay Values		Sp. Gr.	Sample Weight
Actual			Recovery																
From (m)	To (m)	Run (m)	Length (m)	%	Core Size	Rock Type	Description	Structure & Texture	Economic Mineralization	%Zn	%Pb	No.	From	To	Width	%Zn	% Pb		(kg)
198.40	201.45	3.05	3.05	100	NX	Amphi-bolite Calc-Silicate	198.4–200.35 amph-35; fels-35; qtz-25; Other-5 (in %) 200.35- 201.45 Diop-35; fels-30; qtz- 25; gt-5; Others-5	Medium to course gr., gneissic occasionally disseminated. Structure is shown by amphibolite gneissic, granulose tex., medium to fine gr., rarely coarse	RQD 77% Foliation 60–70° to Core Axis; with inter-filling layers of calcite	Tr	Tr	SG/1	198	199	1.00	0.04	Tr	3.00	6.99
												SG/2	199	200	1.00	0.03	Tr	3.10	8.30
												SG/3	200	201	1.00	0.03	Tr	3.01	7.25
214.40	216.75	2.35	2.35	100	NX	Garnet Biotite Sillima-nite Gneiss & ORE ZONE	214.40–216.50 RFM: 70% Qtz-20%; fels-20; mica-15; graphite-5; chlo-5; Others-5	-do-		Tr	Tr	SG/13	214	215	1.00	0.06	0.19	2.79	6.10
										Tr	Tr	SG/14	215	216	1.00	0.07	0.02	2.84	6.65
										Tr	Tr	SG/15	216	216.5	0.50	0.02	0.02	2.79	3.10
						Ore Zone	216.50–216.75		Ore minerals,OFM 30%, disseminations of spl, py, gl, with anastomising texture										

GEOLOGICAL LOG SHEET OF THE DRILL CORE

Drilling Detail						GEOLOGIC INFORMATION				VISUAL ESTIMATION		Sampling Details (m)				Assay Values		Sp. Gr.	Sample Weight
Actual			Recovery																
From (m)	To (m)	Run (m)	Length (m)	%	Core Size	Rock Type	Description	Structure & Texture	Economic Mineralization	%Zn	%Pb	No.	From	To	Width	%Zn	% Pb		(kg)
240.81	243.81	3.00	3.00	100	NX	ORE ZONE	RFM 75% Qtz-20; fels-20; Mica-25; chlo-5; graphite-5; Sillimanite-Tr	Medium to coarse grained gneissic texture. In places, fine grained.	OFM: spl, py, gl, pyrr, mostly as veinlets which comprise octahedral crystals of py., in spl; & pyrr. Medium-grained matrix. OFM, as disseminations and inclusions in other OFM.	(Zn + Pb)	10 - 12	SG/31	241	242	1.00	7.45	1.02	3.16	6.80
										-do-	9 - 12	SG/32	242	243	1.00	8.02	0.35	2.98	7.00
										-do-	6–8	SG/33	243	244	1.00	4.83	0.21	3.14	7.05
243.81	246.81	3.00	3.00	100	NX	-do-	-do-	-do-	-do- Spl, py, pyrr, plus gl, veinlets	-do-	6 - 9	SG/34	244	245	1.00	5.40	0.22	3.04	6.40
										-do-	12 - 15	SG/35	245	246	1.00	11.80	0.14	3.09	6.90
										-do-	10 - 13	SG/36	246	247	1.00	13.64	0.75	3.09	7.30

References

ADITC (Australian Drilling Industry Training Committee), 1997. Drilling: The Manual of Methods, Applications and Management, fourth ed. CRC Press, Boca Raton, FL, p. 295.

Annels, A.E., 1991. Mineral Deposit Evaluation: A Practical Approach. Chapman and Hall, London.

Atlas Copco, 2014. Prospecting and exploration for minerals. Talking Technically, pp. 14–19. <http://www.atlascopcoexploration.com/1.0.1.0/354/TS3.pdf>.

Barrett, W.L., 1987. Auger drilling.. Trans. Instn. Min. Metall. B 96, B165–169.

Blackbourne, G.A., 1990. Cores and Core Logging for Geologists. Whittle's Publishing, Caithness, Scotland, p. 113.

Chugh, C.P., 1971. Diamond Drilling. Oxford & IBH Publishing Company, New Delhi.

Devereux, S., 1999. Drilling Techniques. Penwell, Tulsa, OK, p. 337.

Devico, 2010. DeviDrill—Directional Core Drilling and Borehole Survey Tools. http://www.mining-technology.com/contractors/exploration/device.

DMP, 2012. Mineral Exploration Drilling—Code of Practice: Resources Safety. Department of Mines and Petroleum, Western Australia, p. 56.

Dunster, J.N., 2004. Drilling Manual, Northern Territory Geological Survey. Darwin, Australia, p. 64., www.devoco.com: directional core drilling.

Gandhi, S.M., 2003. Rampura-Agucha Zinc-Lead Deposit. Geol. Soc. of India Memoir. 55, 154 p.

GSI, 2010: Standard Operating Procedure for Drilling Operation. Engineering Division, Geological Survey of India, Kolkata, 51 pp.

Hartley, J.S., 1994. Drilling—Tools and Programme Management. AA Balkema Publishers, Rotterdam, p. 272.

Hartman, H.L., 1968. Principles of drilling. In: Pfleider, E.P. (Ed.), Surface Mining. Amer. Inst. Min. Engrs., New York, NY.

Heinz, W.F., 1994. Diamond Drilling Handbook, third ed. Sigma Press, Halfway House (Ganteng province, South Africa).

Huges, M.D., 1965. Geological planning for diamond drilling. 8th Commonwealth Min. Met. Cong., vol. 2. Exploration & Mining Geology.

Kanck, P., 2014. Hidden depths: CT rig set to revolutionize mineral exploration in Australia. Mining Technology Market & Customer Insight News.

Killeen, P.G., Bernius, G.R., Mwenifumbo, C.J., 1995. Surveying the path of boreholes: a review of orientation methods and experiences. Proc. of the 6th International MGLS Symp. on Borehole Geophysics for Minerals, Geotechnical and Groundwater Applications. October 22–25, 1995, Santa Fe, New Mexico. http://www.mgls.org/95Sym/Papers/Killeen/.

Marjoribanks, R.W., 1997. Geological Methods in Mineral Exploration and Mining. Chapman & Hall, London.

McKinstry, H.E., 1962. Mining Geology, second ed. Asian Publishing House, p. 680.

Peters, W.C., 1967. Cost and value of drill hole information. Min. Congr. J. 53 (1), 56–59.

Sandvik Mining & Construction, 2012. Core Drilling Tools and Diamond Products, 34 pp. www.sandvik.com.

CHAPTER

9

Sampling and Analysis

OUTLINE

Essentials of Mineral Exploration and Evaluation
DOI: http://dx.doi.org/10.1016/B978-0-12-805329-4.00016-8

9.1 INTRODUCTION

One of the fundamental requirements for providing adequate estimate of mineral resource is to carry out sampling within a mineral prospect with an objective to understand the characteristics of the mineralization. The methodology should be prudently outlined so that at every phase of the sampling process, the possibility of collecting proportions other than that being representative is reduced to a minimum. The representative fraction, known as "Sample," is collected following certain preparation procedures that make it appropriate for testing and analysis. The kind of testing and analysis carried out are subject to the attributes needed to classify the mineralization. Thus the process of sampling and sample preparation constitute the most vital procedures that add to the exploration expenditure plan.

The technique of sample collection and the accuracy of the analysis results indicating the closeness of true representation rely upon the degree of heterogeneity of the sampled article. An absolutely homogeneous article would need collection of just one small item keeping in mind the end goal to decide its attributes accurately, whereas an uneven heterogeneous article would need collection of several small-sized items at appropriately spaced intervals, which when consolidated, would characterize the entire body or a part of the body with a tolerable level of accuracy. The number and the spacing of the items required to be collected depend on the variability of the whole being sampled. It is of fundamental significance that all small fractions in the body have the same likelihood of being incorporated in the fraction called sample, which is a basic assumption in sampling. The objective of sampling is to obtain reliable measurements, usually chemical analysis of a bulk sample, drill core, etc., from which the analyzed samples are taken and prepared (Gy, 1992, 1998).

9.2 SAMPLING

Sampling is an act or instance of collecting a small fraction of an article in a way that the consistency of the fraction shall be representative of the whole. Sampling should be carried out keeping the geological features of the orebody in view, and should be in the agreement with the broad geological principles. Theory of sampling states that "if enough small portions of an article, appropriately spaced, are taken, then their average value or content shall be representative of the whole." Definition of sample has essentially two stated and inferred elements, namely, (1) typicality and (2) smallness. Collection of sample should be typical of a mineral body, else it will be collection of specimen. Each sample should be appropriately small portion of parent body, else sampling would amount to mining. Sample collection is a mechanical function and depends on sampling method, site accessibility, geology, objectives of the project, and requirements for statistical analysis. Sampling pattern is determined by the nature and geometry of mineralization, that is, size, shape, orientation, and distribution. It could be random or systematic, random during preliminary investigation and systematic during detailed exploration/target investigation. Sample spacing is a mathematical function determined by coefficient of variability (ratio of standard deviation to mean) of mineralization or in other words the degree of homogeneity/heterogeneity of mineralization. Table 9.1 provides a range of coefficient of variability values for different mineralization types.

Sampling pattern is a very important function in all phases from search for minerals to the stage of actual mining and finally to marketing and shipping. Whatever may be the purpose of sampling and whatever may be the aspect, which is proposed to be studied, it is generally agreed that a sample should be representative of the whole from which it is drawn. It is also to be remembered that in addition to representativeness, there will always be some difference between the values revealed by the sample and the actual value, which will be known only after the ore is mined and used. The aim of sampling is to reduce the difference between the two values to the absolute minimum. Sampling of particulate materials and the subsequent preparation and analysis is the most important in the mining and mineral industries. The purpose of exploration is to estimate the grade and tonnage of the various qualities of the content in a given mineral deposit. In samples of drill core, when the moisture content exceeds a few percent, it should be necessary to dry the crushed samples prior to pulverization using oven.

One has to ensure that the analytical results obtained from the assay of a few grams of pulverized samples reflect the contents of the primary sample taken of a bulk sample, core split, or shipment of the concentrate. While the analytical methods are accurate, the initial sampling practices and preparation of those samples to yield small representative portion for analysis are frequently inadequate. The important physical and mineralogical characteristics of each material must be considered carefully in the design of a good sample preparation system. For precious metals, first it has to be finely ground, whereas sulfide minerals are first screened before pulverization. For coarse gold particles, wet screening and then gravity separation are to be done. Ores and products containing precious metals must be given particular attention because of low concentrations, high unit values, and frequently the irregular distribution of those metals or metal-bearing minerals. Random errors are introduced which cause the final assay to differ from the true assay of the original material. The error depends upon the physical and mineralogical characteristics of the sample.

TABLE 9.1 Coefficient of Variability for Different Mineralization Types

Sl	Mineralization Type	Coefficient of Variability	Types of Deposit
1	Regular (homogeneous)	5–40 (av. 30)	Deposits of coal, building stone, limestone, sulfur, phosphorites, iron and manganese ores
2	Irregular (nonhomogeneous)	40–100 (av. 80)	Most copper and complex ore deposits, some tungsten, molybdenum, and also a few primary gold deposit
3	Very irregular (very nonhomogeneous)	100–150 (av. 130)	Some complex ore deposits, most tin, tungsten, molybdenum deposits, a few gold deposits
4	Extremely irregular (extremely nonhomogeneous)	Over 150 (av. 200)	Many rare metals and gold deposits

9.3 GEOLOGICAL SAMPLING METHODS

It is done to study the various valuable attributes of a mineral deposit. Those attributes may be grade, specific gravity, or any other characteristic, which gives the ore its commercial value. Sampling can be from deposit, which is in situ or from mined ore, which will be a more heterogeneous mixture from dumps or other accumulations, drill cores, etc. Sampling may also be done to reveal the pattern of mineralization in a deposit or in order to demarcate the richer and the leaner portions of the ore or the contact between the ore and waste. McKinstry (1962), Kreiter (1968), Peters (1987), Lacy (1983), Moon et al. (2006), and Dowing (2014) have dealt in great detail about the sampling and sample preparation of various geological materials. Sampling method is resorted to various reasons. In the context of geological exploration, sampling is carried out for: (1) establishing the existence of valuable mineral(s) in a mineral prospect; (2) determining the extent and possible grade of mineralization; (3) determining the grade and distribution of ore in a mineral deposit; (4) mine valuation in an operating mine; (5) grade control in an operating mine; and (6) ascertaining final grade in process plant. Depending on the objective, sampling is carried out at all stages of geological exploration. During prospecting, sampling aims to outline a mineral prospect. During exploration, sampling is undertaken for quantitative estimations of ore grades and tonnages. Sampling aids in the preparation of feasibility report necessary for mine development. Technological tests on samples aid in the selection of treatment methods during beneficiation. During production, sampling process is undertaken for grade control operations, and making decisions pertaining to capital investment. Sampling is also undertaken to maintain a balance between productivity and quality before marketing and shipping.

The processing of sampling is generally divided into three stages, viz. (1) sample collection, that is, the taking of representative fractions; (2) sample preparation, that is, processing of the samples collected through reduction in size and quantity; and (3) sample testing, that is, sample analysis.

Several methods of sampling are practiced depending on the type of the parent body to be sampled and the objective of sampling. In geological exploration, the following types of sampling are recognized. They are:

- Talus Debris (Float) sampling
- Trench and Pit sampling
- Grab sampling
- Chip sampling
- Channel sampling
- Placer sampling
- Drill sampling: core/cuttings/sludge
- Bulk sampling
- Dump sampling
- Car sampling.

9.3.1 Talus Debris (Float) Sampling

This type of sampling is undertaken in areas of scanty outcrops, which are usually under a thin cover of rock talus or float. It is presumed that this material is a product of in situ weathering of parent bedrock and as such has not undergone transportation to considerable distance. The purpose of obtaining float samples is to analyze whether it would be useful in the detection of concealed mineralization with a thin cover.

9.3.2 Trench and Pit Sampling

The trenches and pits are openings in mineral deposits directly or under shallow cover for generating data. Trench sampling is undertaken by cutting channels along the floor of the trench. The trench should be subdivided into sections based on the lithological and/or mineralization pattern. The sections may further be divided into subsections of 0.5–1.0 m length. Dimensions of channel are generally 10 m in width and 3–5 m deep. Sample intervals vary depending upon the nature of geology, mineralization, bedrock, etc.

Pit sampling is undertaken by cutting channels on the walls of the pit employing general principles of sampling. Choice of sampling one wall, opposite wall, or all four walls is dependent upon the type and nature of mineralization. In heterogeneous type of mineralization, all four walls may be sampled to obtain a representative sample.

9.3.3 Chip Sampling

This method is used for sampling hard dense outcrops during prospecting and face on which hard ore is exposed during mining. In chip sampling, first the outcrop or face to be sampled is cleaned properly and a regular rectangular or square pattern is made by drawing lines along and across the outcrop at fixed intervals. Then small pieces of ore are broken loose either from the center or the grid or at the intersection points of the lines. The ore pieces should have approximately the same chip size and weight. Chisel and hammer or moil are used for collection of chip samples. After collecting the rock pieces from each center or intersection point of the grid, the rock pieces are mixed together to form one composite sample.

9.3.4 Grab Sampling

Grab sampling is undertaken to form a quick general idea of the mineralization. If composed of sufficient fragments taken over a wide area, grab sampling may reveal the grade of mineralization for that location. Care should be taken to collect materials of varying sizes according to its proportion by weight. Several such grabs are mixed together to form one sample. Grab samples are a bit more random but still useful as they can provide a basic, overall understanding of the area and the possibility of higher grade ores in the vicinity. The quantity of material to be collected depends on the size of the largest piece present in the materials to be sampled and the scale of nonhomogeneity of the material.

The size of the sample is governed by the "Richards Chichette formula":

$$Q = Kd^2$$

where Q is the reliable weight of the worked down (also initial) sample in kg; d is the size (dia) of the largest grain in the samples in mm; and K is the factor depending on the homogeneity of the mineral.

The value of K, which may be used in most cases, is given below:

Ore Type	Value of K
Homogeneous	0.05
Nonhomogeneous	0.10
Very nonhomogeneous	0.20–0.30
Extremely nonhomogeneous	0.40–0.50

The "K" factor is determined on the basis of the irregularity of distribution of the principal constituent of the ore.

9.3.5 Channel Sampling

This is the most frequently used and accepted method of sampling in most exploration projects and operating mines. The method consists of marking an outline and cutting uniform grooves into the rock about 100 mm wide and 20 mm deep across the exposed mineral body, usually spaced across the strike, collecting all the fragments and dust from the channel, and combining together to constitute one sample. While marking the channel outline, it must be taken into consideration that channel outline, as far as possible, should be along the direction of maximum variability. Various tools ranging from a hammer and moil to a pneumatic chisel can be used to cut channel sample. Before attempting to collect a sample, the exposed rock surface must be cleaned thoroughly, with a wire brush, water, or chipping a fresh surface. The cross section of channel being cut should be kept uniform so that the sample cut becomes representative. The samples can be chiseled out and collected carefully on a clean sheet of canvas placed directly below the channel. The sides and floor of the channel should be smooth and uniform to avoid any overcutting/undercutting that may overrepresent/underrepresent. The channels may be divided into 1–2 m sections or their multiples in case of massive or more homogeneous ore bodies or intersection of 50 cm in case of more heterogeneous distributions. It is a good practice to divide larger samples into smaller intervals according to the structure, change in rock types, or differences in rock hardness or strengths.

9.3.6 Placer Sampling

Placer deposits being characteristically stratified with layers of mineral concentration, particularly in bedrock, sampling becomes a troublesome job. In the event that the placer concentrations are coarse in size, the impact of nugget effect can be significant on the determination of the average (mean) concentration. Thus placer sampling requires adequate collection of materials from bedrock as larger-sized bulk samples. Because the value of a placer deposit depends on the recoverability of the minerals by gravity methods, the sample volume is measured and a part of the sample is ordinarily concentrated in a sluice box and panned for a preliminary estimate of the minerals and values.

9.3.7 Drill Sampling: Core/Cuttings/Sludge

In a drilling campaign three kinds of sampling are done, viz. (1) core samples from diamond drills or from the casing drills, (2) dry cuttings from air-flushed diamond, rotary, auger or percussion drilling, and (3) wet cuttings or sludge from churn drills, diamond drill, or wet rotary or percussion drilling. Core samples drawn mainly from ore zones are collected. Wherever disseminations of minor amount of mineralization noticed, complete host rock was sampled with large sample interval. In order to simplify the computation, for statistical analysis, evaluation, etc., equal core length samples (about 0.5 or 1.0 m) are made.

In dry drilling program cyclones are used to draw cuttings away from drill collar and settle the sample into bottom of the apparatus where it can be dropped out into storage containers. The sample is to be run through a splitter, and the samples are collected in plastic bags and sealed tightly. Care is to be exercised in the handling and transport of dry cuttings because the fine and heavy material has a tendency to settle to the bottoms of containers. The sludge from diamond, churn, rotary, and percussion drilling is caught in various devices, ranging from simple overflow buckets hung under the drill collars to elaborate multiple sludge tanks where all cuttings carefully settled out of drilling water. After siphoning off water for recirculation, the sludge is air-dried or dried on stream tables. Extreme temperature and direct contact with flame during drying are deleterious, as for example, partial roasting of sulfides. One should be careful against gravity settling of heavy minerals since they often are the very object of interest and of high unit volume and great error can result from small involuntary enrichment or impoverishment of samples.

9.3.8 Bulk Sampling

Bulk sampling may be made by collecting a sample from different parts of ore stockpile of trial pit of surface mine, cross-cuts from underground operation or run-of-mine ore (either every blast continuously, or from the shovels or cars) of regular operating mine. The prime purpose of any bulk sampling is to confirm grade indicated by prior sample data from drill holes and establishment of metallurgical criteria to collect more data on geological aspects and investigation of rock mechanics properties. Bulk samples may also be collected from a series of pits or number of trenches, adits, or underground drives in the case of prospects. Bulk sample is normally of large volume (hundreds to thousands of tonnes) representing all the characteristics of the orebody. Thorough mixing of the collected sample is carried out to ensure homogeneity. For carrying out bench scale beneficiation studies, bulk samples are collected from different ore zones of mine developments having different physical and mineralogical characteristics. Samples are drawn from respective zones in proportion to their reserves from a place representing the average grade of that zone determining the physical and mineralogical characteristics of orebody.

9.3.9 Dump Sampling

Dumps are accumulation of waste and inferior grade or fine material generated during mining. Sampling of such dump materials are carried out to examine the suitability for beneficiation studies or its marketability. The sampling of representative sample and estimation of the metal present in an ore dump generated by pitting, trenching, mine production, or a tailings heap are matters which are of considerable importance to most men connected with the mining industry. Dump sampling can be carried by systematically driving the augur into the dumps (up to the base) and collecting the augured material. Benches may be prepared on the dumps and from the benches pits can be driven to collect samples. A composite sample is prepared from the samples collected from various points covering the dump area.

9.3.10 Car Sampling

Car sampling involves collection of samples from the broken ore material transported by mine-car from underground operations, dumpers/trucks, or aerial ropeway tubs that transport ore from surface mining operations for further processing in milling plant/ smelter. The collection of samples is done at random from cars/trucks at periodic intervals to get a quick idea of run-of-mine grade that can be compared with the corresponding mill-head grade.

9.4 CRITERIA FOR THE SELECTION OF A SAMPLING PROCEDURE

Sampling procedure depends on the shape and type of mineralization of the deposit:

- When the orebody is thick and the values of mineralization are uniformly distributed, sampling can be done by chip or grab sampling.
- When the orebody is of medium size and mineralization is uniform, a combined chip and channel sampling will give the best results.
- Where the orebody is too thin but occurs in benches or layers, sampling of various layers can be done by chip sampling.
- When a deposit is of very large dimensions, it is essential to collect a large number of chip samples that would provide reliable results
- With mineral like gold, rare metals, etc., where values are too spotty and irregular, bulk sampling would give the best results.
- Wherever the ore is banded, channel sampling would give the best results and very hard ore, particularly massive types of iron ore would require to be sampled by blast hole cuttings.

9.5 COLLECTION OF SAMPLES

The collection of samples is a job-requiring skill and experience. All chips, blocks, and powder coming from a groove should be gathered irrespective of the size of concentration of mineral value. No extraneous material should get mixed up with the sample. The sample should be collected in a clean canvas bag. After the completion of a groove, the collected chips/blocks, etc., should be put in a bag with a proper reference tag inside the bag. A proper register showing the location of the sample coordinates, channel logs, samples weight, time taken for sampling, etc., should be maintained by the sampler. The register should show the serial numbers of the samples. It is always preferable to complete the register as soon as a sample is collected.

9.6 ERRORS IN SAMPLING

Sampling is subject to certain limitations due to two types of errors: (1) random and (2) systematic. The random errors tend to cancel out each other, whereas the systematic error accumulates to create gross errors, which are easily recognizable because of their magnitude. The errors may also accumulate due to four factors: (1) when check samples are taken from the same spot, there would be a natural divergence between the principal sample and the check sample in terms of the grade, which cannot be overcome, (2) errors accumulated due to measurement errors, poor facilities and equipment, and poor eye judgment of the sampler, (3) errors due to mistakes of calculations, misprints, and poor numbering, and (4) limitations of the assay technique itself. In addition, errors may crop up because of intentional or unintentional salting of the sample itself. All these errors have to be avoided to the maximum extent possible to get a reliable estimate of the orebody. Check sampling and repeated sampling help in avoiding some of the mistakes like salting.

Errors due to several factors may occur in the sampling process and thus affect the result of sampling. These errors can be identified broadly under three categories, viz. (1) human errors; (2) analytical errors; and (3) inherent errors. Of these, human and analytical errors can be reduced to an optimum level, whereas the inherent errors are the ones intrinsic to the type of mineralization.

A list of various factors that are responsible for the errors, which should be taken care of, is given below:

1. *For Human Errors:*
   (a) improper choice of sampling method
   (b) improper collection of samples including "salting"
   (c) improper location of samples
   (d) insufficient number of samples
   (e) lack of inadequate knowledge
   (f) improper mental condition
   (g) improper sample preparation.

2. *For Analytical Errors:*
   (a) improper analysis of sample
   (b) contamination.

9.7 PREPARATION OF SAMPLES

Any sampling method yields large quantities of material but for laboratory analysis only a few grams of ground material is required. The process of sample preparation ensures that the small quantity of material analyzed represents the large quantity collected during sampling. In other words, sample collected from various sample points are again sampled for the purpose of analysis in the laboratory. A geological sample is generally of a size, which is not readily handled by a laboratory for chemical or instrumental analysis. Besides, the individual chips and blocks range in such sizes that they do not mix easily. In order to overcome these, it is essential to decrease the sample to a convenient size ensuring at the same time a proper admixture of the various fractions. The operations are achieved by the process of sizing, coning, and quartering. Sizing should be done on the basis of sample weight to particle size ratio which can be determined by the formula $Q = Kd^2$, which has been explained in Section 9.3.4. Progressive reduction of sample fragment size and quantity is done by crushing, grinding, and pulverizing, the product of which will yield homogeneous sample/subsample (GSI, 2012; Dowing, 2014).

Even with the utmost care in sample preparation, a bias can be generated, that is, a difference between the actual grade of a few grams of powdered material and the large quantity of collected sample. The sample preparation is usually carried out in field or laboratory conditions by trained samplers. However, samplers being unaware of the importance of their work and not conscious of the mistakes are to be carefully addressed to. It is necessary for the exploration geologist to evolve a set procedure which should be logical and simple and ensure preparation of representative sample from the larger quantity drawn. Therefore, the sample preparation is a succession of reductions in weight and grain size. Sampling reduction formula given by Gy (1992) is given at the end of the chapter as Appendix.

In the case of cores, before starting the sample preparation, core is split into two halves along the longitudinal axis with one half preserved and the other half taken for sampling. The preparation of sample usually involves crushing, grinding, sieving, and working down or reducing volume. The most dependable way to obtain a laboratory sample is to crush the initial sample very finely to mix the mass and work it down to the required final weight. In conventional treatment, samples are crushed successively to the next half size and at each stage 50% of the material is discarded employing coning and quartering method till about 300–500 g material is obtained as final sample. The optimum weight for reliable results to which an ore sample crushed to a definite size may be worked down with the reduction error within allowable limits. To facilitate taking minimum and reliable sample weight, following factors need to be considered:

1. the structure and texture of the ores
2. the size characteristics of the mineral grain
3. the proportion of ore mineral grains in the sample
4. the unit weight of the useful mineral
5. the average metal content in the ore.

After obtaining reliable sample weight, the entire material is further ground to finer size and working down the sample, required quantity is kept for laboratory studies.

9.7.1 Drying

All geological samples must be dry, prior to sample preparation. The procedure for drying of mineral sample would differ depending upon the type of sample due to mass, moisture, matrix, and requirement of the types of analysis. It is important that all the moisture content of samples are removed to ensure that particles do not adhere to the preparation equipment.

9.7.2 Comminution

Comminution is the process in which the rock/ore is reduced to desired size where liberation of the minerals are maximum without changing the chemical and physical properties of the rock. Pulverization of a sample plays a very important role in the chemical analysis because of the fact that most of the grains of ore minerals are so closely associated with the waste, hence they must be liberated otherwise the chemical results obtained may not be correct and reproducible. There are several methods of comminution.

The comminution is done in two stages: (1) Crushing and (2) Pulverizing/grinding.

9.7.2.1 Crushing

Crushing of samples is required when grain sizes are too large for pulverizing equipment. This volume reducing process is normally applied to large sample types such as rocks and core. Reduction of ore lumps to a size desired for grinding is achieved through crushing. Choosing the best option for crushing can depend on both the sample type and the mineral that hosts the element of interest within the matrix—a finer crush might be appropriate for heterogeneously distributed precious metals. There are two types of crushers: (1) Primary crushers and (2) Secondary crushers.

1. *Primary crushers:* They are (a) Jaw crushers, (b) Gyratory crushers.

   Jaw crushers (Fig. 9.1 left) are used where the crusher gap is more important than the capacity but gyratory crushers are used when high capacity is required. These produce a product of approximately 10 mm size. Rock and core samples are reduced in size to

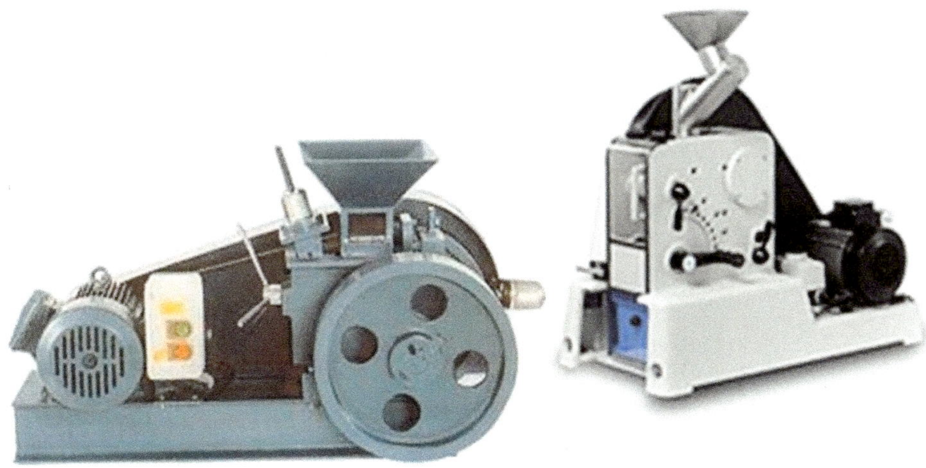

FIGURE 9.1 Lab scale: (left) Jaw crusher and (right) pulverizer. *Sources: Courtesy: (left) Steve and (right) Fritsch.*

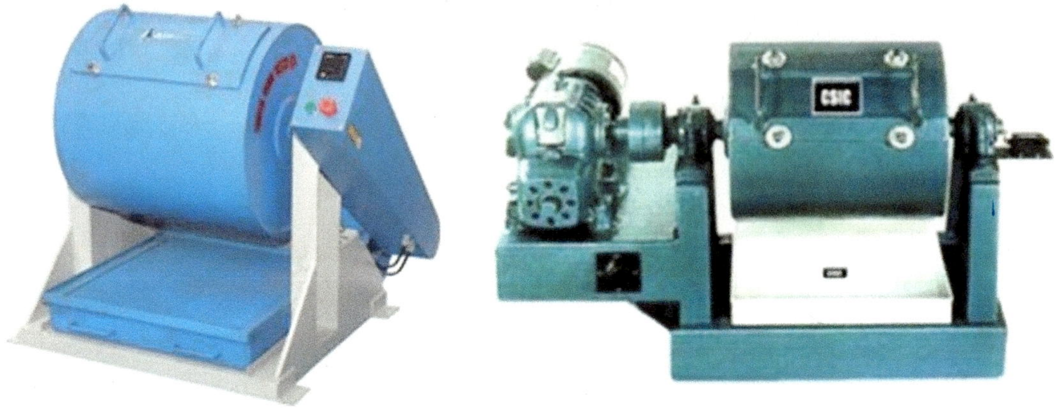

FIGURE 9.2 Laboratory scale: (left) Ball and (right) rod mills. *Source: Courtesy: Steve.*

70% passing 2 mm (Tyler 9 mesh), then a representative split (about 1000 g) is taken and pulverized to 85% passing 75 μm (Tyler 200 mesh). This method is appropriate for rock chip or drill samples.

2. *Secondary crushers:* The maximum feed size will normally be less than 10 cm in dia. in these crushers. The crushers used are (a) Cone crusher, (b) Roll crusher, which normally produce a product of approximately 2–3 mm.

9.7.2.2 *Pulverizing*

Pulverizing of samples creates a fine homogeneous powder which allows for a representative subsample to be taken for analysis. The grains are decreased in size by a mix of impact and abrasion, utilizing different pulverizing mills (Fig. 9.1 right). Rotational cylindrical vessels, known as tumbling mills (Fig. 9.2), are used for the mixed action with media as steel bars, balls, agate materials, or tungsten carbide. In the process of grinding, grains with around 5 and 250 mm size are decreased to around 10 and 30 mm.

9.7.2.3 *Tumbling Mills*

It includes generally (1) Rod mills, (2) Ball mills, and (3) Autogenous mills.

- *Rod mills:* Rod mills are used for selective grinding.
- *Ball mills:* Ball mills are generally used for fine grinding.
- *Autogenous mills:* Grinding in autogenous mills takes place due to the material present in the mill.

9.7.3 Splitting

The samples which have been crushed will require splitting to obtain an appropriate sample size for further processing (pulverization). Splitting is a cost-effective method of reducing sample volume and splits the sample into representative subsamples. To ensure a proper representative sample is obtained, careful consideration is taken when choosing the size of the splitter and its contact with the sample in order to split the rock without

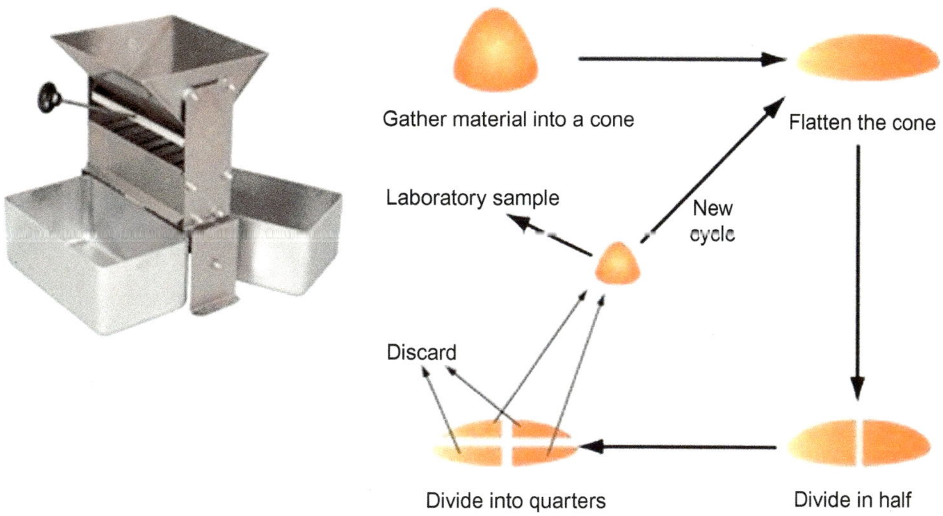

FIGURE 9.3 (Left) Jones sample splitter, (right) sketch of coning and quartering method of sample preparation. *Source: Courtesy: Jacob.*

bias. Coning and quartering is another method to take a representative subsample for further analysis/studies, etc. (Fig. 9.3).

9.8 SCREENING AND PARTICLE SIZE DISTRIBUTION

Screening is performed to decide the distribution of different size divisions. Mesh sizes can be customized for multiple screening processes. Determining the particle size distribution of geological material can be utilized in determining the effectiveness of grinding processes on rock samples. This is done by either the laser diffraction or the mechanical sieve shaker method. In "Laser Diffraction," the degree to which light is scattered from the sample relates to the particle size distribution. A continuous scale can be obtained from submicron size to millimeter size. In "Mechanical Sieve Shaker Method", the sample is passed through a series of screens with different aperture, progressively becoming smaller at the bottom layers. The samples are weighed at each level and a particle size distribution can be determined. Sample treatment procedures for soil and floodplain samples, for humus samples, and for stream sediment samples are given in Figs. 9.4–9.6 (Sandstorm et al., 2001).

9.9 SAMPLE PREPARATION METHODS FOR ANALYSIS

The selection of the appropriate method of sample decomposition is very important for correct analysis. To achieve this, an analyst should carefully take into account the chemical properties of various minerals present in the sample. Owing to the variable characteristics of the ore, it is not possible to have any one set procedure applicable to all samples. It is often necessary to combine two or more methods to effect the dissolution of an ore

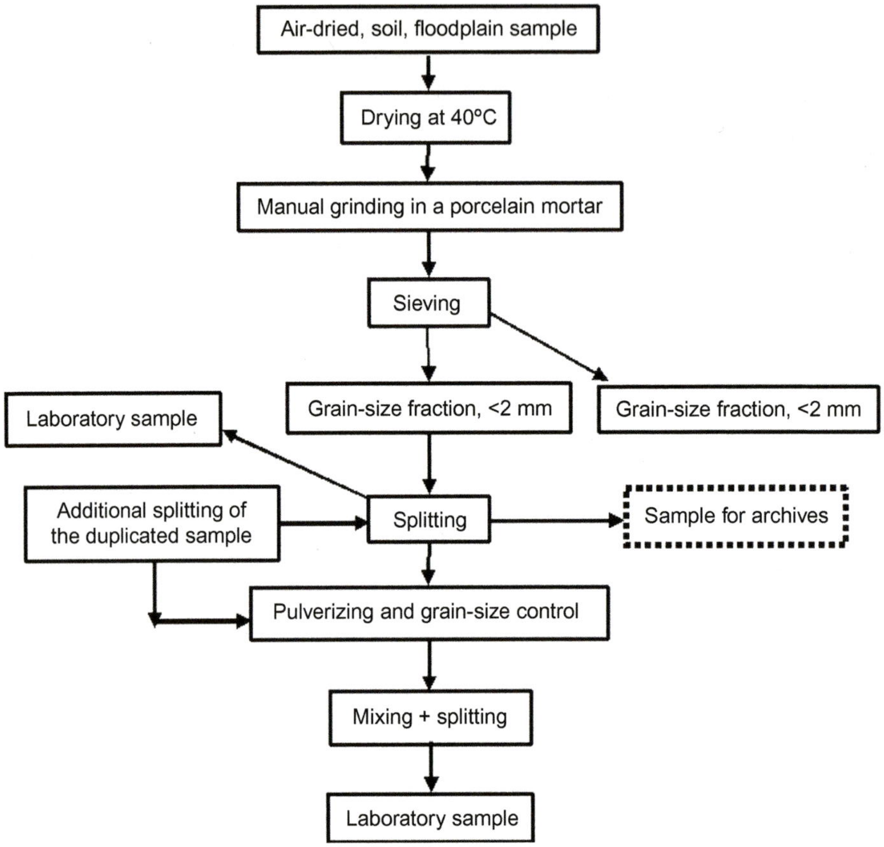

FIGURE 9.4 Pretreatment procedure for Soil and Floodplain samples.

(Levinson et al., 1980). There are mainly two methods to open an ore. They are (1) wet method and (2) dry method.

9.9.1 Wet Method

Wet method constitutes the digestion of an ore sample with acid or combination of acids to convert the complex ore minerals into simple chemical compounds. The action of an acid or a mixture of acids is supplemented by heat and pressure. Heat and pressure hasten the dissolution rate. This is normally preferred over dry method due to its simplicity and it requires less attention; moreover, the vessel used is less affected by the dissolving media.

9.9.2 Dry Method

Dry method constitutes the fusion of an ore with a solid or mixture of solids to convert complex minerals compounds into simple chemical compounds. The reaction is effected

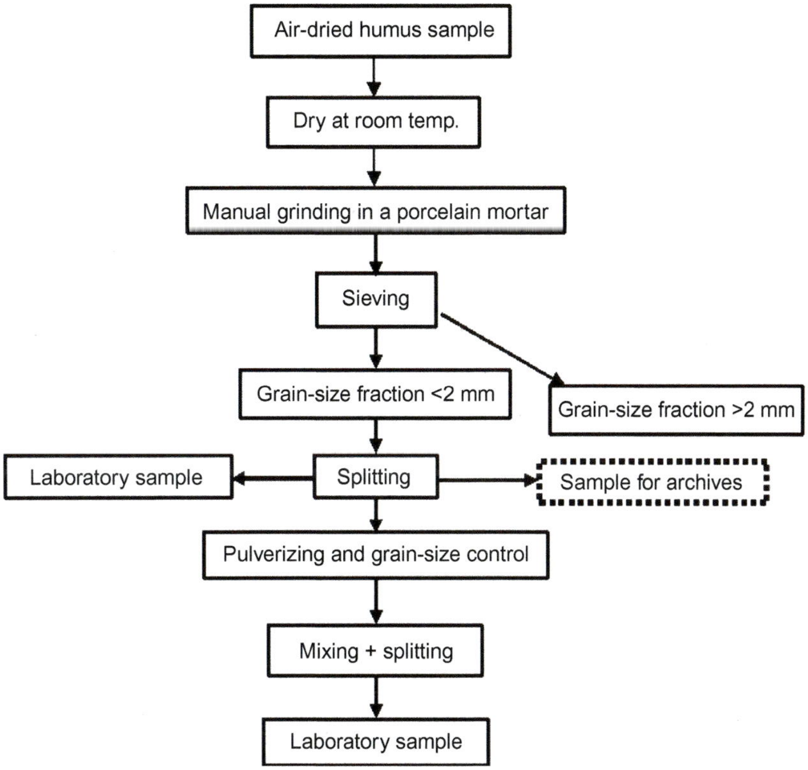

FIGURE 9.5 Sample treatment procedure for Humus samples.

by melting the mass at elevated temperature. The fusion is carried out in variety of crucibles made of platinum, gold, nickel, iron, zirconium, palladium, silica, porcelain, etc. There are number of chemical compounds used for fusion. The choice of fusion mixture depends on the nature of the ore to be fused and the nature of the final product needed for further analysis.

9.10 ANALYSIS OF GEOCHEMICAL SAMPLES

There has been a significant change in the use of analytical methods in exploration geochemistry over the past two decades. The simple colorimetric methods which were in wide use in the geochemical exploration campaigns, in the 1950s and 1960s, were pushed aside with the advent of atomic absorption spectrometry (AAS). AAS method of analysis has been widely accepted and predominantly used in exploration geochemistry surveys of the 1970s, because of its versatility. The uses of colorimetric methods had petered out compared to other techniques like Emission spectrometry, X-ray methods, etc. Calorimetry was the earliest and less expensive method widely adopted for the field analysis of geochemical samples

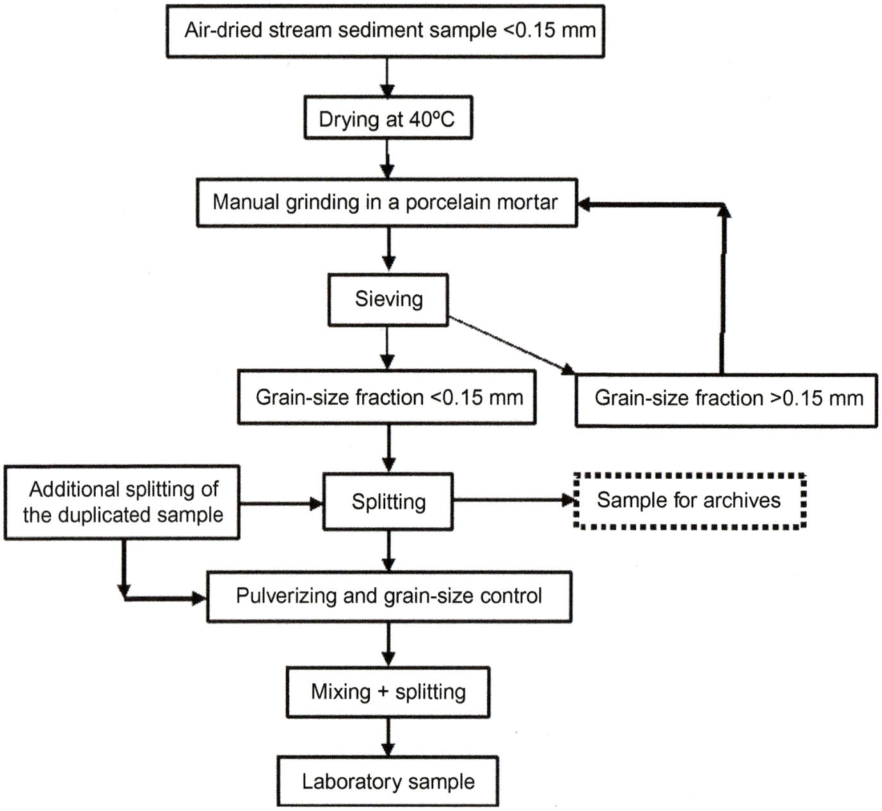

FIGURE 9.6 Sample treatment procedure for stream sediment samples.

in Russia and Western world, by the field parties. Emission spectroscopy was the first instrument widely used for geochemical exploration, since it had great utility in the determination of trace amounts of many elements. In Russia, this was used much. It is semiquantitative in nature and a general scan of all metals could be done. In instrumental techniques of analysis, experienced persons are needed (Levinson et al., 1980).

Owing to rapid changes in exploration objective, analytical methods which are used changed from year to year. With the portable, handheld, X-ray analytical systems (table top and bigger lab-scale XRF unit setups), wet chemical analytical techniques are becoming less and less. In AAS, with one sample digestion, many elements could be analyzed. Whole rock analysis with multielement analysis could be done using XRF techniques. There are many manufacturers of these analytical equipment with different configurations, suiting to different budgets.

Portable field analytical equipment presently available, which, under certain circumstances can have great value for exploration purposes like on-site decision making during geochemical surveys, during evaluation or follow-up studies of recognized anomalies and for directing drilling campaign. Mobile labs are the order of the day in exploration camps. Many private mobile labs are also used to service the needs of exploration groups in areas of high activity, so that immediate direction of a program is possible, especially in remote areas.

9.10.1 Precision and Accuracy

Accuracy describes the nearness of a measurement to the standard or true value, that is, a highly accurate measuring device will provide measurements very close to the standard, true, or known values and is calculated by the formula:

$\%\text{Error} = (\text{MV} - \text{TV}) \times 100 \div \text{TV}$, where MV is your measured value and TV is the true value

Precision is the degree to which several measurements provide answers very close to each other. It is an indicator of the scatter in the data. The lesser the scatter, the higher the precision. It depends on the measuring tool and is determined by the number of significant digits. Accuracy and precision is demonstrated in Fig. 9.7; *Accuracy* is represented by striking the bulls eye (the accepted value) and *Precision* is represented by a tight grouping of shots (they are finely tuned).

In exploration geochemistry precision (the ability to reproduce and replicate the same result) is usually *more important* than accuracy (the approach to the true content) in the initial stages of the project. If the results appear geologically or geochemically improbable, erroneous, they should be checked in another laboratory using identical sample.

Limit of detection (sensitivity) is the minimum content of an element that can be measured by a specific analytical method. This value varies from one method to another and from one matrix to another.

The analytical procedure for geochemical exploration, samples should be:

- sensitive enough to detect elements present in small concentrations
- reliable enough that the chances of missing an anomaly are negligible
- economical enough that huge number of samples be processed as a routine survey
- simplicity of technique, that could be entrusted to relatively untrained personnel
- portability of the equipment, so that lab facilities could be set up near field operations.

In every trace element analysis procedure, four steps are involved as follows.

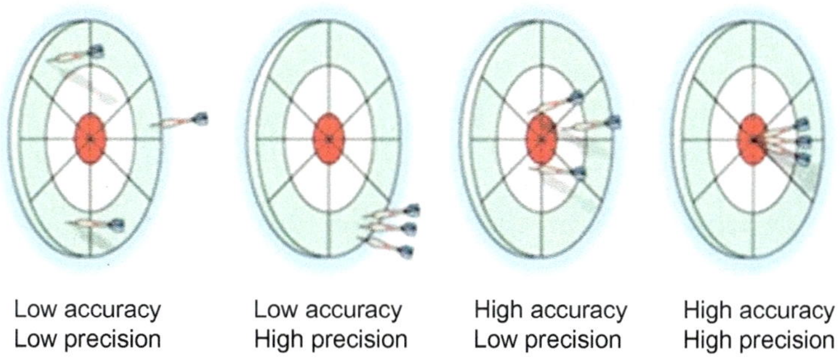

FIGURE 9.7 Illustration explaining accuracy and precision.

9.10.1.1 Preparation of Sample

1. Removal of water by (a) drying of organic and clastic samples and (b) evaporation of water samples
2. Pulverization
3. Sizing
4. Mineral separations according to differences in (a) density, (b) magnetic susceptibility, and (c) electrical properties.

9.10.1.2 Decomposition of Sample

1. Volatilization
2. Fusion with (a) acid flux, (b) alkali flux, and (c) oxidizing flux
3. Vigorous acid attack with (a) oxidizing acid, (b) dehydrating acid, and (c) hydrofluoric acid
4. Weak attack by (a) dilute acids, (b) solution of complexing agents, and (c) pure water
5. Sample oxidation by (a) ignition and (b) wet oxidation.

9.10.1.3 Separation of Element

1. Separation in vapor phase by (a) distillation and (b) sublimation
2. Separation in liquid phase by (a) solvent extraction and (b) complex formation
3. Separation in solid phase by (a) ion exchange and (b) precipitation.

9.10.1.4 Estimation of Element

1. Optical measurements: (a) colorimetry, (b) chromatography, and (c) visible fluorescence
2. Radiation measurements: (a) flame spectrometry, (b) emission spectrometry, (c) absorption spectrometry, (d) X-ray spectrometry, and (e) radiometry
3. Electrical measurements, etc.

The combinations of component of analytical procedures most commonly used in geochemical prospecting are amply described by Rose et al. (1979).

9.11 HIGH-QUALITY ANALYSES FOR EXPLORATION

In geochemical analyses and assaying, various sample decomposition/pretreatment methods are used (labtium.fi):

- aqua-regia digestion for geochemical analyses and assays
- multiacid digestions for geochemical analyses and assays
- sodium peroxide fusion for the assay of refractory and high-grade ores and concentrates
- XRF-analysis for petrology, industrial minerals, and assays
- phase and mineral-selective leaches
- individual methods for additional elements/components and physical properties.

To achieve the best economic and superior quality analysis in mineral exploration, it is imperative to choose the system of analysis by deciding on the suitable analytical methodology for appropriation into the goals of the exploration campaign. The method selected depends upon the element(s) which is (are) being analyzed and the expected amount.

For the success of any exploration program, it is crucial to select proper analytical methodology and element selections. Suitable choice of analytical methodology is essential and attention should be given to the dissolution strengths and constraints of the selected methodology to make sure that all pre-requisite conditions are satisfied. Fletcher (1981), Potts (1987), and Thompson and Walsh (1989) have given detailed description on the various analytical methods used. The range of analytical techniques for elemental analysis that are in use for mineral exploration include:

1. Atomic spectroscopy: absorption, emission, and fluorescence techniques
2. Flame atomic absorption spectroscopy
3. Inductively coupled plasma optical emission spectroscopy
4. Inductively coupled plasma mass spectrometry
5. X-ray fluorescence spectrometer
6. Electron probe micro-analyzer
7. Instrumental neutron activation analysis
8. X-ray diffraction
9. Secondary ion mass spectrometer
10. Scanning/transmission electronic microscope.

9.12 SOURCES OF ERROR

In all mineral exploration program, invariably sources of error would result from various activities including sampling process, analysis, laboratory, reliability of technique, etc. These errors get compounded and produce erroneous values. Errors are reflection of imprecision of information coming about because of the constraints of collecting small representative articles from larger whole and from the sensitivity of analytical methods. Every individual phases from sampling to analysis can be recognized as individual source of error. Adopting of a quality assurance/quality control (QA/QC) plan would aid in reducing the errors to a minimum without expecting that there will be no error. Suitable sampling and QA/QC is vital in all parts of the mineral assessment procedure for accomplishing maximum confidence in estimation of mineral resource and reserve. Comprehensive accounts on QA/QC program have been given by Cavanagh et al. (1997), McQuaker (1999), USEPA (2000), McDonald and LeClair (2004), and Mitchell (2006).

APPENDIX: GY'S SAMPLING REDUCTION FORMULA

Gy (1992) is believed to be the first to devise a relationship between the sample mass (M), its particle size (d), and the variance of the sampling error (S^2). This variance is that of the fundamental error (FE): a best estimate of the total sampling error (TE) is obtained by doubling the FE.

$$M \geq \frac{Cd^3}{S^2}$$

where M is the minimum sample mass in grams and d is the particle size of the coarsest top 5% of the sample in centimeters. Cumulative size analyses are rarely available and from a practical viewpoint d is the size of fragments that can be visually separated out as the coarsest of the batch. S^2 is the fundamental variance and C is a heterogeneity constant characteristic of the material being sampled and = cβfg, where:

c is the a mineralogical constitution factor $= \frac{(1 - a_L)}{a_L}[(1 - a_L)\delta A + \delta G, a_L]$,
a_L is the amount of mineral of interest as a fraction,
δA is the specific gravity of the mineral of interest in g/cm^3,
δG is the specific gravity of the gangue,
β is a factor which represents the degree of liberation of the mineral of interest. It varies from 0.0 (no liberation) to 1.0 (perfect liberation) but practically it is seldom less than 0.1. If the liberation size (d_{lib}) is not known it is safe to use β equal to 1.0. If the liberation size is known $d \geq d_{lib}$ then $\beta = (d_{lib}/d)^{0.5}$ which is less than 1. In practice there is no unique liberation size but rather a size range,
f is a fragment shape factor and it is assumed in the formula that the general shape is spherical in which case f equals 0.5,
g is a size dispersion factor and cannot be disassociated from d. Practically g extends from 0.20 to 0.75 with the narrower the range of particle sizes the higher the value of g but, with the definition of d as above, g equals 0.25.

References

Cavanagh, N., Nordin, R.N., Pommen, L.W., Swain, L.G., 1997. Guidelines for Designing and Implementing a Water Quality Monitoring Program in British Columbia. Resources Inventory Committee, British Columbia Min. of Environment, Lands and Parks, Canada, <http://www.gov.bc.ca/risc/pubs/aquatic/design/index.htm>.

Dowing, B., 2014. ARD Sampling and Sample Preparation. <http://www.technology.infomine.com/enviromine/ard/sampling/intro.html>.

Fletcher, W.K., 1981. Analytical methods in geochemical prospecting. In: Govett, G.J.S. (Ed.), Handbook of Exploration Geochemistry, vol. 1. Elsevier, Amsterdam, 255 pp.

GSI, 2012: Analytical methods. Analysis of Geological Materials, Training Course Notes. Training Division, Hyderabad, 120 pp.

Gy, P.M., 1992. The Sampling of Heterogeneous and Dynamic Material Systems. Elsevier, Amsterdam, 653 pp.

Gy, P.M., 1998. Sampling for Analytical Purposes. John Wiley, New-York and London, 150 pp.

Kreiter, V.M., 1968. Geological Prospecting and Exploration. Mir Publishers, Moscow.

Lacy, W.C., 1983. Mining Geology. Hutchinson Ross Publishing Co, Stroudsburg, PA.

Levinson, A.A., Mc Common, R.B., Hitchon, B., 1980. Introduction to Exploration Geochemistry, second ed. Applied Publishing, Calgary, Canada, 924 pp.

McDonald, D., LeClair, D., 2004. Methods and quality assurance investigations for trace metals data from the long-term river network, 2003. Env. Monit. and Eval. Br., Alberta Environment, 77 pp.

McKinstry, H.E., 1962. Mining Geology. Asian Publishing House, Bombay, 680 pp.

McQuaker, N., 1999. Technical evaluation on sample collection, handling, analysis and interpretation for trace level contamination in water. Aq. Effects Technology Evaluation Program Project 3.1.1. Natural Resources, Canada, Ottawa.

Mitchell, P., 2006. Guidelines for Quality Assurance and Quality Conrol in Surface Water Quality Programa in Alberta. Alberta Environment, July 2006, 67 pp.

Moon, C.J., Whateley, M.E.G., Evans, A.M., 2006. Introduction to Mineral Exploration, II ed. Blackwell Publishing, Oxford, 481 pp.

Peters, W.C., 1987. Exploration and Mining Geology, second ed. John Wiley and Sons, New York, NY, 696 pp.

Potts, P.J., 1987. A Handbook of Silicate Rock Analysis. Blackie, Glasgow, 622 pp.

Rose, A.W., Hawkes, H.E., Webb, J.S., 1979. Geochemistry in Mineral Exploration, second ed. Academic Press, London.

Sandström, H., Reeder, S., Bartha, A., Birke, M., Berge, F., Davidson, B., et al., 2001. Sample preparation and analysis. Special Operating Procedure Report. 14 pp.

Thompson, M., Walsh, N., 1989. Handbook of Inductively Coupled Plasma Spectrometry. Blackie, London.

USEPA, 2000. Guidance for the Data Quality Objectives Process. EPA QA/G4. EPA 600/R-096/055. Office of Environmental Information, Washington, DC.

CHAPTER

10

Geographic Information System and Common Earth Model

OUTLINE

10.1 GEOGRAPHIC INFORMATION SYSTEM

10.1.1 Introduction

Successful and valuable management of a country's mineral resources requires collection and integration of a wide range of topographical, geological, geophysical, geochemical, and other related information. Geographic Information System (GIS), an important development in the field of Information Technology, has made remarkable effect in the field of mineral exploration owing to its capabilities of organizing and analysis of individual layers of spatial data, and of providing tools to analyze and model the interrelationships among various spatial data layers. It is used to provide support for wide variety of

DOI: http://dx.doi.org/10.1016/B978-0-12-805329-4.00017-X

decisions based on data that are in the form of maps, images, reports, and field observations. It is a spatial tool that integrates hardware, software, data, methods, and people. A full range of functional capabilities of a GIS include data capture, input, manipulation, transformation, visualization, combination, query, analysis, prediction, modeling, and map output with respect to all types of geographically referenced data (Moore, 2009; Escobar et al., 2010; ESRI, 2006). The role and use of GIS in geology are very broad and diverse. GIS made its appearance commercially in the market during the mid-1980s. Toward the end of the 1980s, GIS with the ability to manage vector, raster, and tabular data made its appearance in the market. The GIS can be viewed in three ways: database, map, and model views. Major application areas in exploration and mining of minerals include geological and mineral potential mapping; geophysical exploration modeling; mineral resources inventory and evaluation; and mine environmental monitoring, mine design, and operation; and landscape restoration (Tew and Osborne, 1997; Rahimi, 2002; Ayachi, 2003; Kumar et al., 2007; Dobson and Durfee, 2012).

GIS is the best platform to bring all types of data images (topography and geological map, multispectral satellite images, geophysical images, and geochemical images to databases in different format) together. A schematic sketch diagram showing the various fields involved in GIS analysis is given in Fig. 10.1. There are two main types of GIS data: (1) vector GIS, where the basic data are vector-based and represented in the form of coordinates and (2) raster GIS, which involves the use of spatial data, not in the form of points but expressed as a matrix of cells or pixels. There are many similarities between GIS and CAD software, with CAD focused on engineering applications with high precision and accuracy and the GIS works on a much larger scale and is map focused. These similarities have led to the integration of both GIS and CAD software.

The fundamental instrument for earth scientists in all spheres is the map of geology of the area portraying the spatial coverage and name of the rock unit exposures, which has been traditionally the key means for data management in earth science. An organized combination of these maps aids an economic geologist in locating mineral or oil deposits, a groundwater geologist in locating ground water, and structural geologists in identifying

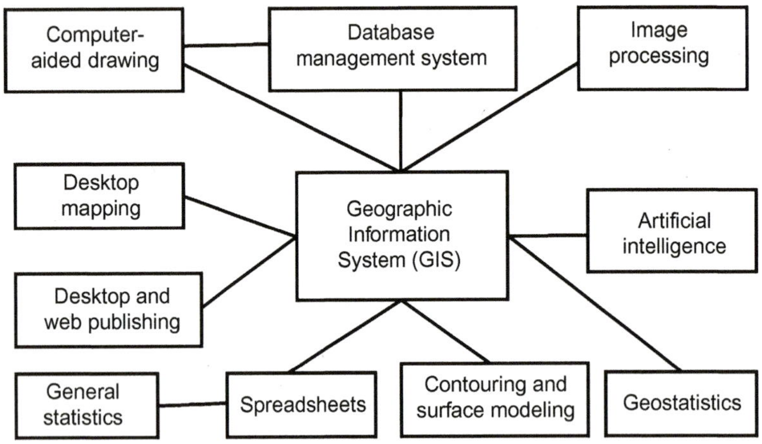

FIGURE 10.1 A sketch showing various fields involved in GIS.

faults, folds, and unconformities. When a customary analog article is transformed to a spatial database, manipulation and analysis of geological information is distinctly reformed. ASTER with its high spatial resolution and bands covering a larger portion of the electromagnetic range gives information that enhances capabilities of geologists to yield more precise geologic maps at a much smaller expense in comparison to traditional ground-based techniques. It is surely understood that ASTER data have attributes that are especially valuable for geologic studies, where good exposures of rocks exist.

GIS can assist geologists in various fields related to their activity domains, viz., data capture, manipulation, analysis, and reporting. It has become likely for field geologists to acquire field data electronically using Global Positioning System (GPS) receivers. The Internet also can act as a source of data sets that can be downloaded directly from it. All spatial and nonspatial data can be overlaid, integrated, and analyzed with the help of a GIS. It can integrate with other specialized programs for image processing. Raster images can be exhibited in GIS and overlaid with vector data such as geology, faults, and geochemical samples.

10.1.2 Types of GIS

There are a number of GIS (or GIS softwares) available today. They range from high powered analytical software to visual web applications, and each of these is used for a different purpose. There are common features in all the available GIS. Understanding these basic features will give a person confidence with any GIS system that a person might use in future. The three types of GIS are: (1) web-based GIS, (2) geobrowser, and (3) desktop GIS.

- The "Web-based GIS" is an online GIS application, which, in most cases, is an excellent data visualization tool. Its functionality is limited compared to the software stored on the computer, but it is user-friendly and particularly useful as it does not require "data download."
- The "geobrowser" is an explorer for seeking and accessing geographic information from the Internet. Like other web browsers, it allows the combination of many types of geographic data from many different sources. The biggest difference between the "world wide web" and the "geographic web" is that everything within the latter is "spatially referred." Google Earth is the most popular "geobrowser" available.
- A "desktop GIS" allows one to interactively work with spatial data. It is a mapping software that needs to be installed onto and runs on a personal computer by using "ArcGIS" (developed by ESRI). It can cover both commercial and educational uses.

10.1.3 GIS and Data Integration

One of the significant favorable circumstances of GIS is its capability to combine remotely sensed information and other spatially referenced computerized information, keeping in mind the end goal to decide how the examples of various variables are connected. Furthermore, GIS adds to the data obtained around a specific range with the utilization of advanced databases. Data preparing and choice making are turning out to be progressively concise. Innovative advances in remote sensing, geographic data frameworks, and reproduction displaying are at the cutting edge of this pattern. With respect to this, remotely sensed information ought to be viewed as only one information set within GIS. It is likewise imperative to note that remotely detected information gives dynamic data about the biosphere, rather than auxiliary information sets, which are static. However, auxiliary information has a fundamental value in expanding the information substance of satellite data.

The technique of integrating remote sensed data into GIS for the most part incorporates a host of methods like data capture, processing, managing, data analysis, data conversion, and presentation of final output in the form of maps (Lunetta et al., 1991).

10.1.4 GIS and Remote Sensing

GIS is a software that allows (1) spatial data capturing from diverse sources (eg, remote sensing), (2) interrelating spatial and tabular data, (3) analysis of tabular and spatial data, and (4) scheming the layout of a final map. Both vector and raster data are managed by GIS. Raster data such as remote sensing need special data manipulation techniques. Nevertheless, subsequent to a remote sensing analysis, its consequences are usually combined in a GIS or into a database, for further analysis, viz., overlaying, buffering, combining, etc. Progressively increasing vector capabilities are being incorporated within a remote sensing software.

10.1.5 GIS and GPS

Integration of new technologies, like that of the GPS with GIS, provides a significant progression in the advancements of spatial databases. GPS is utilized to note latitude, longitude, and elevation of a given location. With respect to reference data, the use of GPS may be made for any or all of the below-mentioned purposes:

- to help in the analysis and explanation of remotely sensed data,
- to standardize a sensor, and
- to corroborate information obtained from remote sensing data.

GPS essentially decreases the expense of controls required for photogrammetry and precision appraisal. The use of GPS to perform customary ground control studies is as of now far cutting edge. The likelihood of giving all the control by method for a GPS receiver is illustrated. In the past couple of years the accuracy of the GPS has enhanced altogether, and more organizations and administrative offices have tackled the GIS usage issue through the incorporated use of GPS. When combined with differential software, GPS/GIS information gathering conveys information with precision.

The enormous amount of multiparametric data obtained from satellites require accurate location on earth. This is achieved by a combination of GIS with GPS and geostatistics. Data processing and interpretation are done with the high-speed computers and the dedicated software. The GIS technology is a combination of the popular database system and highly sophisticated computer graphics. With the inputs from geological, geophysical, and geochemical surveys, the aim of GIS-based geostatistical application is to look for zones of higher probability or "favorability index" for mineral prospecting (Bonham-Carter, 1994, 1997; Heywood et al., 2006).

10.1.6 GIS for Mineral Exploration

Most of the data related to the operations of mineral exploration have a spatial nature. Geologists can apply GIS in mineral exploration and mining to present data in an integrated platform by using traditional cross-sections and graphical strip logs in conjunction with planning map views. Since the business of exploring and extracting for minerals is completely spatial, it can make easier for GIS. The risks of evolving mineral resources should be referred to as precisely as could be allowed. This procedure ought to begin at the exploration stage and proceed through feasibility to the development stage. Numerous mining organizations have developed broad exploration geological databases, which speak to critical resources for the organizations that have taken a large sum of money to compile. Lately, projects of computerized information compilation have been attempted to permit the utilization of more probabilistic data analysis procedures, moving far from the customary master framework techniques. The GIS is the ideal management instrument for compiling spatial information.

The process of mineral exploration is composed of several stages, which starts with small scale and develops into larger scale. In every stage, geological, topographical, geophysical, and geochemical data are collected, processed, and integrated. Finally, having completed every stage, mineral potential map is produced and the study area becomes smaller. GIS has the capability for storing, updating, revising, displaying, retrieving, processing, manipulating, and integrating different geospatial data and information to be used as a policy-making framework and deliver fruitful results.

In mineral exploration, geologists deal with various types and sources of data to explore new economical mineral deposits. Exploration processes require integration of data layers to deliberate a wide range of combinations and to underline different hypotheses. All types of geological data layers, viz., geological surveys, geochemical maps, geophysical maps, boreholes, radius measurements, and mineral accumulations, can be shown, revealed, and interpreted instantly by making use of a GIS. A key application of any GIS is the ability to add multiple data layers for definition and demarcation of mineral prospectivity maps. The use of GIS may be made to integrate survey data with comprehensive models resulting from various mine software like that of GeoSoft, Surpac Range, Mine Sight, Vulcan, and Mining Visualization System.

10.1.7 Mineral Potentiality Mapping

Methods of spatial data analysis used for prospectivity mapping include (1) empirical (data-driven) approach (suitable for mature "brownfields" exploration terrains with abundant data available) and (2) conceptual (knowledge-driven) approach (suitable for "greenfields" exploration terrains with limited number of deposits available for statistical assessment).

GIS also can be used to generate a mineral potential map. In the absence of comprehensive systematic mineral exploration programs, it is important to develop alternative methodologies of mineral potential classification. Then, it is possible to use spatial data that are relevant to mineral potential, such as lithology and topography, which are available for most areas. The importance of such data can be realized by their incorporation in GIS (Bonham-Carter, 1994, 1997; Coulson et al., 1991). Then, it is important to add exploration criteria. Those criteria provided by conceptual mineral deposit models are invaluable bases for the generation of mineral potential information.

Prospectivity mapping is carried out to locate promising sites for mineral exploration. It can be applied in various scales from global to local scale exploration targeting. Definition of the exploration model can be based on a genetic ore deposit model or a mineral system model. This gives the framework for the data used for creating a predictive prospectivity analysis for mineral exploration targeting. Essential part of the procedure is the preprocessing of the raw data into meaningful map patterns for the given task. These preprocessing techniques include data interpolation, classification, clustering, rescaling, filtering, image processing, raster calculation, etc. (Figs. 10.2 and 10.3). A common expression for describing these methods in GIS platform is geoprocessing. After creating the map patterns indicating the vectors toward a mineral deposit type, we can apply various data integration techniques in GIS to create a final potential map delineating target areas for mineral exploration (Nykänen, 2011).

Earth models utilized for mineral investigation or other underground examinations ought to be steady with all accessible geological and geophysical data. Geophysical inversion gives the way to incorporate geological data, geophysical review information, and physical property estimations carried out on rock samples. Integration of geological data into inversions is dependably an iterative procedure. One starts with the geologists' best conceptualization about the Earth (ie, the geological model), and the schemes obtained from geophysical inversion might demonstrate that the geological model ought

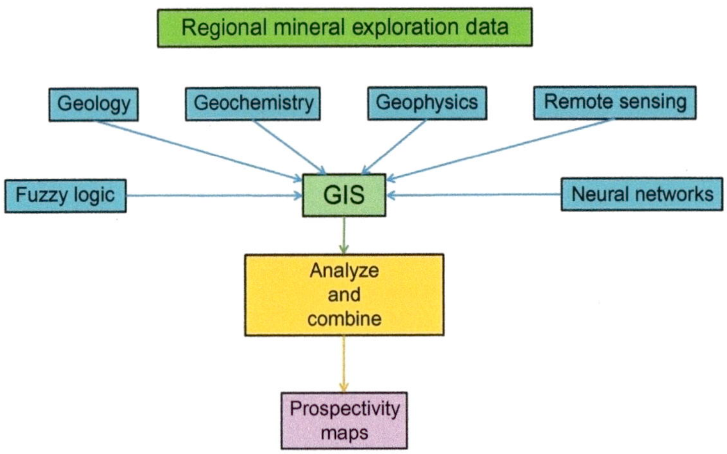

FIGURE 10.2 Integration of critical parameters into prospectivity maps.

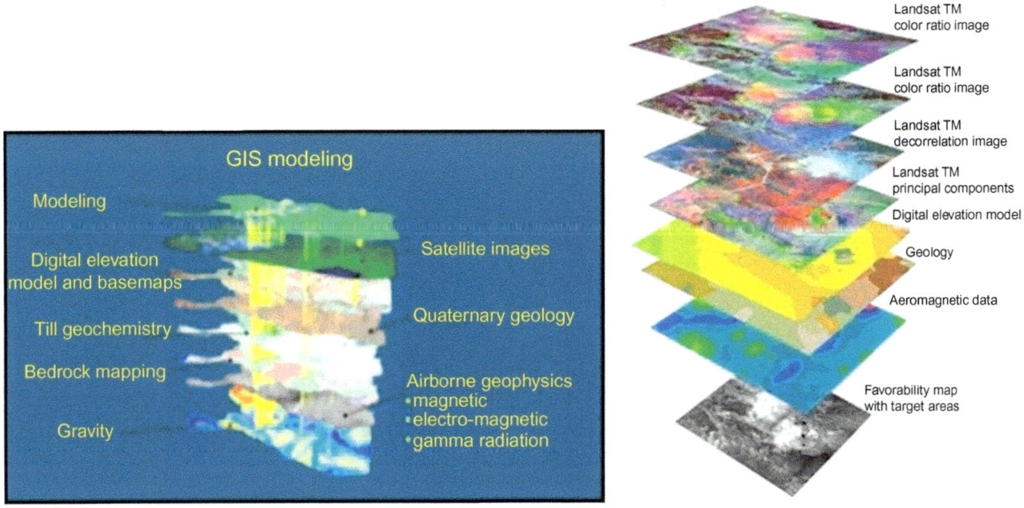

FIGURE 10.3 Left: GIS modeling incorporating various sets of data. Right: Generation of favorability map with target areas (favorability mapping of Satellite data in N. Chile for exploration of Copper (Gold) deposits). *Source: Left: www.satimagingcorp.com, Right: geosphere.gsapubs.org/content/2/4/236.*

to be changed somewhat before the subsequent iteration of the methodology. Furthermore, geological and geophysical information can be consolidated through inversion and progress made for forming a typical Earth model predictable with all the available information. As more data are integrated, the nonuniqueness of the inverse issue is reduced.

10.1.8 Sources of Error in GIS

Normally an error is attached to any geographic information and these are of two types: (1) error inherent to information, viz., source, observation, sampling bias, methods, and devices of collecting information, classification, etc., and (2) error created and accumulated by the processing and interpretation, viz., digitization, Raster/Vector, numerical precision, interpolation, overlay combination, plotting, information used, and/or understood not correctly, etc. The objective of GIS application is how to manage and minimize these errors.

10.1.9 3D GIS Technology

Exploration techniques is advancing into a more thorough subjective science. 3D GIS gives backing to this action through a situation in which a rich and different arrangement of exploration-related perceptions can be dissected and translated. Applying more broad geological mapping has a minimal effort to advantage proportion, with ability to produce quick 3D maps, simplicity of approval against 3D information, and scale freedom of the displaying process. The capacity to demonstrate various geological zones quickly and viably will permit customers to produce regional scale exploration models, lessening risks intrinsic in estimation of resources on inadequately domained mineral deposits. Exploration for deep situated mineral deposits requires incorporation of extensive and heterogeneous spatial information sets. Geological, geophysical, and geochemical perceptions are gained, prepared, and analyzed independently. Geophysical forward modeling and inversion algorithms typically work with rectilinear meshes when parameterizing the subsurface because this simplifies the development of numerical methods (Lelièvre et al., 2012).

The variability of geological features and factors had led to several efforts to decipher complex geological issues using multivariate and geostatistical methods. This eventually resulted in the development of specialized mining packages in the late 1970s and 1980s, for geological data analysis, from simple data interpolation/extrapolation to 3D visualization. However, the costs of the packages were prohibitive, and thus were not easily accessible to the general geological community. GIS, which was essentially developed for the geographical community has, therefore, become attractive to the minerals industry because of its recent advancements involving the incorporation of multivariate geostatistical modules and powerful 3D analysis. The more recent applications of GIS integrate intelligent systems to enhance its capabilities for complex modeling, simulation, and predictive decisions often encountered in mineral exploration and exploitation.

10.1.10 Recent Trends and Future Directions

The geologic modeling of the subsurface is a complex task for GIS technology. But, the spatial nature of the geologic objects always makes GIS to be an important part of the modeling and mapping systems. The conventional activities in which GIS is involved in geosciences are data display, data administration, area analysis, and support-making decision. But, now the evolution allows the GIS capabilities on just a desktop computer with full possibility to integrate the industry-standard relational database management system, 3D visualization, geostatistical functionality, 3D geoprocessing capabilities, and web-based mapping. All these advantages make GIS the most useful and attractive tool for comprehensive and essential geoscientist's needs. The latest GIS software made on an open-standard technology ensures that GIS can be easily integrated with advanced solid 3D geosciences software to gain a solution for complex geologic problems. But until recently, although it's become easier, visualizing the subsurface of the earth within GIS has been a complex technical challenge. Recently, GIS technology is able to help geoscientists and geologists in mapping and modeling the subsurface by a modern tools and software packages and can be integrated with other specialized 3D programs to get precious results. It can be used to advance the mineral exploration process while saving time and money and increasing profitability and efficacy (Moore, 2009; Geosoft, 2010; Nykänen, 2011).

10.2 COMMON EARTH MODEL

The center undertaking of a mineral explorationist is to discover economic mineral deposit. It is consequently fundamental that new focuses for where the deposit may be, are created, organized, and tried. Mineral exploration is developing into a more thorough quantitative science. Achievement of an exploration organization is the capacity to explore and assess mineral deposits rapidly and precisely. This relies upon an exact and auspicious comprehension of various data sets from which 3D geology is built. All that one can derive from available data turns into a "model" of the earth, which is utilized to rationalize vital economic decisions, for example, fixing of exploration priorities, drill targeting, and forthcoming exploration costs. Earth models are images of the subsurface conceived in minds limited by existing data, understanding, and conceptualized hypothesis as opposed to quantitative testable forecasts of the project geology. As a mental build, an earth model combines qualitative views, judgment, and inferences drawn on the subsurface that may be portrayed, conveyed, and distributed within exploration group or to organization. However, a "common earth model" is a quantitative scheme of the earth with data consistency, testable by boring, and constrained by refinement as new data come in. Common earth modeling is a creation that integrates diverse databases and permits for adequate exploration decisions. As a consequence of quantitative refining, a common earth model is a necessity for maximizing the worth obtainable from huge investment in data collection.

Common earth model as a concept was first in use in the petroleum business for interdisciplinary approach to data integration and work procedures. According to Garrett et al. (1997), *"The advent of 3-D earth modelling computer systems suggests there is potential to transform the work processes in cross-disciplinary asset teams. By sharing common digital 3-D representations of the subsurface, the team can iterate between disciplines more easily, rapidly incorporating new information into existing models. Up to now, many cross-disciplinary teams have emphasized the importance of software communication, 3-D visualization and data access. From now on, we believe that earth modelling issues will assume greater significance in the business of these teams"* (www.mirageoscience.com).

3D GIS has developed as the new platform in which all the subject specialists along with organization controlling group can upgrade its worth through particular interpretative and identification activities. It is presently conceivable that multifaceted 3D spatial query tools can express adequately the local exploration scheme and analyze and corroborate the query outputs through drilling, 3D mapping, or by exercising validation tests (McGaughey, 2006; McGaughey et al., 2009).

In spite of the fact that it took numerous years to be formalized as standard practice, quantitative integration of map data used for exploration by 2D GIS innovation is by and large recognized to have generated considerable value through the making of integrated maps. New programming innovation stretches out to 3D all the advantages of 2D GIS mapping by facilitating the making of common earth models as displayed in Fig. 10.4. The common earth model innovation permits the express representation of 2D and 3D spatial exploration data. The outcome is a predictable spatial earth model in terms of geology, geophysics, and geochemistry, which individually constitutes subsets of the common earth

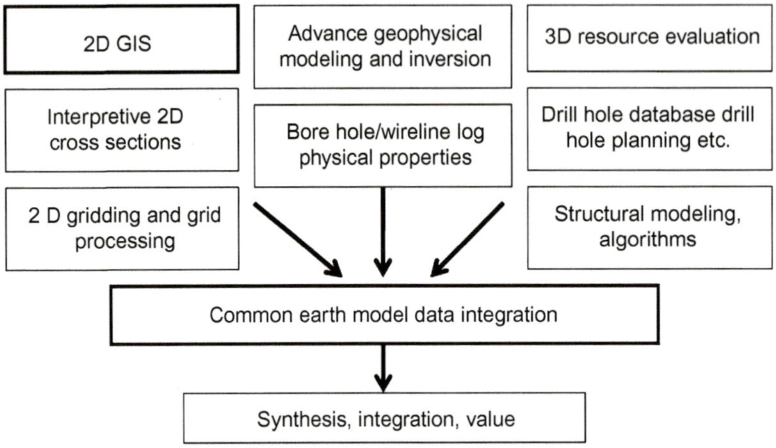

FIGURE 10.4 The common earth model takes all conventional, spatial exploration data as input and provides explicit integration of all components into a single model of the earth consistent with all input data.

model. Significant advancement has been made in integrating differing data sets. 3D geological models have been utilized as exploration and interpretation tools, and for delineation of mineralized bodies. The resultant models can be pictured, shared, altered, and queried. These models can be utilized in outlining drill course (ie, drill length and direction) and as background geological framework for geophysical data interpretation. Constructing such models requires some persistence and the will to apply new innovation.

An important advancement in the field of mineral exploration is the concept of common earth model, in which the expert understanding of a mineral deposit is captured and stored in a 3D GIS with an aim that the same can be utilized by the explorationist. A 3D GIS offer potential environment to combine such diverse data sets to imagine and describe geological associations in a more stable interpretive setting. In other words, it provides a multiparameter spatial model that offers multiple realizations of target corroboration. Since mineral deposits along with host geology can be defined as 3D objects, it is important that exploration for mineral deposits should be viewed as a 3D spatial framework to examine key spatial and characteristic associations that will govern exploration criteria for locating mineral at the deposit scale.

Generally, common earth model is an instrument for ascribing the normally imparted 3D space to all the significant bits of data that are expected to make a 3D mineral targeting activity important (Fig. 10.5). Information of a mineral deposit or obvious geological potential area is extracted and stored in 3D GIS format that permits data integration, visualization, overlay, and final output as mineral potential map (gsc.nrcan.gc.ca).

It must be emphasized that spatial query and analysis are data dependant; in other words, where data are sparse or not adequately dependable, they do not mystically make new bits of knowledge of mineral genesis or produce exploration targets. Like all spatial technologies, they are essentially an expansion of the exploration tool kit and should be utilized suitably with all necessary cautions. Mineral targeting is currently a completely advanced 3D procedure for generating mineral predictive maps. With knowledge on ore genesis and acquisition of progressively increasing relationships of geological setting with geographical locations of mineral deposit, 3D GIS innovation aids in finding concealed and deep-seated orebodies (McGaughey, 2006; de Kemp, 2006, 2007).

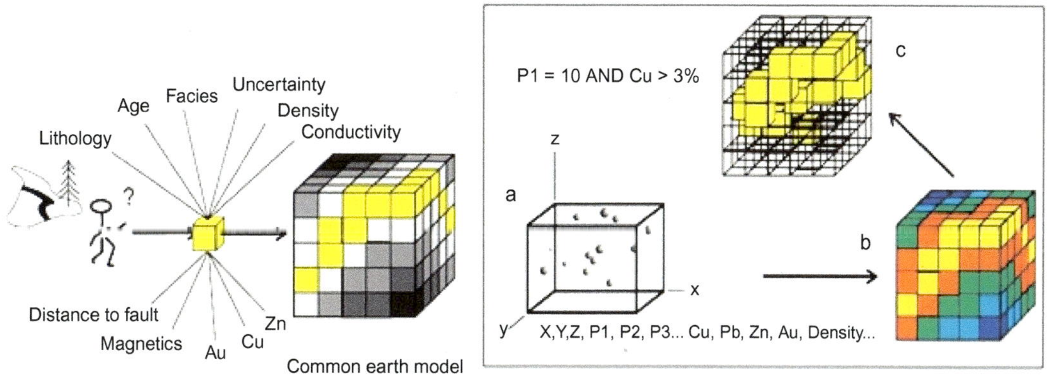

FIGURE 10.5 (Left) Common Earth Model concept: Implemented regular partitioned multiparameter voxel model. Implementation varies depending upon exploration. Requirements with cell sizes varying in dimension, regularity and size (Right); Geospatial query using a property P1, a lithology code, and Cu abundance >3 wt%. (a) The starting data set, which is a set of samples from drill core, surface, or drifts that support a model estimation, (b) Combines all these estimated parameters into the common earth model, (c) A selected set, resulting from a query for which the criteria is true. *Source: From de Kemp (2007).*

10.3 3D & 4D GIS GEOMODELING

The 3D GIS provides a means of key innovation in which geological modeling can be attempted in a thorough and quantitative way. It can provide knowledge-driven and data-driven understandings that can be tried through cross-acceptance tests/direct boring of targets. It is a domain that, in the coming years, will see a wealthier set of devices and work processes for particular kinds of exploration tasks or deposit types. Quantitative geology suggests interconnecting geological, geochemical, and geophysical data. Various thematic geoinformation captured at different scales, ranging from regional to deposit, leads to development of geological models that are able to visualize different spatial parameters (including economic) of mineral provinces along with mineral deposits in 3D and 4D.

A 3D GIS exploration gives a way to use every single accessible parameter to infer 3D coordinated targets. The approach provides a way to target at depth and transforms to resourceful and comprehensive exploration practices. The methodology can combine geology and mineral deposit models using geophysics to incorporate the physical qualities of the rocks and their relationship with the geology and rocktypes of an area prior to subsurface imaging.

4D geomodels can be utilized interactively as a part of visualization and conceptualization of exploration environment defined in terms of geographical location and geological setting of the mineral belt/deposit. The models will likewise encourage the investigation of likely natural and societal effects of mineral extraction all through the whole life cycle from mineral disclosure to conclusion, and money-related and administrative issues, all of which will maintain a strategic distance from clashes in area use.

10.4 COMMON EARTH MODEL AT EXPLORATION STAGES

3D earth models are produced from map information prior to any drilling in the area being modeled. On building a 3D geological model even with exceptionally scanty information, the model exists to hold all future information with a structure for the understanding of that information. The earth model finally attained is with no or little subsurface information showing that an intelligible conceivable model might be worked out. Such a model, even at the earliest stage of exploration, gives a handy mechanism to correspondence and, in addition, premise for drill targeting. The way to the convenience of the model lies in its adaptability as to type of information and the straightforwardness with which it can be edited. The first one certifies that the model can give a far reaching visual exploration data archive in a reliable coordinate framework, while the second ensures the continuous helpfulness of the model as a dynamic and economically profitable exploration resource. The 3D picture permits visualizing, analyzing, and interpreting consistently including extremely complex tectonic settings. Information from various investigation techniques, for example, geophysical studies (seismics, gravimetrics, geoelectrics, well logs, etc.), drilling, or remote sensing can be combined into one information model. It underpins the geologist's assignment to accomplish an exhaustive assessment of the underground by decreasing uncertainties.

Uncertainty and risk are part of the earth model. There is uncertainty in data and interpretation, conceptual geological model, structural model, and petrophysical model. One has to investigate the complete range of uncertainty to understand, which are the relevant uncertainties, to assess the impact of "key uncertainties," and to assess necessary actions to reduce key uncertainties. The capacity to model and cross-examine geophysical, geological, geochemical, and geotechnical information decidedly affects targeting capabilities and diminishes geological uncertainty. Positive execution requires the capacities, among others, of creating 3D models of mineral situations, demonstrating their expected geophysical response, and upgrading existing models to authorize consistency with geophysical information. Such common earth models, predictable with all lines of proof, should likewise be liable to 3D spatial query or expert-system–assisted examination to recognize, rank, and prioritize targets. 3D earth models facilitate to make effective decisions by reducing the risk and uncertainties in exploration, resource assessment, and production. Common earth model is the foundation for decision support, and 3D-GIS capabilities are for investigation and understanding.

References

Ayachi, A.K., 2003. Advance Mineral Resources Information System (M.R.I.S v 2.0) Using Remote Sensing and GIS Inputs. Map Asia (2003).

Bonham-Carter, G.F., 1994. Geographic Information Systems for Geoscientists: Modelling with GIS. Pergamon, Oxford, p. 398.

Bonham-Carter, G.F., 1997. GIS methods for integrating exploration data sets. In: Gubins, A.G., Ed., Proceedings of Exploration 97: Fourth Decennial Conference on Mineral Exploration, pp. 59–64.

Coulson, R.N., Lovelady, C.N., Flamm, R.O., Spradling, S.L., Saunders, M.C., 1991. Intelligent geographic information systems for natural resource management. In: Turner, M.G., Gardner, R.H. (Eds.), Quantitative Methods in Landscape Ecology. Springer Verlag, New York, USA, pp. 153–172.

de Kemp, E.A., 2006. 3-D interpretive mapping: an extension of GIS technologies for the geoscientist. In: Harris, J.R. (Ed.), GIS for the Earth Sciences, vol. 44. Geological Association of Canada Special Publication, pp. 591–612.

de Kemp, E.A., 2007. 3-D geological modelling supporting mineral exploration. In: Goodfellow, W.D. (Ed.), Mineral Deposits of Canada—A Synthesis of Major Deposit Types, District Metallogeny, The Evolution of Geological Provinces and Exploration Methods, vol. 5. Geological Association of Canada, Mineral Deposits Division, Special Publication, pp. 1051–1061.

Dobson, J.E., Durfee, R.C., 2012. ORNL and the geographic information systems revolution. http://www.ornl.gov.

Escobar, F., Hunter, G., Bishop, I., Zerger, A., 2010. Introduction to GIS. http://www.sli.unimelb.edu.au.

ESRI, 2006. What is GIS? The Guide to Geographic Information Systems. http://www.gis.com.

Garrett, S., Griesbach, S., Johnson, D., Jones, D., Lo, M., Orr, W., Sword, C., 1997. Earth model synthesis. First Break 15 (1), 13–20.

Geosoft, 2010. FRESH PERSPECTIVE: Technology Trends and Solutions for Mineral Exploration. Ebook, 44.

Heywood, I., Cornelius, S., Carver, S., 2006. An Introduction to Geographic Information Systems. 3rd Edition, Pearson-Prentice Hall, Harlow, England, p. 426.

Kumar, L., Viswakarma, J.K., Singh, V., 2007. Use of GIS in Underground Mining for Optimization. Abstract, www.httplinks.co/site/beta.gisdevelopment.net. Map World Forum (2007).

Lelièvre, P.G., Carter-McAuslan, A., Tycholiz, C. Farquharson, C.G., Hurich, C.A., 2012. Refining 3D earth models by unifying geological and geophysical information on unstructured meshes. Geophysical Research Abstracts v.14, EGU2012-1306, 2012.

Lunetta, R.S., Congalton, R.G., Fenstermaker, L.K., Jensen, J.R., McGuirre, K.C., Tinney, L.R., 1991. Remote sensing and geographic information system data integration: error sources and research issues. Photogrammetric Engineering and Remote Sensing 57 (6), 677–687.

McGaughey, J., 2006. The Common Earth Model: A Revolution in Mineral Exploration Data Integration. In: Haris, J.R. (Ed.), GIS for the Earth Sciences, Vol. No. 44. Geological Association of Canada Special Publication, pp. 567–576.

McGaughey, J., Perron, G., Parsons, S., Chalke, T., 2009. GOCAD workflows for expert-system decision making. Proceedings of the 26th GOCAD Research Conference, Nancy, France.

Moore, P., 2009. GIS—a Mine of Information. Mining Magazine, Jan–Feb 2009, pp. 42–44.

Nykänen, V., 2011. Prospectivity mapping in GIS: integrate geochemistry data with geophysics and geology Workshop 25th International Appl. Geochem. Symp. Aug. 2011, Rovaniemi, Finland. 82.

Rahimi, M., 2002. GIS in Exploration with Special Viewpoint to Mineral Potential and Defining the Exploration Priority Maps. Map India (2002).

Tew, B.H., Osborne, W.E., 1997. Digital Geologic Mapping Program at the Geological Survey of Alabama, U.S. G. S. Open-File Report 97-269 (1997).

CHAPTER

11

Conventional and Statistical Resource/Reserve Estimation

OUTLINE

Essentials of Mineral Exploration and Evaluation

DOI: http://dx.doi.org/10.1016/B978-0-12-805329-4.00018-1

11.1 INTRODUCTION

A mineral resource can be estimated principally on the basis of geoscientific information with some input from other disciplines. A mineral reserve, which is a modified subset of a measured or indicated mineral resource, needs consideration of all factors affecting extraction, including mining, metallurgical, economic, marketing, legal, environmental, social, and governmental factors, and should in most instances be estimated with input from a range of disciplines.

The purpose of mineral resource/reserve estimation is to first assist in determining if a property is worth mining, and, if so, to guide its later development. The ore deposit models are the underlying foundation for numerous consequent economic decisions and the correctness of those decisions will be directly dependent upon the accuracy of resource/reserve estimation. The ability to provide desirable resource/reserve estimates rests on the development of geological, geometrical, and mathematical–statistical techniques of mineral deposit model construction. Resource/reserve estimates remain an art, requiring practice and judgment in its application. Even with the power of techniques (both present and future), this will remain so, due to the inherent geological complexity of mineral deposits. From the very limited data, well-conceived geological inferences must be drawn, subject to frequent review as new information becomes available. Any resource/reserve estimation must begin with the collection and treatment of samples (drill hole and geological) drawn from and defining the ore body. At the preliminary resource/reserve estimation stage, the sample may represent as little as 10^{-6} or 10^{-9} of the bulk of the mineral deposit. Considering the cost of collection and importance of the sample data, on which all further computations rest, accurate information about each sample must be obtained.

How long a mine can continue to operate depends on how much ore is there in the ground. The calculable tonnage, including some that is believed to exist, even though not conclusively proved, is known as "reserves." The amount of ore that will be available for future mining cannot be always calculated accurately because no one can be sure of how much ore is there in a mineral deposit/mine, until it has been mined out. Hence, reserve figures are "estimates" rather than "certainties."

A reliable estimate of the resource/reserve depends on an adequate number of reasonably spaced drill holes, representative and unbiased sampling, reliable assays, reasonable and coherent geological interpretation of the deposit geometry and continuity, and appropriate resource estimation methods. The reliability of the estimate increases with the level of geological knowledge of the deposit, and depends on the amount of data available and the characteristics of the deposit. The number of drill holes and their spacing depend on the type of mineralization, deposit geometry, geological complexity and continuity, assay variability and sensitivity to nugget effect, reliability of sampling and mining method considered. The resource/reserve estimation must incorporate all aspects of the geological framework and should always be based on the geologist's best judgment of reality.

11.2 CONVENTIONAL RESOURCE/RESERVE ESTIMATION

Calculation of grade and tonnage in a mineral deposit by the conventional methods is usually made by the analysis of sample data found in a polygonal, triangular, cross-sectional, random stratified grid (RSG), contour, and longitudinal vertical sectional (LVS) pattern (Fig. 11.1). The choice of method depends upon the shape, dimensions, and complexity of the mineral deposit and the type, dimensions, and pattern of spacing of the sample information. The conventional methods used for calculating reserves assume that the area of influence of an assay in any one drill hole is a function of the distance to any other adjacent drill hole (Popff, 1966; Roscoe, 1997). Geometric areas are assigned grades for reserves based on averages of assays from adjacent drill holes. In using these methods, no consideration is given to the mineralization which actually exists between the drill holes. These methods are a function of geometry which simplifies the calculation (Sarkar, 1988).

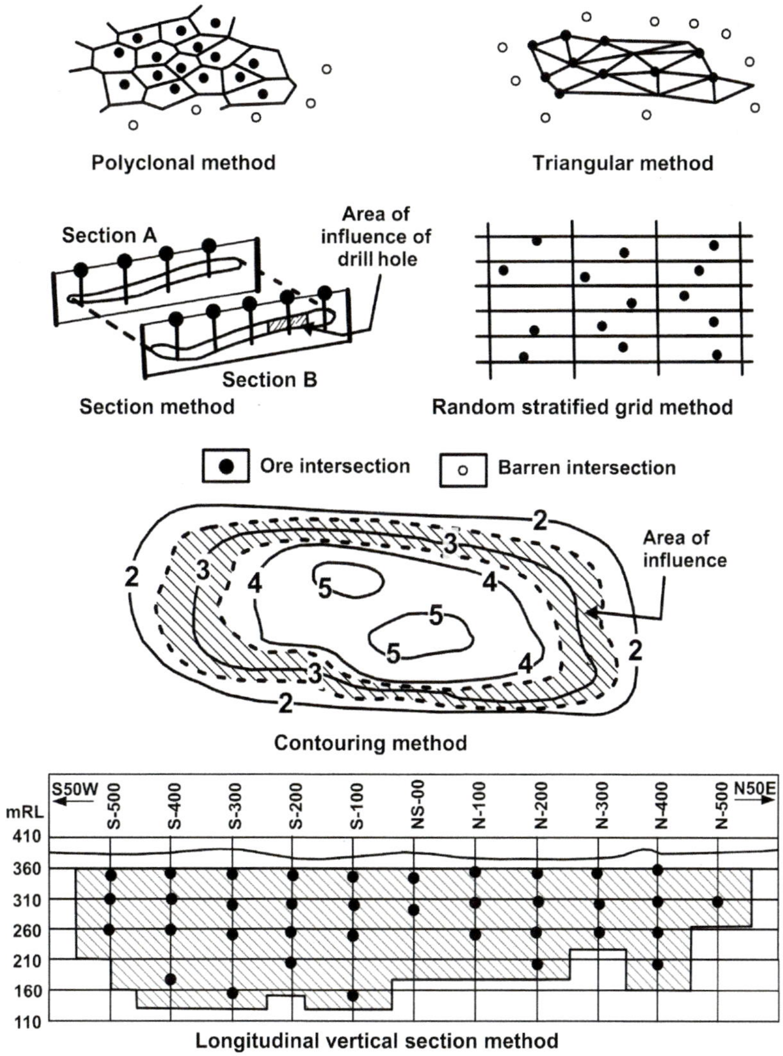

FIGURE 11.1 Conventional methods of resource/reserve estimation.

11.2.1 Polygonal Method

Polygons may be constructed on plans, cross sections, or longitudinal sections by drawing perpendicular bisectors of lines connecting sample points. The polygons are then planimetered to define the area of mineral body. The thickness of above cutoff grade mineralization is applied to the entire polygon to establish the volume estimate. The individual volumes are then summed and converted to tonnage on the application of an appropriate tonnage factor. The average grade of mineralization encountered by the sample point lying within the polygon is considered to accurately represent the grade of the entire volume of material within the polygon. The global grade estimate is determined by summing the sample grades weighted by the area of the relevant polygons. The major drawbacks of this method are: (1) in an irregularly sampled area, the zone of influence of a drill hole will vary inversely with the density of sampling and not according to the mineralization and (2) the problem of closure at the boundary of a deposit depends on the interpretation and the experience of the interpreter.

11.2.2 Triangular Method

In this method, a series of triangles are constructed with the drill holes at the apices. The areas of the triangles are calculated by geometry or by coordinates. The thicknesses above cutoff grade and average grade are calculated for each drill hole. The average grade of each triangle is estimated from the mean of the three corner samples as a thickness-weighted mean or by weighting as a function of their distance from the center of the triangle. The problems with this method are: (1) it is possible to construct the triangles in a number of ways on the samples depending on the interpretation and (2) similar to that in the polygonal method, problem of closure at the boundary of a deposit exists here too.

11.2.3 Cross-Sectional Method

The method estimates a block of ground that is bounded between regularly spaced sections. The mineralization outline of each bounding section is divided into areas of influence based on the drill hole or other sample data. These individual areas assume the grade of the composite drill intersection. These areas are then planimetered or calculated geometrically. The individual areas are totaled for each section and the volume is calculated based on the spacing between the sections. While volume-to-tonnage conversion is carried out by the application of the appropriate tonnage factor, the average grade is calculated by weighting the area of the respective sections. Merits of this method are that they portray graphically the geology of a deposit, while their general procedure is simple and rapid. On the other hand, poor selection of section lines may necessitate interpolation over unmerited distances and where drilling is irregular, nonparallel sections must be constructed causing an increasing in complexity of calculations. The method is best suited to assess well-defined and large deposits that are uniform in thickness and grade, that is, sedimentary-bedded deposits.

11.2.4 RSG Method

A regular grid of a suitable size and an orientation is adjusted on a set of sample values distributed in space by trial and error until, as far as possible, at least one sample value falls per grid panel. The grid assumes the value of one sample or the mean of the sample values, in case of more than one sample. An estimated global mean grade of the deposit is the thickness-weighted average of the individual grid grades.

11.2.5 Contour Method

Isolines are constructed by interpolation between points of known values that assume a gradual uninterrupted change from one point to another. When constructing thickness contours (ie, isopach), the mineral body, which is confined in nature by irregular surfaces, is transformed by computations into an equivalent body limited by a planer base and a complex surface above, which then becomes suitable for volume calculations. A second map is produced for accumulation contours. The areas lying between each successive pair of contours are planimetered on both the maps, each of these areas is then multiplied by the average value of its confining contours. While the global tonnage is computed by summing the volumes derived from thickness contours greater than a specified minimum mining height, the average grade is equal to the sum of the accumulation volumes above a certain minimum mining height times the cutoff grade divided by the total volume. The isopach map gives an idealized likeliness of the deposit next to a three--dimensional model. The use of this method is limited to deposits showing gradual change and assumes that the values can be contoured.

11.2.6 LVS Method

The limits of mineralization (ore body) along with surface profile, drill hole intersection data (projections at different levels), exploratory UG working, if any, correlatable details and other data as they occur within the plan of mineralized body are plotted on LVS. The mineralized envelope on LVS is divided into subblocks around the positive intersection taking into consideration of different influence factors. The subblocks/blocks are measured with a planimeter and their average is multiplied by the height to give block volumes. By applying appropriate tonnage factor to the volume, tonnage estimate is obtained. To obtain grade of each block, assays of all drill hole intersections within the block outline are averaged in proportion to their widths. Block tonnages are totaled; the overall grade is obtained as a weighted average of block grades (Gandhi, 2003).

11.3 DRAWBACKS OF CONVENTIONAL RESOURCE/RESERVE ESTIMATION

Large errors in estimation of thickness can be made when assuming that the thickness/grade of a block is equal to the thickness of grade of a single point about which the block has been drawn. If the area and shape of ore body on adjacent sections vary considerably, significant errors in grade and tonnage could be introduced. These methods do not take into account any correlation of mineralization between sample points nor quantify any error of estimation.

When assay values are interdependent at intervals less than the drill hole spacing, the best estimate of grade of ore for the area between drill holes at a distance beyond the area of influence of the assay values is average assay values surrounding the area, unless geologic considerations decide otherwise. Computing the average of assays surrounding an area requires some form of weighting/interpolation based on distance (linear, exponential, two-dimensional, and vector) for combining and averaging assays. Sampling and reserve calculations are usually simpler and more reliable for coal and iron ore deposits (with less variation of assay range) than for copper and gold deposits. In the latter, the range of assays may be considerable, demanding statistical evaluation to approximate the reliability of the estimation of grade of ore. The observed factual data, which indicate areas of influence of assays in drill holes more accurately than standard geometric pattern, should improve the accuracy of reserve calculations. Thus, in many cases, geological information should be used to modify the outlines of blocks used in estimating reserves (Hazen, 1968).

Geometric methods for estimating the grade and tonnage of ore, although simple to use, are not realistic models for complex ore bodies. These methods might be good for calculating tonnage rather than the grade of ore. Conventional methods of estimation are deterministic and these methods *do not provide any objective way of measuring the reliability of estimates*.

11.4 STATISTICAL RESOURCE/RESERVE ESTIMATION

Application of statistics to problems in geology, mining as well as to hydrology dates back to a considerable time. The use of statistical techniques provides a better basis for developing models to produce additional information about an ore deposit. Statistical methods assume that samples taken from an unknown population are randomly selected and are independent of each other. In the context of an ore body, this implies that the relative positions of samples are ignored and it is assumed that all sample values in a deposit have an equal probability of being selected (Sinclair and Blackwell, 2002). The likely presence of trends, zones of enrichment, or pay shoots in the mineralization, is ignored. The fact that two samples, taken close to each other are more likely to have similar values than if taken far apart is also not taken into consideration. Detailed exploration campaign begins with surface drilling. The drill holes are widely spaced in the initial stages which provide broad knowledge of a deposit. It is in this early stage of exploration, the quality of the deposit is examined by estimating mean (average) grade, m of the deposit. For this purpose, n samples of same support (size, shape, and orientation) are taken at points X_i. The drill hole sample values are used to estimate m of the population mean, μ and the confidence limits of the mean. The estimator for this purpose would vary according to the probability distribution of sample values (Barnes, 1980). In classical statistical analysis, since it is assumed that all sample values are independent (ie, random), the location X_i of the sample is ignored. The parameters estimated from a classical statistical model refer to variables of mineral deposits.

11.4.1 Statistics and Probability

Statistics is defined as "mathematics applied to observational data" that enables to analyze and interpret such observed information effectively and efficiently. It involves making statements about a larger population on the basis of measurements made on a relatively small sample. It deals with collection, organization, analysis, interpretation of data, and drawing of inferences from the data. The phase of statistics dealing with conditions under which inference drawn is valid is called *inductive statistics or statistical inference*. Because such inference cannot be absolutely certain, probability is often associated in stating such inference. On the other hand, the phase of statistics which seeks only to describe and analyze a given group without drawing any inference about a larger group is called *deductive or descriptive statistics*. There are two branches of statistics, viz., (1) parametric and (2) nonparametric. *Parametric statistics* is the branch of statistics concerned with data measurable on interval or ratio scales so that arithmetic operators are applicable to those data enabling parameters such as mean and variance of the distribution to be defined. *Nonparametric statistics* is the branch of statistics that studies data measurable on a nominal scale or an ordinal scale to which arithmetic operators cannot be applied directly.

Probability is a numerical measure of likelihood of occurrence(s) of random process(es). The theoretical foundation for interpretations and inferences that can be made from statistics is *probability theory* which is the mathematical structure devised for providing models of chance happenings. A variable whose value is determined by a chance experiment and assumes to each of its possible values with a definite probability is called a *random variable*. A random variable which can only assume only integer values is called *discrete random variable* (eg, number of mineral deposits in a mining district), while a random variable whose values may range continuously over an interval is called *continuous random variable* (eg, mineral sample values). An *event* is simply the outcome of a random process or a statistical experiment. The probability of occurrence of a given event, "A" lies between 0 and 1. If it is absolutely certain that event "A" cannot occur, then the probability of occurrence of "A" is 0, that is, $P(A) = 0$. If, on the other hand, it is completely certain that it will occur, then $P(A) = 1$. All other probabilities, however, would have a fractional value between "0 and 1."

11.4.2 Probability Distribution

Possible outcome of a random selection of a sample is expressed by its probability distribution that may or may not be known. In the case of a *discrete distribution*, which can only assume integer values, the distribution would associate with each possible value X, a probability P(X). The individual value of P(X) will be positive and the sum of all possible P(X) will be equal to 1. The function f(x) is a mathematical model that provides the probability that the random variable X would take on any specified value x, that is, $f(x) = P(X = x)$. This function f(x) is called the probability distribution of the random variable X and describes how the probability values are distributed over the possible values, x of a random variable X. In the case of a *continuous distribution*, to each possible value x, a density of probability f(x) is associated so that the probability of a value lying between x and $x + dx$ is $f(x)\,dx$, where dx is infinitesimal. This serves as a mathematical model for describing the uncertainty of an outcome for a continuous variable. The probability of x lying between lower limit (a) and upper limit (b) is expressed as:

$$\text{Prob}(a \leq X \leq b) = \int_a^b f(x)\, dx.$$

The individual probability density value will be positive and the sum of all such values extending from $-\propto$ to $+\propto$ will be 1. The probability of X being smaller than or equal to a given value x is called the cumulative probability distribution function F(x):

$$\text{Prob}(X \leq x) = \int_{-\alpha}^{x} f(x)\, dx = F(x);\; F(-\propto) = 0;\; \text{and } F(+\propto) = 1.$$

The following holds true for the cumulative distribution function, F(x):

(1) $0 \leq F(x) \leq 1$ for all x;
(2) F(x) is nondecreasing.

11.4.3 Frequency Distribution

Frequency distribution of sample data is an estimate of the probability distribution for the population from which the samples are drawn. In other words, sample is a statistical image of a population that enables deductions about the population to be made. A frequency distribution obtained from n samples can be transformed into a probability distribution simply by dividing each frequency by n, the total number of observations. Frequency distributions may either be symmetrical or asymmetrical.

11.5 CHARACTERIZATION OF A DISTRIBUTION

11.5.1 Parameters of Central Tendency

Mean (or *average*) of a series of independent measurements is the sum of the values of all the measurements divided by the total number of such measurements. The computation of mean assumes that all measurements x_i are of the same size of sampling unit (ie, of same support). For ungrouped data, Mean is estimated as:

$$\overline{X} = (1/n)\sum X_i; \quad \text{and for grouped data}, \overline{X} = 1/n\sum f_i x_i,$$

where f_i is the frequency and x_i is the midpoint of class interval.

Median for a series of n independent measurements, X_i arranged in order of magnitude, is the value which divides it into exactly two equal halves. For ungrouped data, it is the middle value in case n is an odd number or the mean of the two middle values in case n is an even number. For grouped data, *median* $= L_1 + [((n/2) - (\sum f)_1)/f_{median}]\, C$, where L_1 is the lower limit of median class, n is the number of items in the data (ie, total frequency), $(\sum f)_1$ is the sum of frequencies of all classes lower than the median class, f_{median} is the frequency of median class, and C is the size/width of median class interval.

Mode is the value that occurs most frequently, that is, the value with the greatest frequency. A distribution having only one mode is called *unimodal*; if having two modes, it is called *bimodal*, in particular and *polymodal*, in general. When measurements are grouped, $mode = L_1 + [(f_1 - f_0)/\{(f_1 - f_0) + (f_1 - f_2)\}]\ C$, where L_1 is the lower limit of modal class, C is the width of modal class, f_1 is the frequency of the corresponding modal class, f_0 is the frequency preceding modal class, and f_2 is the frequency succeeding modal class.

11.5.2 Parameters of Dispersion

It is natural that the sample values are not all located at the central value but are dispersed around it. In some cases, they are closely packed around the central value, while in other cases, they are widely scattered away from it. In order to understand the nature of a distribution, it is thus necessary to know the dispersion characteristics. The spread of values around the mean is measured by estimating the sample *standard deviation*. It is a measure of the square root of the mean squared deviation of the individual value x_i from the mean. For a series of *n* sample values X_i, the standard deviation, S is expressed as:

$$S = \sqrt{\sum (X_i - \overline{X})^2/(n-1)}.$$

With n sample values, there will be *n* squares of deviation from the mean of which only (n − 1) are independent. The unit of expression is same as the sample values. The square of the standard deviation is the parameter called *variance*. If the sample values are expressed in percentage, then variance is expressed in square percentage. *Coefficient of variation (cv) or relative standard deviation* is another useful measure of dispersion used to compare the relative variability of values around mean, among different distributions. It is defined as the quotient (σ/μ). The parameter, being independent of unit measurement, can be used to compare the relative variations of two or more data sets regardless of the units involved. The coefficient of variation is a parameter that in the early stages of a mineral exploration is very suitable for providing a quick indication of the variability of the sample grades and the block grades by comparing the coefficient of variation with known values derived from other deposits of same type. This information from other deposits of same type also helps as *a priori* information in the first order of magnitude estimation of statistical parameters.

11.5.3 Parameter of Symmetry

Skewness (Sk) is a measure of lack of symmetry. It is a shape parameter that characterizes the degree of asymmetry of a distribution. A distribution is said to be positively skewed with degree of skewness greater than 0 ($Sk > 0$, usually observed in low-grade mineral deposits) when the tail of a distribution is toward the high values indicating an excess of low values. Conversely, it is negatively skewed with degree of skewness less than 0 ($Sk < 0$, usually observed in high-grade mineral deposits) when the tail of the distribution is toward the low values indicating an excess of high values. The degree of skewness, Sk is given by:

$$Sk = [1/(n-1)] \sum_{i=1}^{n} (X_i - \overline{X})^3 / S^3.$$

11.5.4 Parameter of Peakedness

Kurtosis (Ku) is a measure of relative peakedness of a distribution. It is a shape parameter that characterizes the degree of peakedness. A distribution is said to be leptokurtic when the degree of peakedness is greater than 3, it is mesokurtic when the degree of peakedness is equal to 3, and it is platykurtic when the degree of peakedness is less than 3. The degree of kurtosis, Ku, is given by:

$$Ku = [1/(n-1)] \sum_{i=1}^{n} (X_i - \overline{X})^4 / S^4.$$

11.6 PROBABILITY MODELS

The theoretical models of probability distributions, which commonly conform to in most mineral deposits or geological situations to represent sample value frequency distribution, are either Normal (Gaussian) or Lognormal. Various other distributions are known but the assumption of either normality or lognormality can be made for most deposits and the use of more complex distributions may not be justified (Rendu, 1981).

11.6.1 The Normal Distribution Theory

A symmetrical bell-shaped probability distribution with asymptotic tails is described as a normal distribution or a Gaussian distribution. It is known as the Gaussian distribution since it was derived by C.F. Gauss in his work on the theory of measurement of errors. The theory of normal distribution is of fundamental importance in the evaluation and treatment of various geological and mineral deposit data. If independent random samples of the same size *n* are repeatedly collected from a population, which can have any distribution, meaning different sample series but with the same sample size *n*, then the distribution of the mean of the samples for a sample size *n* is approximately normal. A normal curve occurs in sample distribution that is subject to chances in which the outcome depends on a large number of causes, each of which has a fifty-fifty chance. The total area under the standard normal distribution curves from $-\propto$ to $+\propto$ is 1 (Fig. 11.2). The size of the area under the standard normal distribution curve between defined limits is related to the probability density with which the value, Z_i, is located between the defined limits. About 68.27% of the total area under the normal curve lie between $-\sigma$ and $+\sigma$ limits; 95.45% lie between $+2\sigma$ and -2σ limits; and 99.73% lie between -3σ and $+3\sigma$ limits (Davis, 2002).

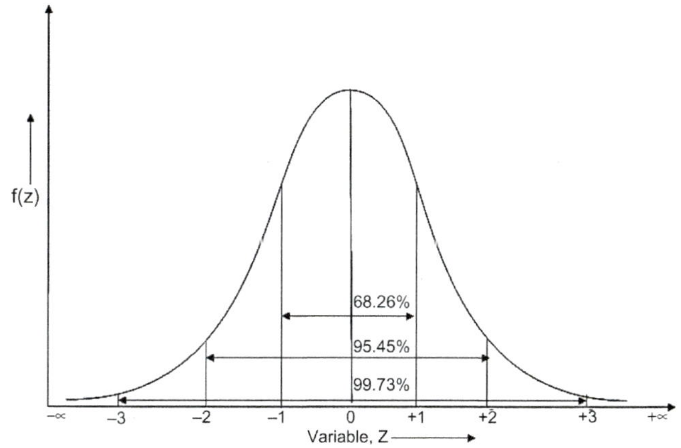

FIGURE 11.2 Standard normal distribution.

The probability density function (pdf) of a normal distribution given by f(X) is expressed as:

$$\text{pdf, } f(X) = [1/(\sigma\sqrt{2\pi})]\exp[-(1/2)((X-\mu)/\sigma)^2] \quad \text{for } -\propto \leq X \leq \propto,$$

where μ is the population mean which is estimated by the sample mean, $\overline{X}$, and σ is the population standard deviation which is estimated by the sample standard deviation, S. The distribution can be standardized by expressing $[(X-\mu)/\sigma]$ equal to Z:

$$f(Z) = [1/\sqrt{2\pi}]\exp[-{}^{1}\!/\!_{2}Z^2].$$

This standard normal distribution has a zero mean and unit standard deviation, that is, N(0,1). The cumulative probability density function (cdf), F(X) of a normal distribution has the expression:

$$\text{cdf, } F(X) = [\,1\,/(\sigma\sqrt{2\pi})\,]\int_{-\alpha}^{X_i}\exp[-(1/2)((X-\mu)/\sigma)^2]dx,$$

which can be standardized to:

$$\text{cdf, } F(Z) = [1\,/\sqrt{2\pi}]\int_{-\alpha}^{Zi}\exp[-(1/2)(Z)^2]\,dz.$$

The normal probability distribution function does not have a simple integral and therefore the areas under a normal distribution curve have been tabulated extensively. These areas provide the probabilities of certain interval values. Because a normal distribution is completely characterized by its mean and standard deviation, it is possible to tabulate its areas using a standardized normal distribution and to calculate probabilities for a normally distributed random variable.

11.6.1.1 Fitting a Normal Distribution to Sample Distribution

Suppose one has n sample values, X_i, i = 1,2,....., n. The first step in the analysis of these values consists of grouping them in classes and counting the number of samples which

TABLE 11.1 Histogram Table

%Zn (in Class)	Frequency (f)	Percentage Frequency (%f)	Cumulative Frequency (cf)	Percentage Cumulative Frequency (%cf)
0–1	4	$(4/72) \times 100 = 5.56$	4	5.56
1–2	7	$(7/72) \times 100 = 9.72$	11	15.28
2–3	15	$(15/72) \times 100 = 20.83$	26	36.11
3–4	21	$(21/72) \times 100 = 29.17$	47	65.28
4–5	15	$(15/72) \times 100 = 20.83$	62	86.11
5–6	8	$(8/72) \times 100 = 11.11$	70	97.22
6–7	1	$(1/72) \times 100 = 1.39$	71	98.61
7–8	1	$(1/72) \times 100 = 1.39$	72	100.00

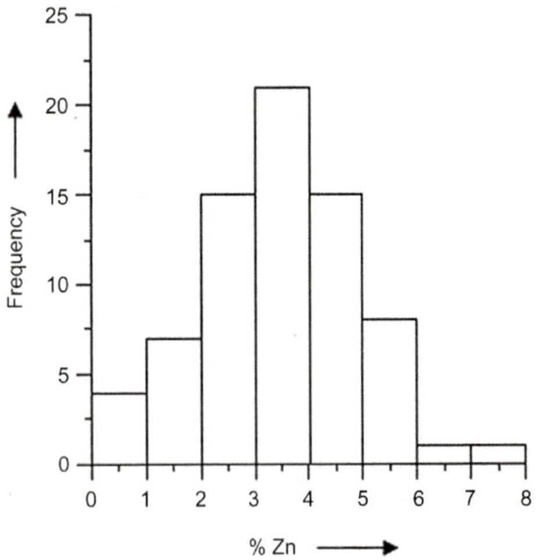

FIGURE 11.3 Histogram plot of zinc values.

fall within each class. The number of samples per class is the frequency for that class. This frequency when divided by the total number of samples gives the percentage or relative frequency (Table 11.1). The results of such an analysis enable to construct a histogram (Fig. 11.3). The histogram is the first-hand tool in determining whether or not the sample distribution is reasonably symmetrical, and in detecting visually possible outliers, if any. The shape of a histogram is affected by the class interval chosen to group the sample values and the starting value of the class interval. To check the assumption of normality, or in other words, to fit a normal distribution to an experimental histogram, a convenient graphical method, known as the probability paper method can be used. Cumulative frequency distribution of the values is calculated and plotted in an arithmetic probability paper against the upper limits of the class values. From the definition of arithmetic probability scale, the cumulative distribution of a normally distributed variable will plot as straight line on arithmetic probability graph (David, 1977). If the graphical plot points obtained by this approach can be considered or closely approximated as distributed along a straight line, the assumption of normality can be accepted, and the theory of normal distribution to estimate the mean, variance, and confidence limits of mean can then be applied.

11.6.1.2 Numerical Estimation of Mean, Variance, and Confidence Limits of Mean

The mean and variance for a normal distribution are estimated as follows:

$$\text{Mean, } \overline{X} = [1/n] \sum_{i=1}^{n} X_i,$$

$$\text{Variance, } S^2 = [1/(n-1)] \sum_{i=1}^{n} (X_i - \overline{X})^2,$$

where $S = \sqrt{S^2}$ which is an estimate of the population standard deviation. The mean value, "m," of the mineral deposit is estimated by: $m = \overline{X}$; with variance of error, $v = S^2/n$.

Three confidence terms associated with the estimate of mean are confidence level, confidence interval, and confidence limits. The confidence level is the desired level of probability assigned to the confidence estimates about the mean. The confidence interval is the range associated with the mean estimate of a normal population at a specified confidence level. The confidence limits are the two value bounds, viz., lower and upper about the mean estimate of a normal population. If m_p be confidence limits of the true mean "m" such that the probability of "m" being less than m_p is p, then m_{1-p} is the confidence limit such that the probability that "m" is larger than m_{1-p} is $1-p$. The probability that "m" falls between m_p and m_{1-p} is 1_2p confidence limits of the mean. The following equations can be used to calculate m_p and m_{1-p} for the mean value, "m," of a mineral deposit (Rendu, 1981):

$$\text{Lower limit, } m_p = m - t_{1-p}(S/\sqrt{n}); \text{ and Upper limit, } m_{1-p} = m + t_{1-p}(S/\sqrt{n}),$$

where t_{1-p} is the value of student's t-variate for $f = n-1$ degrees of freedom, such that the probability that "t" is smaller than "t_{1-p}" is $1-p$. The t-statistics value can be obtained from statistical table of student's t distribution by selecting the desired probability level of confidence across the top and the degrees of freedom on the left-hand column.

11.6.1.3 Graphical Estimation for Normal Distribution

Graphical estimation of mean and standard deviation can be made from arithmetic probability plot of the cumulative frequency distribution of sample values, provided the number of samples is large enough. Value corresponding to the 50% cumulative frequency provides an estimate of the mean, and difference in values corresponding to 50% and 84% cumulative frequencies or in values corresponding to 50% and 16% cumulative frequencies provides an estimate of the standard deviation, that is

$$X_{84\%} - X_{50\%} = +S,$$

or,

$$X_{50\%} - X_{16\%} = -S.$$

Alternatively,

$$S = \frac{1}{2}[(X_{84\%} - X_{50\%}) + (X_{50\%} - X_{16\%})].$$

11.6.1.4 Measures of Degree of Skewness and Kurtosis

Besides graphical methods, other methods to test the fit of a normal distribution include (1) measures of degree of skewness and kurtosis, and (2) χ^2 (Chi-squared) goodness-of-fit test. *For a normal variate, the degree of skewness is zero and that of kurtosis is 3, and the calculated value of χ^2 must be less than or equal to the table value of χ^2 at "α" level of significance and "ν" degrees of freedom.* Degrees of skewness and kurtosis for a normal distribution are given by the expressions:

$$\text{Skewness, Sk} = [1/(n-1)]\sum_{i=1}^{n}(X_i - \overline{X})^3/S^3,$$

$$\text{Kurtosis, Ku} = [1/(n-1)]\sum_{i=1}^{n}(X_i - \overline{X})^4/S^4.$$

11.6.1.5 Chi-Squared (χ^2) Goodness-of-Fit Test

Once the optimum solution for "m" has been determined, it is desirable to check for the goodness of fit of a normal distribution to the sample distribution. Chi-squared is a distribution to measure the goodness of fit. It is a means of testing the agreement between the observation and hypothesis. It tests essentially whether the observed frequencies in a distribution differ significantly from the frequencies that might be expected according to an assumed hypothesis. Chi-squared test provides a robust technique for the fit. The test statistics is given by:

$$\chi^2_{\text{calculated}} = \sum_{i=1}^{n}(O_i - E_i)^2/E_i,$$

where O_i is the observed frequency in group, i and E_i the expected frequency in group, i.

In a χ^2 test, three degrees of freedom are lost because the total, the mean, and the standard deviation of expected frequencies are made to agree with that of the observed frequencies.

Conditions of a χ^2 test include (1) all the individuals in the samples should be independent; and (2) the differences between small observed and expected frequencies at the ends of a distribution have a great effect upon χ^2-calculated value. As suggested by a noted statistician Fisher, no group should contain fewer than five expected frequencies. Groups, containing less than five expected frequencies, may be clubbed with the one following it (at the lower end of the distribution) and/or clubbed with the one preceding it (at the upper end of the distribution).

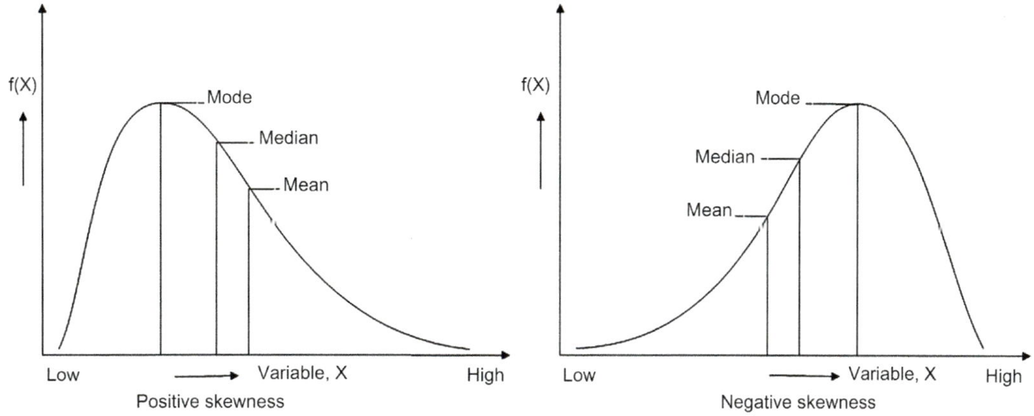

FIGURE 11.4 Skewed distributions.

11.6.2 The Lognormal Distribution Theory

In many mineral deposits as well as geochemical exploration programs, especially in the case of low-grade deposits (eg, gold, molybdenum, etc.) or high-grade deposits (eg, iron ore, manganese, etc.), where the distribution of the sample values is asymmetrical, either positively or negatively skewed (Fig. 11.4), it has been observed that this skewed distribution can be represented either by a two-parameter or a three-parameter lognormal distribution. If $\log_e(X_i)$ has a normal distribution, it is called a two-parameter lognormal distribution, and if $\log_e(X_i + C)$ has a normal distribution, it is called a three-parameter lognormal distribution (where C is the additive constant).

The value of the additive constant, C, is usually positive for a positively skewed distribution, that is, a distribution showing an excess of low values with tail toward high values; usually negative for a negatively skewed distribution, that is, a distribution showing an excess of high values with tail toward low values; and equal to zero for a two-parameter lognormal variate. In other words, C is positive for low-grade mineral deposits and negative for high-grade mineral deposits with exceptions to marginally skewed distributions. The pdf of a lognormal distribution is given by the expression:

$$f(X) = \left[1/(x\beta\sqrt{2\pi})\right] \exp\left[-(1/2)\left(\frac{(\ln x-\alpha)}{\beta}\right)^2\right];$$

where α is the log mean and β^2 is the log variance.

The probability distribution of a three-parameter lognormal variate, X_i, is defined by (1) the additive constant, C; (2) the log mean of $(X_i + C)$; and (3) the log variance of $(X_i + C)$.

11.6.2.1 Fitting a Lognormal Distribution to a Skewed Sample Distribution

For n samples with values X_i $(i = 1,2,\ldots\ldots, n)$, the cumulative frequency distribution of a two-parameter lognormal variate will approximate to a straight line on a logarithmic probability scale. If the variate is three-parameter lognormal, the cumulative curve will show either an excess of low values for positively skewed distribution and or an excess of high values for negatively skewed distribution. In such cases, plot of $(X_i + C)$ will approximate to a straight line on the logarithmic probability scale conforming to a lognormal distribution (Sarkar, 1988).

11.6.2.2 Estimation of Additive Constant (C)

If a large number of samples are available, the cumulative distribution may be plotted on a log-probability paper. Different values of "C" can then be tried until the plot of $(X_i + C)$ is reasonably assumed to be a straight line. Alternatively, the value of "C" can be estimated using the following approximation:

$$C = \frac{M_e^2 - F_1F_2}{F_1 + F_2 - 2M_e},$$

where M_e is the sample value corresponding to 50% cumulative frequency (ie, the median of the observed distribution) and F_1 and F_2 are sample values corresponding to "p" and "100 − p" percentage cumulative frequencies, respectively. In theory, any value of "p" can be used, but a value between 5% and 20% gives best results.

11.6.2.3 Proof of the Equation

If $\log_e (X_i + C)$ is normally distributed, then because of the symmetry of the normal distribution about the mean, it can be expressed as:

$$\log_e(F_1 + C) + \log_e(F_2 + C) = 2 \log_e(M_e + C),$$

or,

$$(F_1 + C)(F_2 + C) = (M_e + C)^2,$$

or,

$$C = \frac{M_e^2 - F_1F_2}{F_1 + F_2 - 2M_e}.$$

11.7 GRAPHICAL ESTIMATION OF LOGARITHMIC MEAN AND LOGARITHMIC VARIANCE

log mean $(\alpha) = \log_e(X_{50\%})$, that is, $\log_e$ value corresponding to 50% cumulative frequency for the straight line plot on a log-probability scale;

log standard deviation (β) = difference in the $\log_e$ values corresponding to 84% and 50% cumulative frequencies or 50% and 16% cumulative frequencies for straight line plot on a log-probability scale, that is:

$$\log_e(X_{84\%}) - \log_e(X_{50\%}) = \beta; \text{ or } \log_e(X_{50\%}) - \log_e(X_{16\%}) = \beta.$$

Alternatively,

$$\beta = \frac{1}{2}\left[(\log_e(X_{84\%}) - \log_e(X_{50\%})) + (\log_e(X_{50\%}) - \log_e(X_{16\%}))\right].$$

11.8 NUMERICAL ESTIMATION OF LOGARITHMIC MEAN AND LOGARITHMIC VARIANCE

Let, $Y_i = \log_e(X_i + C)$

$\log_e$ mean, α or $\overline{Y} = [1/n]\sum_{i=1}^{n} Y_i$;

$\log_e$ variance, β^2 or $v(y) = [1/(n-1)]\sum_{i=1}^{n}(Y_i - \overline{Y})^2$.

11.9 ESTIMATION OF AVERAGE OF A MINERAL DEPOSIT

$$m^* = e^{\overline{y}+(v(y)/2)} = e^{\alpha+(\beta^2)/2} = e^{\alpha}.e^{\beta^2/2}$$

$e^{\beta^2/2}$ value can also be approximated by using $\gamma_n(v)$ factor read from statistical table (where n is the number of samples, and v is the log variance), that is

$m^* = e^{\alpha}\gamma_n(v)$.

Average value, $m = (m^* - C)$; variance, $S^2 = m^2[\exp(v) - 1]$.

11.10 ESTIMATION OF CENTRAL 90% CONFIDENCE LIMITS OF MEAN OF A LOGNORMAL POPULATION

The lower and upper limits for the estimation of Central 90% confidence interval of the mean of a lognormal population can be obtained using factors $\psi_{0.05(v,n)}$ and $\psi_{0.95(v,n)}$ from Sichel's table (David, 1977; Rendu, 1981):

$$\text{Lower limit} = (\psi_{0.05(v,n)}m^*) - C; \text{ and}$$

$$\text{Upper limit} = (\psi_{0.95(v,n)}m^*) - C.$$

11.11 NUMBER OF SAMPLES

Number of samples required in sampling a mineral deposit is decided by the required level of precision in the estimate of the mean value. Precision estimate is expressed as:

$$\text{C.I.} = \pm t_{(\alpha,\nu)}\frac{S}{\sqrt{n}},$$

where C.I. is the confidence interval at a desired confidence level, "S" is the sample standard deviation, and t is the student t value as a function of degrees of freedom (ν) and desired level of significance (α). Assuming that standard deviation estimator S remains same, one can estimate a sample size n that would provide the required precision:

$$\text{C.I.} = \pm t_{(\alpha,\nu)} \frac{S}{\sqrt{n}} \text{ or, } \sqrt{n} = \frac{t_{(\alpha,\nu)}S}{\text{C.I.}} \text{ or, } n = \left(\frac{t_{(\alpha,\nu)}S}{\text{C.I.}}\right)^2.$$

The use of statistical techniques provides an improved basis for developing models to produce additional information about a mineral deposit. However, an important concept of mineral deposit statistics is that all symmetrical distributions either in its raw or log-transformed form are not necessarily a normal distribution, but a normal distribution either in its raw or log-transformed form is essentially a symmetrical distribution (Sarkar, 2014). The inadequacy in using statistical techniques is not in the techniques, but in the lack of pertinent data measurements during exploration. Classical statistical methods are generally restricted to a global estimation of volume/grade within the mineralization envelope.

11.12 DEMERITS OF STATISTICAL RESOURCE/RESERVE ESTIMATION

1. Although it estimates the mean and produces *an error of estimation stated by confidence limits but ignores the spatial relations within a set of sample values,* in other words, the spatial position of sample values is ignored. If the sample interchanges their position, there is no effect on the estimates—a big drawback.
2. Thereby, the methods are unable to define which sample lies nearer or distance away with respect to another sample.

References

Barnes, M.P., 1980. Computer-Assisted Mineral Appraisal and Feasibility. Society of Mining Engineers, New York, NY, p. 167.

Davis, J.C., 2002. Statistics and Data Analysis in Geology, Third Edition, John Wiley & Sons, New York, p. 656.

David, M., 1977. Geostatistical Ore Reserve Estimation. Elsevier Scientific Publishing Co., Amsterdam, p. 364.

Gandhi, S.M., 2003. Rampura-agucha zinc-lead deposit. Geol. Soc. India., Mem. 55, 154 p.

Hazen, S.W., 1968. Ore Reserve Calculations in "Ore Reserve Estimation and Grade Control", CIM Special, vol. 9, pp. 11–32.

Popff, C., 1966. Computing Reserves of Mineral Deposits—Principles and Conventional Methods, USBM Information Circular No. 8283.

Rendu, J.-M., 1981. An introduction to geostatistical methods of mineral evaluation. SAIMM Monograph, Johannesburg, 84 p.

Roscoe, W.E., 1997. Delineation of resources after the discovery. In: Proc. of Short Course. Prosp. Dev. Assn. of Canada.

Sarkar, B.C., 1988. An Integrated System for Geology-Controlled Geostatistical Evaluation (Ph.D. thesis). Imperial College of Science and Technology, London, 218 pp.

Sarkar, B.C., 2014. Geostatistics for Natural Resources Modelling, Edited Book Volume. Indian School of Mines, Dhanbad, p. 178. , ISBN 978-93-5156-635-9.

Sinclair, A.J., Blackwell, G.H., 2002. Applied Mineral Inventory Estimation. Cambridge University Press, UK, p. 381.

CHAPTER

12

Geostatistical Resource/Reserve Estimation

OUTLINE

Essentials of Mineral Exploration and Evaluation
DOI: http://dx.doi.org/10.1016/B978-0-12-805329-4.00019-3

12.1 BACKGROUND

Classical statistics, based on the theory of random variable, does not take into account of the position of samples and thus not able to define the spatial relationship(s) among the sample values. The fact that two samples if taken at smaller spacing are likely to have similar values than at wider spacing is not considered in classical statistics. Geostatistics, on the other hand taking into consideration of the spatial positions of samples, emerged as a geomathematical subject in France in the early 1960s from the work of Prof. Georges Matheron and was based on the empirical studies of Prof. D.G. Krige dealing with the estimation of block values in gold and diamond mines in South Africa (Krige, 1962). The techniques are not merely an amalgamation of the geological sciences with probability theory and classical statistics, but rather an entirely new methodology based on "Regionalized variable" theory. The fundamental tool of geostatistical analysis, which permits quantification of geological parameters is called the semivariogram (Barnes, 1980). Most regionalized variables (in resource/reserve estimation) display a random aspect consisting of highly irregular and unpredictable variations, plus a structured aspect reflecting the spatial characteristics of the regionalized phenomenon.

12.2 GEOSTATISTICS

Geostatistics recognizes that the samples in a mineral deposit are spatially correlated with one another, and that the samples at closer spacing are probably not independent. The techniques are based on a set of theoretical concepts known as the "Theory of Regionalized Variable" and involve a study of the spatial relationships between sample values, thickness, or any geological phenomena showing intrinsic dispersion (Matheron, 1971). From this theoretical basis, a range of practical methods gradually developed known by the general term "kriging" for estimating unsampled point values or block averages from a finite set of observed values sampled either on a regular or an irregular sampling pattern.

Geostatistical methods utilizing an understanding of the interrelations of sample values within a mineral deposit provide a basis for quantifying the geological concepts of:

- an inherent characteristic of the deposit type;
- a change in the continuity of interdependence of sample values according to the type of mineralization;
- a range of influence of the interdependence of sample values.

12.3 RANDOM FUNCTION

A random function is a probabilistic model of spatial distribution of a variable. Various complex attributes that one studies in geology can be considered as random functions, which is a concept that may be thought of in two different forms:

- First, a set of correlated random variables with one such variable for each sample location. In other words, it has a structured component, consisting of the regionalized variable that exhibits some degree of spatial autocorrelation.
- Second, it may be considered as an independent random variable where values are functions rather than numbers. In other words, it has a local random component, consisting of the random variable, showing little or no spatial autocorrelation.

12.4 REGIONALIZED VARIABLE

A variable can be considered "Regionalized," if it is distributed in space and/or time and exhibits some degree of spatial correlation. Variables such as grade of ore, thickness of formation, and elevation of the surface of the earth, are few examples of regionalized variables. Most regionalized variables in resource/reserve estimation display a random aspect consisting of highly irregular and unpredictable variations, plus a structured aspect reflecting the spatial characteristics of the regionalized phenomenon. The main objectives of the regionalized variable theory are: (1) to express the spatial properties of regionalized variables in adequate form and (2) to solve the problem of estimating regionalized variable from sample data. To achieve these, Matheron introduced a probabilistic interpretation to regionalized variables. In order to make estimation possible, characteristics can be obtained from the available sample data if certain "stationarity" assumptions about the function are made. According to Chiles and Delfiner (1999), "Geostatistics aims at providing quantitative descriptions of natural variables distributed in space or in time and space. Examples of such variables are:

- ore grades in a mineral deposit;
- depth and thickness of a geological layer;
- porosity and permeability in a porous medium;
- density of trees of a certain species in a forest;
- soil properties in a region;
- rainfall over a catchment area;
- pressure, temperature, and wind velocity in the atmosphere;
- concentrations of pollutants in a contaminated site.

These variables exhibit an immense complexity of detail that precludes a description by simplistic models such as constant values within polygons, or even by standard well-behaved mathematical functions".

12.5 WHY GEOSTATISTICS

Among the available methods, *the conventional ones do not provide any objective way of measuring the reliability of estimates*. Classical statistics *produces an error of estimation stated by confidence limits but ignores the spatial relations within a set of sample values*. These limitations point to the need for an estimation technique that is capable of producing estimates with minimum variance. Geostatistics consists essentially of a set of theoretical ideas known as "Theory of Regionalized Variable" and a variety of practical descriptive and estimation techniques derived from them. The key purpose is to reliably estimate the resource/reserve and the distribution of *in situ* and recoverable tonnage and grades throughout a mineral deposit. A meaningful geostatistical reserve study with careful attention to geological controls on mineralization would provide not only an adequate global resource/reserve estimate but also a more reliable block by block mineral inventory with an indication of relative confidence in the block grade estimates. Understanding the geological character of a mineral deposit as thoroughly as possible is an essential base on which to build an estimate of mineral inventory (Sinclair and Blackwell, 2002). Geostatistical methods utilize an understanding of the interrelations of measurement (sample) values and provide a basis for quantifying the geological concepts of (1) an inherent variability characteristic of a deposit; (2) a change in the continuity of interdependence of measurement (sample) values according to the spatial variability; and (3) a range of influence of the interdependence of measurement (sample) values.

12.6 SEMIVARIOGRAM FUNCTION

The underlying assumption of geostatistics is that the values of samples located near or inside a block of ground are most closely related to the value of the block. This assumption holds true if a relation exists among the sample values as a function of distance and orientation. The function that measures the spatial variability among sample values in pairs is known as the semivariogram function, $\gamma(h)$. Comparisons are made between each sample of a data set with the remaining ones in pair (Fig. 12.1) at a constantly increasing distance, known as the lag interval.

Thus a semivariogram function numerically quantifies the spatial correlation of mineralization parameters (eg, grade, thickness, accumulation, bedrock elevation, etc.). If $x(z)$ be the value of a regionalized variable x at a location z in a mineral deposit and $x(z + h)$ be the value of x at "h" distance away from z location, that is,

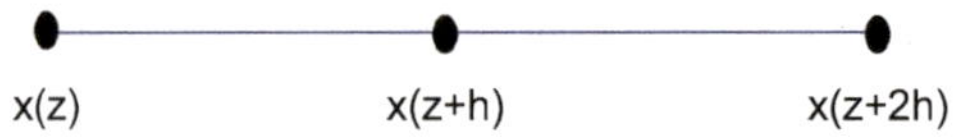

a semivariogram function, $\gamma(h)$ is expressed mathematically as:

$$\gamma(h) = (1/2n(h)) \sum_{i=1}^{n(h)} [x(z_i) - x(z_i + h)]^2,$$

where, n(h) is the number of pairs of samples at h distance apart.

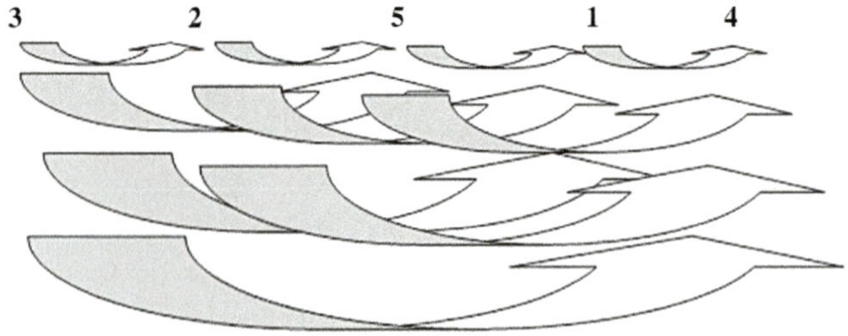

FIGURE 12.1 Comparison of sample value at different lags for semivariogram computation.

Consider following set of values that are spatially distributed in two different arrangements (Arrangement I: most ordered and Arrangement II: most disordered):

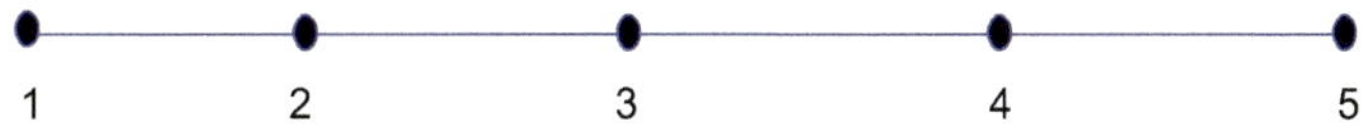

Arrangement I (Most ordered) has mean = 3, $s^2 = 1.58^2$, $s^2{}_{\Delta h}$ (spatial variance) = 0.50;

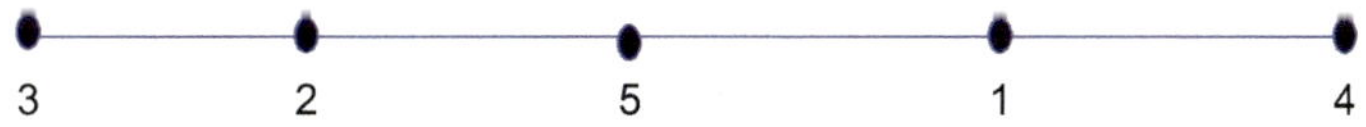

Arrangement II (Most disordered) has mean = 3, $s^2 = 1.58^2$, $s^2{}_{\Delta h}$ (spatial variance) = 4.37.

Spatial variance calculated for various lag distances will change from arrangement to arrangement, which in this example has a range of 0.50–4.37 for a total of 120 possible arrangements. Statistical variance remaining the same for all possible arrangements reflects its inability to measure the variability of a spatially distributed variable.

The function $2\gamma(h)$ is known as variogram function. However, rather than using variogram function $2\gamma(h)$, it is the semivariogram function $\gamma(h)$ that is used because the relation between semivariogram and covariogram is straightforward. According to Rendu (1981), "if the second-order stationarity conditions are satisfied, variogram and covariogram are related as follows:

$$\begin{aligned} 2\gamma(h) &= E\{[(x(z) - x(z+h))^2]\} \\ &= E\{[(x(z) - \mu) + (\mu - x(z+h))]^2\} \\ &= E\{[(x(z) - \mu)^2\} + E\{[x(z+h) - \mu]^2\}\text{-}2E\{[(x(z) - \mu)][x(z+h) - \mu]\} \\ &= 2[\sigma^2 - \sigma(h)]] \end{aligned}$$

Hence the relationship between the semivariogram and the covariogram is given by

$$\gamma(h) = \sigma^2 - \sigma(h)".$$

The relationship can be graphically represented as shown in Fig. 12.2.

An experimental semivariogram permits the interpretation of several characteristics of the mineralization (Fig. 12.3) as follows:

1. *The Continuity* (C): The continuity is reflected by the rate of growth of $\gamma(h)$ for constantly increasing values of "h."

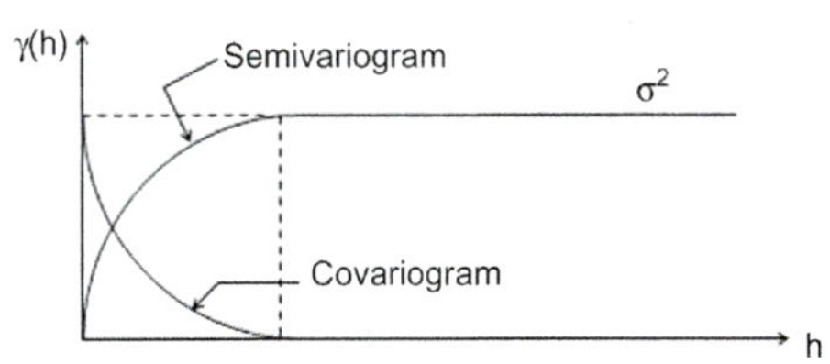

FIGURE 12.2 Relation between semivariogram and covariogram.

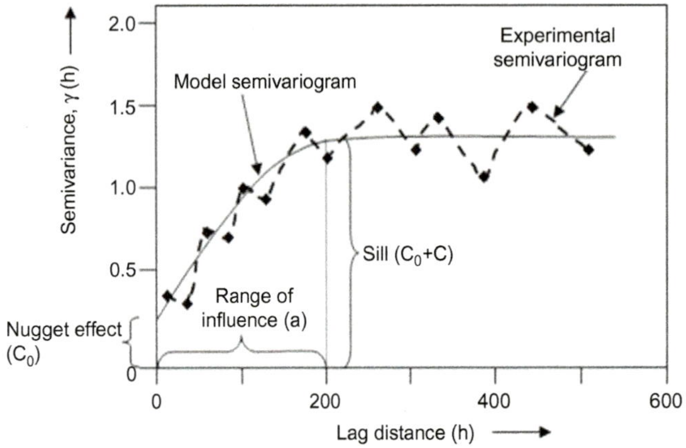

FIGURE 12.3 An experimental semivariogram with fitted spherical model.

2. *The Nugget Effect* (C_o): This is the name given to the semivariogram value, $\gamma(h)$ at $h \rightarrow 0$. It expresses the local homogeneity (or lack thereof) of the deposit. The nugget effect represents an inherent variability of a data set which could be due to both the spatial distribution of the values together with any error encountered in sampling.
3. *The Sill Variance* ($C_o + C$): The value where a semivariogram function $\gamma(h)$ plateaus is called the sill variance. For all practical purposes, the sill variance is equal to the statistical variance of all sample values used to compute an experimental semivariogram.
4. *The Range* (a): The distance at which a semivariogram levels off at its plateau value is called the range (or zone) of influence of semivariogram. This reflects the conventional geological concept of an area of influence. Beyond this distance of separation, values of sample pairs do not correlate with one another and become independent of each other.
5. *The Directional Anisotropy*: This denotes whether or not the mineralization has greater continuity in a particular direction compared to other directions. This characteristic is analyzed by comparing the respective semivariogram ranges computed along different directions. Where the semivariograms in different directions are very similar, it is said to be isotropic.

In practice, since sampling grids are rarely uniform, semivariograms are computed with a tolerance on distance (ie, $h \pm dh$) and a tolerance on direction (ie, $\alpha \pm d\alpha$) to accommodate sample pairs not falling on the grid. The tolerances on distance and direction should be kept as low as possible in order to avoid any directional overlapping.

12.7 MATHEMATICAL MODELS OF SEMIVARIOGRAM

In practice, $\gamma(h)$ is not known and is estimated from the available samples. A series of experimental semivariogram function values, $\gamma^*(h)$ is obtained for constantly increasing values of "h" from available sample pairs. Next, a mathematical scheme is fitted to the experimental semivariogram values that would represent the true underlying semivariogram. Different mathematical models of semivariogram are described in Sections

12.7.1–12.7.9 and graphically shown in Fig. 12.4. For each of the models of semivariogram, γ(h), there is an equivalent covariogram model, CV(h) given by the relation:

$$CV(\mathrm{h}) = \sigma^2 - \gamma(\mathrm{h}).$$

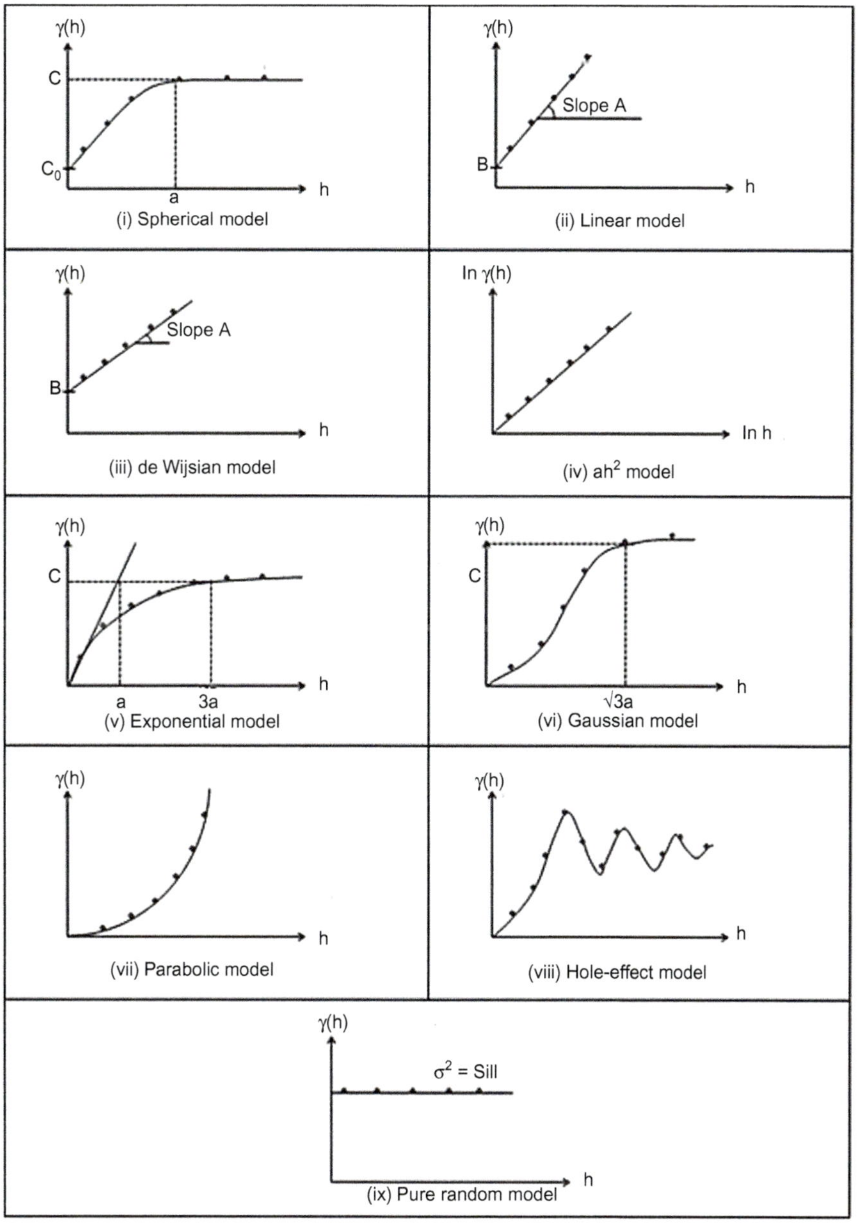

FIGURE 12.4 Different models of semivariogram.

12.7.1 The Spherical Model

This model is encountered most commonly in mineral deposits where sample values become independent once a given distance of influence (ie, the Range) "a" is reached. The equations are given by:

$$\begin{aligned}
\gamma(h) &= C_0 + C[3/2(h/a) - 1/2(h^3/a^3)] && \forall h < a;\\
\gamma(h) &= C_0 + C && \forall h \geq a;\\
\gamma(h) &= C_0 && \forall h \text{ tends to } 0;\\
\gamma(h) &= 0 && \forall h = 0.
\end{aligned}$$

This model is common in most sedimentary and porphyry deposits. Deposits as different as iron, copper, lead–zinc, gold, bauxite, nickel, uranium, phosphates, and coal have been found to have their grade distribution adequately represented by this model (David, 1977). This model also known as Matheron model is said to describe transition phenomena as it is the one which occurs when one has geostatistical spatial structures independent of each other beyond the range but within it, sample values are highly correlated.

12.7.2 The Linear Model

It is the simplest model encountered where there is no existence of the range. $\gamma(h)$ continuously increases as h increases. It shows a moderate continuity, observed sometimes in iron ore deposits. It is described by a linear equation.

$$\gamma(h) = Ah + B,$$

where A (slope) and B (intercept) are constants.

12.7.3 The de Wijsian Model

In some hydrothermal deposits, semivariogram plots as a straight line when $\gamma(h)$ is plotted against ln(h). The equation is given by:

$$\gamma(h) = A \ln(h) + B.$$

12.7.4 The ah^{λ} Model

In some cases, semivariogram can be made linear by plotting it on a log–log scale. The equation is: $\gamma(h) = ah^{\lambda}$; where λ is a power factor and "a" is intercept. This model is frequently encountered in elevation semivariogram or in the study of mill feed variability.

12.7.5 The Exponential Model

This model is not encountered too often in mining practice since its infinite range is associated with a too continuous process. The equation is: $\gamma(h) = C[1-e^{-h/a}]$. The sill is intersected at a point where "h" equals "a" by the tangent drawn from the origin of the curve with its slope as C/a. The practical range is regarded as 3a.

12.7.6 The Gaussian Model

This model is characterized by two parameters C and a. The curve is parabolic near the origin and the tangent at the origin is horizontal, which indicates low variability for short distances. Excellent continuity is observed which is rarely found in geological environments. The equation is:

$$\gamma(h) = C[1 - e^{(-h^2/a^2)}].$$

The practical range is $\sqrt{3}$a.

12.7.7 The Parabolic Model

The parabolic semivariogram is given by $\gamma(h) = Ah^2$, where A is the slope. This model is observed when there is a linear (drift) "trend."

12.7.8 The Hole-Effect Model

This model has an equation: $\gamma(h) = C[1-(\sin(ah)/ah)]$. It can be used to represent fairly continuous process. The tangent at the origin is horizontal and it shows a periodic/cyclic behavior which is often encountered when there exists, for instance, a succession of alternate rich and poor zones or alternate layers.

12.7.9 The Pure Random Model

No continuity is observed in this model thereby indicating the existence of a very high degree of randomness of the variable distribution. $\gamma(h)$ is then equal to the statistical variance, that is, $\gamma(h) = S^2$.

12.8 KRIGING: CONCEPTS AND APPLICATIONS

Kriging is an optimal spatial interpolation technique. In general terms a kriging system calculates an estimated value, G* of a real value, G by using a linear combination of weights, a_i of the selected surrounding n values such that:

$$G^* = \sum_{i=1}^{n} a_i g_i, \quad \text{where} \sum_{i=1}^{n} a_i = 1 \text{ and } g_i \text{ are the sample values.}$$

If G* is the estimate of a block average grade G by applying straight average method, that is,

$$G^* = \frac{1}{n}\sum_{i=1}^{n} g_i,$$

then equal weight is given to all the sample values, and the error of estimation of G is given as:

$$\sigma_E^2(\text{S to V}) = E[(G^* - G)^2] = \overline{\gamma}(S, S) - \overline{\gamma}(v, v) + 2\overline{\gamma}(S, V).$$

In many cases, however, assigning equal weight to all selected surrounding samples may not provide the best possible estimate. Let us consider the case of a block valued by a center sample and a corner sample as per configuration shown in Fig. 12.5.

Clearly, the center sample should be given a greater weight than the corner one. Say, we give weight a_1 to S_1 and a_2 to S_2. The new grade estimate would be:

$$G^* = a_1 g_1 + a_2 g_2$$

The weights of selected surrounding sample values are so chosen that:

- G^* is an unbiased estimate of G, that is, $E[(G^* - G)] = 0$;
- the variance of estimation of G by G^*, that is, $E[(G^* - G)^2]$ is minimum.

By definition, kriging is known as Best (because of minimum estimation variance) Linear (because of weighted arithmetic average) Unbiased (since the weights sum to unity) Estimator—BLUE.

12.8.1 Practice of Kriging

The technique for 3D block kriging (Sarkar, 1988) within a delineated mineralized boundary entails (1) computation of average sample to sample variability of samples falling within the radius of search; (2) selection of nearest samples lying within the radius of search; (3) establishment of kriging matrices involving setting up of a semivariance matrix that contain expected sample variabilities for each of the neighborhood sample values, and setting up of a matrix that contain the average variability between each of the nearest neighborhood sample values and the block; (4) establishment of kriging coefficient matrix; and (5) multiplication of kriging coefficients by their respective sample values to provide

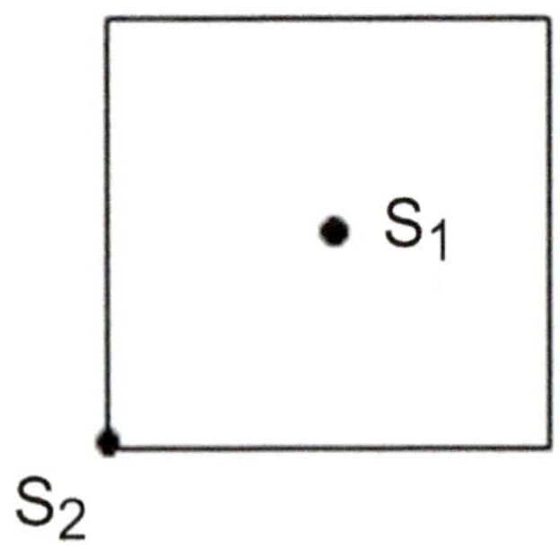

FIGURE 12.5 A square block with one sample at the center and another at corner.

kriged estimates (KE). The kriging variance (KV) is calculated from the sum of the products of the weight coefficients and their respective sample block variances. An extra constant, the lag-range multiplier is added to minimize the KV.

12.9 INTEGRATED GEOSTATISTICAL MODELING PROCESS

12.9.1 Geology and Geostatistics

Application of statistical methods to the analysis of geological data is not applied geostatistics. Understanding the geology of the ore body and other geological information constitutes essential prerequisite to determine which geostatistical method should be employed (Sarkar, 1988). The results obtained by geostatistical study must be interpreted in geological terms. Many a times, it is observed that both are conflicting. Any conflict between geology and geostatistics must be explained. The error can be either in the geological hypothesis or in the geostatistical analysis or both.

One of the important assumptions must be satisfied to complete a geostatistical study is the assumption of "stationarity." The deposit must be divided into areas which present reasonably constant statistical properties. The definitions of such areas must be based on geologic evidence, for example, (1) a supergene enriched zone must be separated from the primary zone; (2) two host rock types with very different fracture densities may be analyzed individually. In some cases, the geostatistical study may indicate the need for division of deposit into zones which are not recognized by the geologists. The geological justifications of these divisions must be thoroughly understood (Sarkar, 2014a).

12.9.2 Exploration Database

A key prerequisite to an integrated geostatistical modeling process is the organization of a realistic and reliable exploration database. Since much of the later work will be based on this data, utmost care is to be taken to compile and organize the database. Raw geological and exploration data are entered in a computer database and processed to a desired format to aid the data processing. The relevant geological and exploration data are stored in the computer database for interpretation of mineralization pattern with the aid of potential geological controls and for generation of an appropriate ore body model. The exploration database contains:

- Geological information pertaining to the geology of a deposit. Different mineralization environments reflect various physicochemical conditions within which different deposits occur. The distribution of economic mineral constituents in any deposit is governed by one or more geological control(s), viz., rock type, mineralogy, stratigraphy, structure, and genesis among other controls. These controls are of special importance in the geostatistical modeling process that aid in the identification of homogeneous zone(s) of mineralization.
- Drill hole information pertaining to *drill identification*, viz., number, type, and date; *survey information*, viz., collar coordinates in XYZ coordinate system (where X represents Easting, Y represents Northing, and Z represents Elevation), and downhole survey

measurements, that is, azimuth and dip at collar, depth of survey down the hole and their corresponding azimuths and dips; and *sample information*, viz., depths to top and bottom of sample and sample grade information.

These are stored in a form that is easily verifiable and can be updated whenever additional data are available.

12.9.3 Integrated Geostatistical Modeling

An integrated geostatistical modeling process must provide for close interaction among the geologist, the computer, and the mathematical estimation procedures in a way that the geologist retains control and has confidence in the final results. With exploration database in hand comprising geological and drill hole information, the integrated geostatistical modeling process entails the following operations (Sarkar, 2014b):

- sample value compositing;
- statistical modeling;
- semivariogram analysis;
- trend analysis;
- semivariogram modeling;
- block kriging;
- mineral inventory; and
- grade–tonnage relations.

12.9.3.1 Sample Value Compositing

The location of individual drill hole samples at given distances of constant interval leads to compositing of sample values wherein the individual drill hole samples are grouped into segments of constant length along the hole, subject to geological homogeneity. The values corresponding to the segments of constant length are the composited weighted values. The location and size of the composites are chosen to coincide with the bench or slice elevation and bench or slice height, respectively. The sample value composites and their respective centroid XYZ coordinates constitute the *composite value database*.

12.9.3.2 Statistical Modeling

Probability modeling of frequency distribution of the composite values of constant support provides a significant insight into mineralization characteristics of a deposit. The type of probability distribution representing the sample values is estimated through statistical modeling. The assumption of a probability model fit to a sample frequency distribution is validated graphically by its probability plot and numerically by its degree of skewness, degree of kurtosis, and chi-squared goodness-of-fit test statistics. Typical averaging of sample values in a mineral deposit without identifying the representative probability distribution may lead to mixing of populations in the case of a multipopulation deposit (Rendu and Readdy, 1982; Rendu, 1985). Geostatistical modeling carried out on such mixed population yields poor estimates. If a sample frequency distribution exhibits multiple modes, individual population components are split based on the knowledge of geological control(s) and analyzed separately. Significant geological controls on the grade of mineralization and zoning of various types that influence the probability distribution modeling and geostatistical properties of a mineral deposit include lithological units, mineralogical composition, folding, faulting, fracturing, genesis, episodes of mineralization, barren zones, intrusion, alterations, grade zones, deformation, and unconformity among many. However, in the absence of a complete knowledge of geological controls, a straightforward overall statistical analysis of the composite sample values and a plot of the percent cumulative frequencies in an arithmetic probability scale would provide indication to the number of population component(s) based on the number of breaks in the probability plot. In case such an analysis reveals a mixed population, contour plots of the overall data set on different planes are made to aid in the identification of possible homogeneous zones of mineralization. Contour plotting may be carried out extensively to improve the identification of population components based on a knowledge of the deposit *type*. The physical split of the data set is carried out on the surface where the contour plot shows sharp change and/or distinct discontinuities in the data set.

12.9.3.3 Semivariogram Analysis

Once a geologic understanding of a mineral deposit is obtained, population component(s) representing distinct homogeneous geological regimes are identified and parameters of probability model estimated, the next step is to quantify the spatial relation between sample values. Experimental three-dimensional (3D) semivariograms are constructed along various directions to take into account of the directional variability (in most cases, along four principal directions, viz., 0, 45, 90, and 135 degrees or along the directions of strike and dip of the mineralized body and along two directions oblique to the strike and dip in the case of a vein type deposit) with a lag distance approximately equal to the average sample interval, an angle of regularization of 11.25 degree and a spread limit of 1.5–2 times the lag distance. The experimental semivariograms express the spatial dissimilarity as a function of distance at different orientations. The consideration of the distance, whether straight line or curved, is dependent on the shape and geometry of the mineral deposit. If no marked difference in the characteristics of the directional semivariograms is observed, a mean isotropic 3D semivariogram is assumed to explain the spatial structure of the mineralization.

12.9.3.4 Trend Analysis

In some instances, experimental semivariograms exhibit a trend in the sample values. To overcome this during semivariogram modeling, a simple surface is fitted to the data set through trend surface analysis. In mathematical terms, the grade x(z) at any point is a function of a deterministic trend m(z) plus a random component known as the residual r(z) with zero mean. If the sample values are freed from the trend, the remaining component is the residual (Sarkar et al., 1988), that is,

$$x(z) = m(z) + r(z),$$

or,

$$r(z) = x(z) - m(z).$$

These residuals, which are at least intrinsic, are employed for normal semivariogram analysis.

12.9.3.5 Semivariogram Modeling

The behavior at the origin for both nugget effect and slope plays a pivotal role during the fit of a mathematical function to an experimental semivariogram. While the slope can be assessed from the first three or four semivariogram values, the nugget effect can be estimated by extrapolating back to the $\gamma(h)$ axis. The choice of nugget effect is extremely important since it has a very marked effect on kriging weights and in turn on KV. There are, at present, three methods for model fitting which are described as follows:

(1) *Hand Fit Method*

The sill ($C_o + C$) is set at the value where experimental semivariogram stabilizes. In theory, this should coincide with the statistical variance. Estimate of nugget effect is achieved by joining the first three or four semivariogram values and projecting this line to the $\gamma(h)$ axis. By projecting the same line until it intercepts the sill provides two-thirds the range. Using the estimates of C_o, C, and "a," calculate a few points and examine if the model curve fits the experimental semivariogram. Although the method is straightforward, and simple to practice, there is an element of subjectivity involved in the estimation of model parameters.

(2) *Nonlinear Least Squares Fit Method*

Like any curve fitting technique, this method uses the principle of polynomial fit by least squares to fit a model with sum of the deviations squared of the estimated values from the real values being minimum. However, polynomials obtained by least squares do not guarantee the positive definite function (otherwise semivariance could turn out to be negative).

(3) *Point Kriging Cross-Validation Method*

Point kriging cross-validation is a technique referred to by Davis and Borgman (1979) as a procedure for checking the validity of a mathematical model fitted to an experimental semivariogram that controls the kriging estimation. The principle underlying the technique is as follows:

A sample point is chosen in turn on the sample grid that has a real value. The real sample value is temporarily deleted from the sample set and value of the sample is kriged using the neighborhood samples confined within its radius of search. The error variance of the estimated value and the real sample value is calculated. The kriging process is then repeated for rest of the known data points.

A crude semivariogram model is initially fitted by visual inspection to the experimental semivariogram. Estimates of the initial set of semivariogram parameters (viz., C_o, C, and "a") are made from the initial model and cross-validated through point kriging empirically. The error statistics such as mean error, mean variance of errors, and mean KV are then computed. The model parameters are varied and adjusted until: (1) a ratio of mean variance of the errors (estimation variance) to mean KV approximating to unity (in practice, a value of 1 ± 0.05 has been observed to be adequate acceptable limits); (2) a mean difference between sample values and estimated values close to zero; and (3) an adequate graphical fit to the experimental semivariogram are achieved. For a good estimate, most of the individual errors should also be close to zero. A model approximated or fitted by this approach eliminates subjectivity.

12.9.3.6 Block Kriging

Once the model semivariogram parameters characterizing all information about the expected sample variability are defined, the subsequent step involves the estimation of block values together with their associated variances through kriging. At this stage, a homogeneous mineralized zone is considered and sliced into a number of regularly spaced horizontal sections by projecting sample data from various transverse and longitudinal sections. Mineralized boundaries are then delineated on each of the horizontal sections based on geological and mining considerations. The spacing of horizontal cross sections is manipulated from constant length at which drill hole samples are composited, generally equaling the height of a proposed mining bench (as in an open pit) or vertical lift (as in underground operation). This involves minimum projection of sample data from transverse and longitudinal sections onto horizontal sections. Each of the horizontal sections (hereinafter termed horizontal slices), with a mineralized boundary delineated on them, is divided into smaller grids equaling the size of a block.

Decision on the choice of a block size, or in other words, an SMU is generally influenced by several factors (David, 1977) such as sampling density, geological structure, precision of sample data, method of mining, equipment capabilities, production target, desired use of block, and capability of manipulating a huge number of blocks. Ideally, height of a block should usually be taken as that of the proposed bench height or vertical lift, since this is the way it would be mined. The other two dimensions should equal at least a quarter of the average drill spacing (David, 1977). Daily production target is another important contributory factor, since the choice of an equipment depends on the tonnage of material it can handle. The individual slices, when divided into smaller grids based on SMU, form a set of X (Easting) and Y (Northing) arrays of blocks with constant Z (Elevation) value. The arrays of blocks are then kriged slice-by-slice, producing KE and KV for each of them and also a slice average. The input parameters that are found to be adequate for block kriging include: (1) a minimum of 4 samples (because of the minimum necessity to define a surface) and a maximum of 16 samples (because of reasonable computational time and cost) with at least one sample in each quadrant (or one sample in each alternate octant) to krige a block; and (2) the radius of search for sample points around a block center to be within the full range of influence.

The individual slice averages are then further averaged to produce a mean kriged estimate and a mean KV in order to provide global estimates. The 95% geostatistical confidence limits are calculated as: $m \pm 1.96 * \sqrt{\sigma^2_k}$, where m is the mean kriged estimate; σ^2_k is the mean KV.

12.10 MINERAL INVENTORY

Each of the slices with regularly spaced kriged blocks is then stacked one below the other from top to bottom thereby giving a 3D array of blocks distributed regularly in space with their kriged mean (KM) and KV and tonnages per block obtained by multiplying the block dimensions by the bulk density of the block constituents. Such a 3D network of blocks is known as the mineral inventory, which provides the *in situ* stock of mineral.

12.11 GRADE–TONNAGE RELATIONS

Subsequent to the development of a mineral inventory, estimates of grade–tonnage at different probabilistic cutoff grade values are obtained. In the process of estimation, it is observed that a greater tonnage is obtained with a relatively low grade. Progressively, higher grades may be obtained by increasing the values of cutoff grade and thus reducing the tonnage. This is known as grade–tonnage relation. A simple numerical procedure is to model the relation statistically involving a progressive integration of the block distribution over various probabilistic cutoff grade values and obtain: (1) tonnages of ore, metal, and waste; (2) grades of ore and waste; (3) ratio of quantity of waste to ore. Plots of these relations provide grade–tonnage curves. These curves together with the mineral inventory provide a sound basis for mine decisions.

The Confidence limit of tonnage at a stated probability is related to three factors:

- Inherent variability of the 3D mineralized outline and its width;
- Inherent variability of *in situ* bulk density of the mineralized body; and
- Frequency and spacing of borehole intersections.

An example displaying calculation of geostatistical assessment of ore grade and tonnage at a range of cutoff grades for an iron ore deposit is given in Table 12.1.

12.12 A STEP-BY-STEP SUMMARY FOR AN INTEGRATED GEOSTATISTICAL STUDY

In summary, a geostatistical study entails the following steps:

- Understanding of the deposit geology and identification of geological controls.
- Stratification or splitting of mineralization into geological domains.
- Compositing of sample values within each geological domain.
- Frequency distribution analysis of the composite sample values.
- Validation of the hypothesis of one population (ie, single mode) through classical statistical modeling.
- Geostatistical structural/spatial analysis of each geological domain individually by constructing experimental semivariograms (at least along four principal directions).
- Detecting the presence of geostatistical anisotropy and trend, if any and corrections thereof.

TABLE 12.1 Calculation of Geostatistical Assessment of Ore Grade and Tonnage at a Range of Cutoff Grades for an Iron ore Deposit

Block Grades (%)	No. of Blocks (f)	Class Average (A)	Expectancy (E = f/n)	CE (High to Low)	ExA	C (ExA) (High to Low)	C/O (%)	Av. Grade g (%)	r_o (mt)	W/O ratio
52–54	2	53	2/20 = 0.10	1.00	5.30	57.90	52	57.90	40	0.00
54–56	3	55	3/20 = 0.15	0.90	8.25	52.60	54	58.44	36	0.11
56–58	6	57	6/20 = 0.30	0.75	17.10	44.35	56	59.13	30	0.33
58–60	4	59	4/20 = 0.20	0.45	11.80	27.25	58	60.56	18	1.22
60–62	3	61	3/20 = 0.15	0.25	9.15	15.45	60	61.80	10	3.00
62–64	2	63	2/20 = 0.10	0.10	6.30	6.30	62	63.00	4	9.00
	Sum = 20		Sum = 1.00		Sum = 57.90					

Assuming total tonnage of ore (t_0) = 40 mt; block dimensions = 100 m × 100 m × 50 m; Total no. of blocks = 20; bulk density = 4 t/m^3.
CE, cumulative expectancy; C(E × A), cumulative product of expectancy and class average; g = C(E × A)/CE, average grade; $r_o = t_o \times CE$, tonnage of ore at a cutoff (C/O) grade; $(t_o - r_o)/r_o$, W/O ratio.

- Semivariogram model fitting and establishment of model parameters.
- Delineation of mineralized boundary on horizontal slices.
- Decision on the choice of a block size (SMU).
- Division of each slice into smaller grids equaling the size of a block.
- Kriging of blocks within the limits of mineralized boundary slice-by-slice, producing kriged estimates and associated KVs.
- Compilation of kriged outputs in three dimensions by stacking each of the slices, with regularly spaced kriged blocks, one below the other from top to bottom to produce a mineral inventory.
- Establishment of grade–tonnage relations at various probabilistic cutoff grade values to provide a basis for choosing an optimum cutoff value (grade) and then estimating reserves from the mineral inventory by applying appropriate cutoff criteria.
- On completion of these steps, block estimates are displayed, assessed visually, and a comparison of block, sample composite, and individual sample values is then made for a reconciliation of the results. Only when the reconciliation process is complete to the satisfaction of all concerned, will the estimation of block values be accepted and used for follow-up exploration and mine decisions.

A workflow for an integrated geostatistical evaluation is given in Fig. 12.6.

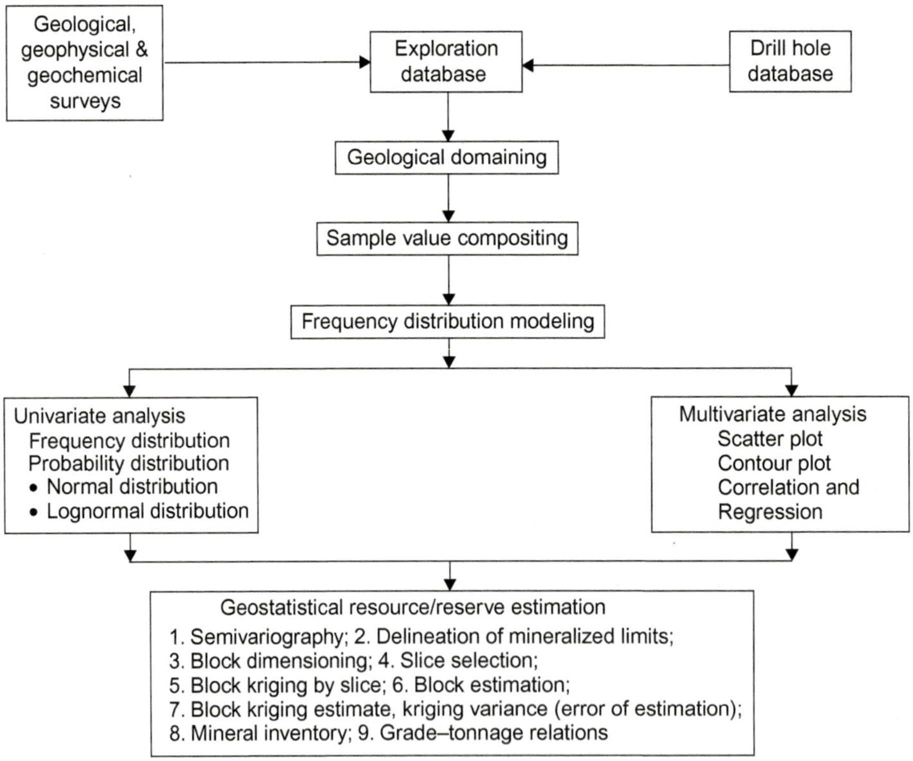

FIGURE 12.6 Workflow for an integrated geostatistical evaluation.

12.13 GEOSTATISTICS IN MINERAL INDUSTRY

Various domains of application of Geostatistics in Mineral industry include:

Database Verification and Preliminary Study of Sample Values: Common statistical methods including frequency distribution and scatter diagrams can be used to detect anomalous (ie, outlier) and potentially erroneous sample values, differentiate between geologic environments, and provide visual and numerical displays of the properties of the sample values. These studies constitute essential prerequisites to determine the optimal method of geostatistical analysis.

Geologic Modeling: Geostatistics aids in understanding the genesis of ore body as well as geological factors that control the mineralization. Mixtures of geologic zones as well as the presence of successive mineralizing events can be detected. Trends and directions along which the mineralization shows the best continuity that can be detected and quantified.

Optimization of Sampling Program: The quality of information supplied by different sampling methods can be compared. Drilling programs can be optimized including optimization of drill hole spacing. The need for additional drill holes and the optimal location of these holes can be assessed.

Reserve Estimation: It is most commonly used to optimize the estimation of reserves. Both geologic and mineable reserves can be estimated geostatistically with respect to a cut-off. The amount of information needed to determine the properties of a deposit at a given precision can be minimized.

Mining and Production Control Method: The influence of various mining methods on the grade–tonnage relations and on the economics of a project can be quantified early in the analysis of the deposit.

Risk Analysis: Risk involved in a project can be quantified and minimized. The need for additional drilling can be assessed by comparing the expected drilling cost with the expected value of the information which will be obtained.

12.14 LIMITATIONS OF USE OF GEOSTATISTICS

Most of the problems encountered in the application of geostatistics to resource/reserve estimation come from:

- misunderstanding of, or insufficient information concerning the geology of a deposit;
- insufficient sample information;
- poor choice of geostatistical method, resulting in inaccurate calculation, interpretation, and modeling of the parameters of the deposit, including the semivariogram;
- misunderstanding of the constraints that must be satisfied to apply a given geostatistical method, including the stationarity condition.

These problems can usually be resolved in parts by ensuring that the geological properties of the deposit are well understood by the geostatistician and that the results of the geostatistical study are interpreted in geologic terms. Close cooperation of exploration geologist and geostatistician right from the early stage of the project would result in a substantial decrease in the cost of the study and increase in the precision of estimates.

Geostatistics provides a new dimension in the exploration of mineral deposits and mineral deposit evaluation and is used invariably by all exploration and mining companies. Geostatistics is a powerful tool, but it must be used with caution. There are many ways of using geostatistics incorrectly, but if it is well-understood and properly applied, it can be of great benefit.

References

Barnes, M.P., 1980. Computer-Assisted Mineral Appraisal and Feasibility. Society of Mining Engineers, New York, NY, p. 167.

Chiles, J.P., Delfiner, P., 1999. Geostatistics—Modelling Spatial Uncertainty. John Wiley & Sons, New York, NY, p. 695.

Davis, B.M., Borgman, L.E., 1979. A test for hypothesis concerning a proposed model for the underlying variogram. Proceedings 16th APCOM. pp. 163–181.

David, M., 1977. Geostatistical Ore Reserve Estimation. Elsevier Scientific Publ. Co., Amsterdam, p. 364.

Krige, D.G., 1962. Statistical applications in mine valuations. J. Inst. Mine Surv., S. Afr. 12 (2 and 3), p. 82.

Matheron, G., 1971. The Theory of Regionalized Variables and Its Applications, Booklet No. 5. Ecole Nat. Sup. Des Mines, Paris.

Rendu, J.-M., 1981. An introduction to geostatistical methods of mineral evaluation. SAIMM Monograph Johannesburg, pp. 12–84.

Rendu, J.-M., 1985. Geostatistical modelling and geological controls. Trans. IMM London. pp. B166–172.

Rendu, J.-M., Readdy, L., 1982. Geology and the semi-variogram—a critical relationship. 17th Int. Symp. on Comp. Appl. in Min. Industry, Colorado. pp. 771–783.

Sarkar, B.C., 1988. An Integrated System for Geology-Controlled Geostatistical Evaluation (Ph.D. thesis). Imperial College of Science and Technology, London, p. 218.

Sarkar, B.C., 2014a. Geostatistics: concepts and applications in mineral deposit modeling for exploration and mining. J. Indian Geol. Congr. 6 (1), 3–26.

Sarkar, B.C., 2014b. Geostatistics for Natural Resources Modelling, Edited Book Volume. Indian School of Mines, Dhanbad, p. 178., ISBN 978-93-5156-635-9.

Sarkar, B.C., OLeary, J., Mill, A.J.B., 1988. An integrated approach to geostatistical evaluation. Mining Mag 159 (3), 199–207.

Sinclair, A.J., Blackwell, G.H., 2002. Applied Mineral Inventory Estimation. Cambridge University Press, UK, p. 381.

CHAPTER

13

Mineral Resources Classification

OUTLINE

13.1 INTRODUCTION

The classification of mineral inventory was necessary, in the 20th century, for several reasons that include: (1) creation of a formal inventory of the principal assets of a mining/exploration company; (2) documentation of assets to demonstrate potential for medium-term or long-term production; (3) to raise development funds in the speculative money market; (4) to provide a reasonable confidence level for senior financing institutions; and (5) to provide governments with a basis for royalties, taxation, land use management, and so on. Formal classifications were suggested in the early 20th century (Table 13.1) and were pertained only to ore. Since that time, there has been increasing recognition that mineralized zones can be either very poorly or very well established with all gradations of certainty possible. This situation has led to a widely accepted trend to describe separately the terms resources and reserves and their relations in context to minerals, both viewed as part of "mineral inventory" (Sinclair and Vallee, 1994; Sinclair and Blackwell, 2000).

Essentials of Mineral Exploration and Evaluation
DOI: http://dx.doi.org/10.1016/B978-0-12-805329-4.00020-X

TABLE 13.1 Early Definitions of Mineral Estimation

	Quality of Estimate		
Source	**Highest**	**Middle**	**Lowest**
Kendal et al. (1901)	Puritan	Cavalier	Others
	Ore in sight, ore blocked out, ore in Reserve, ore available for extraction	Ore reasonably expected, visible ore not blocked out; probable ore in sight; Ore probable but not blocked out	Witwatersrand drill-inferred geologic projection
Hoover (1909)	Proven ore	Probable ore	Possible ore
	Ore without any risk of failure in terms of geological continuity	Ore with some risk, but adequate justification for assumption of geological continuity	Ore which are not included in either proven or probable classes with no known or stated in any terms of tonnage
Leith (1935)	Assured ore	Prospective ore	Possible ore
	Ore blocked out in three dimensions by mining or drilling, risk of failure remote	Extensions near at hand; probability high but extent is less precise	Presumptive evidence of ore; but of indeterminate quantity
Fennel (1939)	Ore blocked out	Probable ore	Prospective ore
	Exposed on three or four sides	Partly exposed on one or two sides	Not exposed
USBM/USGS (1976)	Measured ore	Indicated ore	Inferred ore
	Sampled tonnage from dimensions in workings, trenches, drill holes, tonnage, or grade estimates good within 20%	Tonnage partly from measurements and partly projected	Few measurements but some geological knowledge presumed

From Taylor (1994).

Since quite some years, the mineral industry has been international in scope. Hence it has become necessary to develop and maintain international reporting standards of Resources, Reserves, and Exploration Results in context to minerals. Raising necessary capital with respect to various stages of mining cycle has been usually restricted to home country of a specific mining company. With an undeniably globalized mining industry, the mineral resources of nations attracting potential political consideration and the effect that minerals have on the commercial, accounting and investing groups, there has been a requirement of common terminology for public reporting with global acceptance. It has become a necessity that all exploration companies are required to raise its own capital in several jurisdictions, few being independent of management of a mining company. It may be emphasized that it is a challenge to raise necessary capital knowing fully well that mineral industry is associated with high-risk capital. Because of the associated high risk, the mineral industry should be in a position to evolve means of managing capital and to provide complete transparency to the stakeholders. Fulfilment of these conditions require internationally recognized reporting standards in respect of mineral resources and reserves.

Inside of nations, numerous minerals reporting frameworks have been built up; some by governments for public, others by expert and corporate houses for business use. In any case, with the present day cross outskirt mergers and acquisitions, corporate postings on various universal stock trades and expanding speculation by outside nations in the creating nations of Asia, Africa, and South America imply that a worldwide dialect is expected to depict the benefits of a definitive mineral abundance of nations, all in all and resources of mining organizations, specifically. Experts in such manner have been cooperating for a by and large adequate plan.

Reporting measures give conspicuous benchmarks inside the business and are perceived by others outside the business. Benchmarks pick up quality and validity when they are commonly perfect. They lessen perplexity, expand seeing, ideally keep things as basic as could reasonably be expected and are pertinent under all circumstances. There are obviously issues of similarity between frameworks as we analyze them more. Keeping to the latest advancements and amending national and global benchmarks is a time consuming job.

13.2 HISTORY OF THE DEVELOPMENT OF REPORTING STANDARDS

Historically, most nations followed their own arrangement of classifying and reporting resources and reserves without any standardization. Be that as it may, cases of numerous happenings and makeovers in the mineral business required the initiation of universally acknowledged mineral reporting norms. From a starting point in the United States, Australia had taken the activity in giving a Code and Guidelines to characterizing and ordering Reserves, Resources and Exploration Results in context of minerals. The Joint Ore Reserves Committee (JORC) Code was initially formalized in 1989 with further upgrades in 1992, 1993, 1996, 1999, 2004, and 2012. From that point forward, there was a fast development in the improvement of new models in different nations such as China, Russia, and so forth. Along these lines, reporting models turned into a worldwide phenomenon without there being a universal standard into classes taking into account the level of geological certification of its presence.

Several classification schemes have been proposed, all based more or less on the same principle (Henley, 2004; Weatherstone, 2008; Vaughan and Felderhof, 2002). CRIRSCO published "International Reporting Template" in 2006 with reference to minerals. It represents the most updated reporting standards that have wide recognition and global acceptance with reference to market-based reporting and business investment. Several other national systems of reporting include:

1. USA: SEC Industry Guide 7 or, SME Guide 2014 (revised);
2. Australia: JORC Code, 2012;
3. Europe: PERC, 2013;
4. South Africa:SAMREC Code, 2009;
5. Chile: IIMCh, 2007;
6. Mongolia: Mongolian MRC Code, 2014;
7. Canada: NI 43-101, 2014;
8. Russia: NAEN, 2011;
9. Chinese scheme, adapted from old UNFC system.

It is to be noted that except for the Chinese system, all other systems listed above are CRIRSCO template compliant. The UNFC 2009 code can be mapped to the CRIRSCO template through a bridging document.

13.3 EXPLORATION RESULTS

Exploration results are antecedents to Resources. Exploration data, for example, drill samples or geological samples, constitute basic input to assess a volumetric dimension, quantity (tonnage) and quality (grade) of mineral body. Evaluations of *in situ* tonnage and grade of mineral body with reasonable prospects of resulting in commercial extraction at profit refer to the term Mineral Resources, that is, they are not simply mineral body rather they are prone to be mineable, treatable, and saleable. On the other hand, the known economic part of Mineral Resources that are determined by applying the "modifying factors" are called Mineral Reserves. Fundamentally, this implies that the resources change to reserves through technoeconomic evaluations and are usually called prefeasibility and feasibility studies. While reporting, evaluation techniques should take into consideration of all the modifying factors collectively to justify economic extraction.

13.4 COMPETENT PERSON AND RESPONSIBILITY

Proviso 11 of the JORC Code (2012) Edition characterizes Competent Person and the required abilities and experience an individual must possess to be known as a Competent Person. The person should be a Member of globally Recognized professional body in context to mineral resources. The person should possess at least 5 years practical skill pertinent to the mineralization style and deposit type relevant to the expert's domain of activity (www.iom3.org).

On the basis of relevant experience, a competent person estimates and/or supervises the evaluation process. For a competent person to estimate and supervise such assessment, the person should be conversant with estimation, market pricing, and commercial appraisal of mineral commodity (www.pgi.co.za).

13.5 MINERAL RESOURCE CLASSIFICATION

Mineral resources are classified based on the degree of geological knowledge of continuity and consideration of technoeconomic parameters. The classification is regulated by laws, norms, and industry standards. Numerous schemes have been proposed around the world; however CIM, JORC, UNFC, SEC, SAMREC are in vogue in the worldwide mining industry.

Classification of Mineral resources is the process of categorization of mineral deposits based on their geologic assurance and commercial worth. Characterization of Mineral assets is the procedure of classification of mineral deposits in respect of the mineralization continuity and market value. Mineral deposits can be categorized as (1) resources which could become profitably extractable in foreseeable future and (2) reserves that are associated with high degree of geological confidence and currently extractable legally, technically, and economically. In a similar manner, mining phrasing, an ore deposit essentially has a known economic component known as reserve with or without additional resources. (www.russia.gneizno.pl).

13.5.1 Mineral Resources

Proviso 20 of the JORC Code (2012) characterizes a Mineral Resource as a concentration of minerals of importance for business concern inside or on the Earth's crust in such geometrical shape, size, grade, and tonnage that there are sound prospects for ultimate profitable extraction. The geographical location and geological attributes of a Mineral Resource are identified, calculated, or inferred based on geological information, continuity, and sampling results. Mineral Resources are classified in order of geological knowledge and experience into Inferred, Indicated categories.

Reporting of Mineral Resources must fulfill the prerequisite that there are sound prospects for ultimate profitable extraction, irrespective of the various categories (www.draigresources.com).

13.5.2 Mineral Reserves/Ore Reserves

Proviso 29 of the JORC Code (2012) characterizes Ore Reserve as the economically extractable part of a Measured and/or Indicated resource categories. It incorporates subeconomic components and admissible extent for losses that might happen at the time of extraction and is characterized by prefeasibility or feasibility studies including application of modifying factors. The prefeasibility or feasibility studies reveal sensible justification of extraction during reporting. The reference point at which reserves are characterized generally where the ore is sent to the handling plant should be specified. It is critical that, in all circumstances where the reference point is not the same, an explanatory declaration is incorporated to ensure what is being reported (www.asx.com.au).

Resources are convertible to reserves by applying modifying factors such as geological and mining, metallurgical, economic, environmental, marketing, legal, political, and social. Some multinational mining companies explain Mineral resource/reserve as given below:

Mineral Resource: That part of a mineralized volume of rock that is currently expected to be capable of eventual economic extraction, but is not yet demonstrated to be so.

Ore Reserves: That part of a mineralized volume that is estimated to be capable of legal and economic exploitation. Almost always a smaller subset of Mineral Resources.

Tier One Resource: An exceptional mineral deposit or group of deposits that will ultimately host one or more large, long-life, low-cost mining operations. Often the largest in their class, they have embedded options (eg, future extensions, expansions, and tie-ins) and have a lasting impact on the industry, investors, governments, and societies.

13.6 THE JORC CODE

The JORC Code (2012) gives an obligatory framework to categorization of estimates of tonnage and grade as indicated by increasing degree of geological knowledge of continuity and on the consideration of technoeconomic parameters and sets minimum specifications for Open Reporting of results of Exploration, Resources, and Reserves. Open Reports arranged as per the JORC Code are reports arranged with the end goal of appraising financial specialists or potential speculators and their counselors (www.ausimm.com.au).

13.7 REPORTING TERMINOLOGY

Mineral resources/reserves are reported on the basis of the level of geological knowledge and confidence. Reports providing the results of exploration, resources, and reserves should custom the terms given in Fig. 13.1. Mineral resources convert to mineral reserves on consideration of 'Modifying Factors'. However, it is also possible that mineral reserves can back convert to mineral resources because of the influence of new information on the modifying factors. In few specific cases, measured mineral resources convert to probable mineral reserves because of the improbabilities associated with modifying factors without any decrease in the geological knowledge and confidence.

13.8 CODIFICATION OF UNFC SYSTEM

This provides a system for codification of ore reserve blocks on geological, feasibility, and economic parameters. Classification directly reflects the status of investigation in exploration and feasibility phases along with economics of a mineral block/reserve. The UNFC system is shown in Fig. 13.2. This grouping cases to (1) explain the already existing semantic issues (2) provide correspondence of the diverse coal and mineral deposit types, and (3) offer crisp, explicit documentation of resource/reserve classes that would aid in automated data processing and information exchange.

The Table 13.2 summarizes the codes for various stages of mineral reserves/resources.

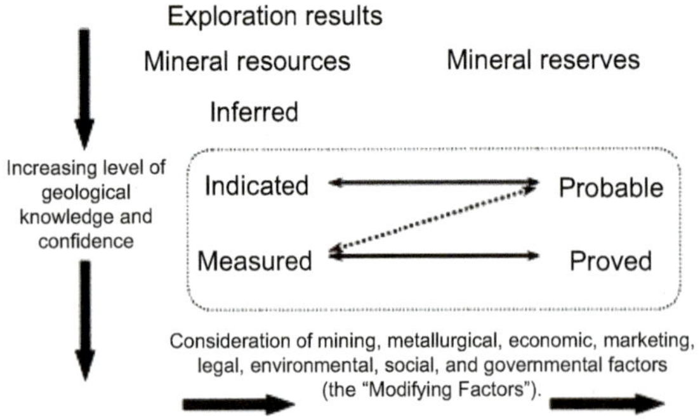

FIGURE 13.1 General relationship between exploration results, mineral resources, and ore reserve.

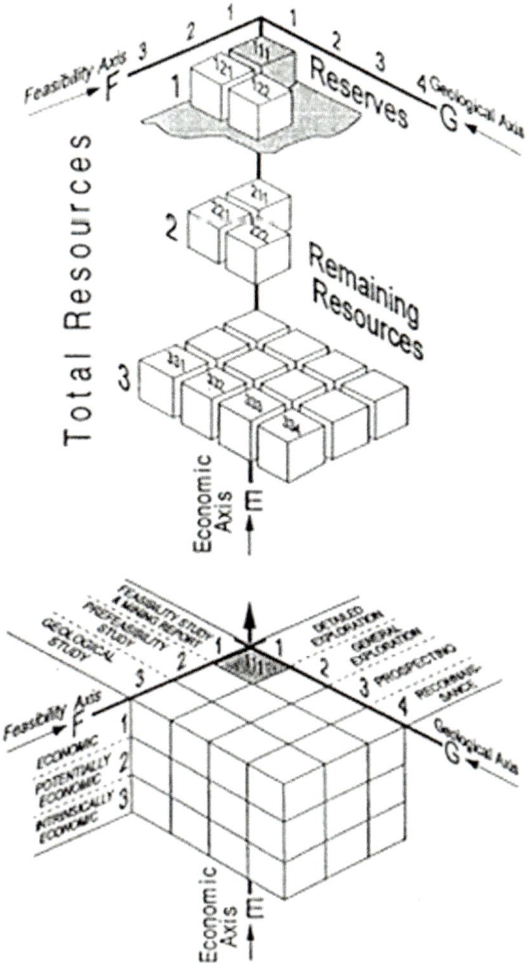

FIGURE 13.2 UNFC classification system.

TABLE 13.2 Example of Translation From UN to CMMI System Using Codes

CODE	CMMI System	UN Proposal
111	Proved reserve	Proved reserve
121 & 122	Probable reserve	Probable reserve
211	Measured resource	Feasibility resource
221& 222	Indicated resource	Prefeasibility resource
331	Measured resource	Measured resource
332	Indicated resource	Indicated resource
333	Inferred resource	Inferred resource
334	Not existing	Reconnaissance resource

This classification is best suited for exploration projects leading to feasibility studies but it does not cover the producing mines/deposits under development as it does not reflect the concept of ore availability. Although the code 111 (Proved Mineral Reserve) is almost equivalent to "Developed Reserves," but the latter is bound by two fully developed haulage levels and is ready for extraction. UNFC does not facilitate production budgeting, scheduling, ore accounting, measure of dilution, and cater to development needs of a mine for tonne to tonne ore replacement.

13.9 THE RUSSIAN FEDERATION CLASSIFICATION SYSTEM

Sutphin et al. (2011) have given a point-by-point account on the Russian Reporting framework of Reserves and Resources. The previous Soviet framework for grouping of resources/reserves, created in 1960 and amended in 1981, is even utilized today in Russia and different CIS republics. Basically, it segments mineral concentrations into seven classes under three main categories, considering the quantum of exploration as fully explored reserves or resources (A, B, C_1), evaluated reserves or resources (C_2) and prognostic resources (P_1, P_2, P_3) (Fig. 13.3). On a fundamental level, these take after a progression of approximations that are connected to different phases of exploration, which implies resources/reserves are relegated to classes in view of the level of confidence and demonstrate their similar significance for the national economy, that is, the classification is not characterized absolutely by exploration certainty levels but rather include some financial criteria) (www.stansenergy.com).

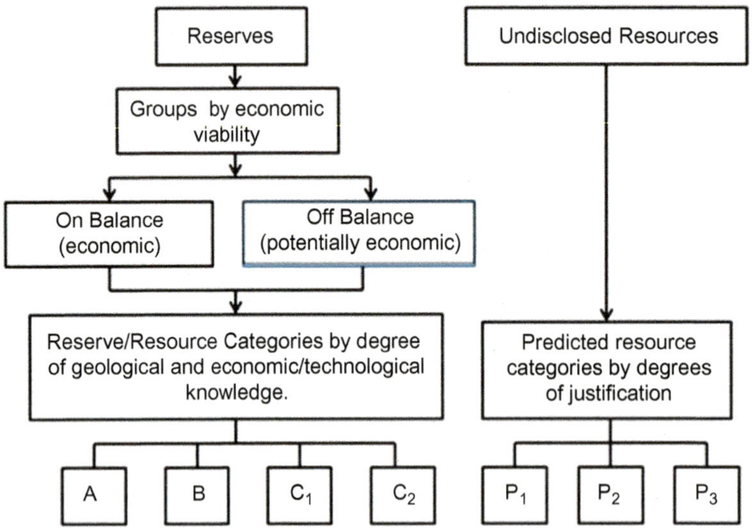

FIGURE 13.3 Russian federation classification system.

According to Henley (2004), a wide proportionality between the groupings might be introduced as:

Russian Reporting Code	JORC and Others
A, B	Proved reserve/measured resource
C_1	Proved or probable reserve/Indicated resource
C_2	Reserve/Indicated resource/Inferred resource
P_1	Inferred resource
P_2	Reconnaissance Mineral Resources (or UNFC code 334)
P_3	no equivalent".

13.10 THE CHINESE RESERVE AND RESOURCE REPORTING SYSTEM

There are important logical and real-world contrasts between the JORC and Chinese resource framework system. The framework for grouping resources/reserves in China has as of late been reformed. The old customary framework, developed from the previous Soviet framework, has been gradually eliminated following 1999, which had used classes A through F for resources, in view of diminishing levels of geological certainty. The present (New) framework utilizes a three-dimensional network taking into account level of certainty in Economic, Feasibility, and Geological assessments. The follow-up arrangement is sorted by a three number code, and is likewise distributed either to Resource or Reserve standing. The new framework fulfills the guidelines requirement of the UNFC framework that was planned for global use in 2004 (www.srk.ru.com). The old and the revised Reserve and Resource Reporting System are given in Figs. 13.4 and 13.5.

Sinclair and Blackwell (2000), after carefully analyzing the various classification systems and the "Qualified Person," opine that with the increasing emphasis on protection of the public against outright fraud and low-quality work by professional earth scientists relative to mineral deposits, require that reports of mineral inventory be of a high standard. Important attributes to high quality include:

- estimates can be made only where indicated by an appropriate geological argument that is clearly documented by the Qualified Person;
- within a zone interpreted by the qualified person to be continuously mineralized, mineral inventory estimates can be classified using clearly documented criteria; and
- the general procedure for classification should be easily reproducible by an independent auditor.

According to them, a report on mineral inventory should also include (www.legalink.ch):

- a brief statement regarding the basis for confidence in the geological continuity of a mineralized zone;

Geological Assurance Classification and Type Degree of Economic Viability	Total Identified Mineral resources			Undiscovered Resources
	Measured	Indicated	Inferred	Reconnaissance
Economic	Proved Extractable Reserve (111)			
	Basic Reserve (111b)			
	Probabale Extractable Reserve (121)	Probabale Extractable Reserve (122)		
	Basic Reserve (121b)	Basic Reserve (122b)		
Marginal Economic	Basic Reserve (2M11)			
	Basic reserve (2M21)	Basic reserve (2M22)		
Sub-marginal Economic	Resources (2S11)			
	Resources (2S21)	Resources (2S12)		
Intrinsic Economic	Resources (331)	Resources (332)	Resources (333)	Resources (334)?

Notes: Of the codes (111-334) used in the table above, the first digital number indicates the degree of economic viability: 1=economic, 2M=marginal economic, 2S=sub-marginal economic, 3=Intrinsic economic, 7=economic interest undefined: the second digital number Indicates phases of feasibility assessment: 1=feasibility study, 2=pre-feasibility study, 3=geological study: the third digital number Indicates geological assurance: 1=measured, 2=indicated, 3=inferred, 4=reconnaissance, b=before the deduction of extractable quantities lost in the process of designing and mining.

FIGURE 13.4 Chinese reserve and resources classification.

Old Classification		A & B		C			D	E & F
New Classification								
'E' Economic Evaluation (100)	Designed mining loss accounted	Recoverable Reserve (111)	Probable Recoverable Reserve (121)		Probable Recoverable Reserve (122)			
	Designed mining loss not accounted (b)	Basic Reserve (111 b)	Basic Reserve (121 b)		Basic Reserve (122 b)			
Marginal Economic (2M00)		Basic Reserve (2M11)	Basic Reserve (2M21)		Resource (2M22)			
Sub-Economic (2S00)		Resource (2S11)	Resource (2S11)		Resource (2S22)			
Intrinsically Economic (300)		.	.	Resource (331)		Resource (332)	Resource (333)	Resource (334)
'F' Feasibility Evaluation		Feasibility (101)	Pre-Feasibility (020)	Scoping (030)	Pre-Feasibility (020)	Scoping (030)	Scoping (030)	Scoping (030)
'G' Geological Evaluation		Measured (001)			Indicated (002)		Inferred (003)	Predicted (004)
JORC							Unclassified or Exploration Potential	
					Inferred			
			Probable Reserve OR Indicated Resource					
		Proved/Probable Reserve OR Measured Resource						

FIGURE 13.5 Modified Chinese Reserves and Resources Classification. *Source: Courtesy from SRK Consultancy.*

- a concise commentary on sample density and the various sampling methods;
- a justification of the block size used as a basis of classification;
- a concise list of criteria used to classify individual blocks including, where possible, reference to levels of confidence for estimates; and
- justification of the cutoff grade used to distinguish waste from ore.

Classification schemes are a function of individual political jurisdictions; consequently, agreement on an internationally acceptable classification system, while highly desirable, might be difficult. A Qualified Person must be aware of the *specific requirements of the jurisdiction* within which he/she is working. A Qualified Person should clearly document the basis on which *continuously mineralized* zones are interpreted to exist.

Establishing a reasonable cutoff grade is required even at the resources classification stage because various jurisdictions indicate the need for classified resources to have economic potential. A Qualified Person should justify the cutoff grade selected.

Within the continuously mineralized zone, resources/reserves can be classified through the rigorous applications of appropriate, *clearly defined criteria*. These criteria should be documented by a Qualified Person in sufficient detail that the classification is easily reproducible by an auditor.

Wherever possible an array of *uniformly sized blocks* is desirable as a basis for resource/reserve classification because local criteria are then used to classify each block and because the block array approach lends itself to reproducible procedures that facilitate audit.

Vallee (2000), in the wake of having contemplated of different classes in the resource classification framework and the part of a Qualified Person, had observed a few inadequacies. A few reviews of mineral project execution in the previous 35 years demonstrate an example of incidental, yet intermittent, mining venture disappointments because of inadequacies in geological information, estimation, mineral handling and metallurgical testing, goals and prerequisites of feasibility study, and venture startup techniques, among others. A promise to quality certification and progressive development techniques and to gathering of adequate information of suitable quality specification is crucial. Current worldwide and Canadian proposition for resource/reserve definitions have taken a non-prescriptive attitude that will undermine plan of necessities, rules and industry norms for effective demarcation and estimation of mineral deposit, and for the project feasibility analysis. In order to reduce the difficulties, each feasibility analysis in support of a mine development and production choice should be on the basis of goals and techniques satisfactory to achieve end targets like the investors, as a decent commercial focus for a gainful task (www.cim.org).

The term "reserve" is applicable to a project that meets required technical, economic, legal feasibility requirements for successful project implementation, and the definitions of these categories should provide explicit limits. Current shortcomings will limit the ability of the Qualified Person to provide "adequate professional practice," fulfill his/her professional responsibilities to his/her employer or client and provide adequate information to the investing public. One must take advantage of present opportunities for "Setting New Standards for the current millennium."

References

Fennel, J.H., 1939. Ore reserves. Trans. Inst. Min. Metall. 49, 316.

Henley, S., 2004. The Russian Reserves and Resources Reporting system. Resource Computing International Ltd., UK, p. 13.

Hoover, H.C., 1909. Principles of Mining. McGraw-Hill Pub. Co., New York, 199 pp.

JORC. 2012. Australian Code for Reporting of Identified Mineral Resources and Ore Reserves, issued by the Joint Ore Reserve Committee (JORC), comprising Australasian Institute of Mining and Metallurgy (AusIMM), Australian Institute of Geoscientists (AIG) and Minerals Council of Australia (MCA), July, 19 pp. (AusIMM, Melbourne), 1996.

Kendal, J.D., 1901. Ore in sight. Trans. Inst. Min. Metall. 10, 143–202.

Leith, C.K., 1935. National mineral policy—purpose of planning committee's recommendations for the mining industry. Mining Cong. J. 21, 10–14.

Sinclair, A.J., Blackwell, C.H., 2000. Resource/reserve classification and the qualified person. CIM Bull. 93 (1038), 29–35.

Sinclair, A.J., Vallee, M., 1994. Reviewing continuity: an essential element of quality control for deposit and reserve estimation. Explor. Min. Geol. 3 (2), 95–108.

Sutphin, D.M., Renaud, K.M., Drew, L.J., 2011. An analysis of the published mineral resource estimates of the haji-gak iron deposit, Afghanistan. Nat. Resour. Res. 20 (4), 329–353.

Taylor, H.K., 1994. Ore reserves, mining and profit. CIM Bull. 87 (983), 38–46.

USBM/USGS, 1976. Principles of the mineral resource classification system of the U.S. Bureau of Mines and the U.S. Geological Survey. Geological Survey Bull. 1450-A, 5.

Vallee, M., 2000. Mineral resource + engineering, economic and legal feasibility = ore reserve. CIM Bull. 93 (1038), 53–61.

Vaughan, W.S., Felderhof, B., 2002. International mineral resource and mineral reserve classification and reporting systems. In: Presented at 48th Annual Rocky Mountain Mineral Law Institute, Lake Tahore, Nevada, July 2002.

Weatherstone, N., 2008. International standards for reporting of mineral reserves and resources—status, outlook and important issues. In: World Mining Congress & Expo, 2008, p. 10.

CHAPTER

14

Valuation of Mineral Properties

OUTLINE

14.1 INTRODUCTION

Mineral property valuation entails assigning of a monetary value to the worth of a mineral deposit. Valuation of mineral properties requires an integrated involvement of the geology, mining, processing, metallurgy, market, society, and the environment. Mineral exploration property is a mineral asset for which economic viability is yet to be demonstrated. Such a property is purchased, traded, and multiparty expressed on the premise of its perceived potential for the existence and discovery of a viable mineral deposit. The inherent worth of an exploration property is valued on the basis of its exploration potential (Roscoe, 2007). With regard to the best performing worldwide equity sectors, although mining and metals happen to be among those, yet clashing issues, viz., price fluctuations; impending retreats; and request demolition to resource shortage, regularly befuddle financial specialists. Of late, the significance of mining has turned out to be seeming, as article of trade and equity costs have surpassed most anticipations. As a result, mineral commodity investments have turned out to be more alluring buying and holding stocks for the long haul as they provide refuge in times of economic crunch and give an insurance against money devaluation.

Essentials of Mineral Exploration and Evaluation
DOI: http://dx.doi.org/10.1016/B978-0-12-805329-4.00021-1

The sum that can be reasonable to expend on exploration for an economic mineral deposit is a procedure to examine exploration potential. The appraised value scheme utilizes a cost approach methodology to assess exploration property. It depends on the reason that an exploration property is valued as the significant previous spending on exploration in combination with justified forthcoming expenses to examine balance exploration potential. Aftereffects of past investigation work are dissected, to hold just those past consumptions that are beneficial as far as recognizing remaining potential (http://www.cim.org). According to Bhappu and Guzman (1995), justified future expenses involve a sensible exploration spending plan to test that potential.

An investment in mineral exploration would be productive, provided its expected value is positive at the commencement of exploration program and expected return is higher than prevailing interest. Mineral developments generally have long lead-times prior to commencement of production. This stage is succeeded by mineral production, and ultimately to the stage of mine closure and rehabilitation (Lilford and Minnitt, 2005). Necessity of mineral resource price assessment can occur at any of these stages, yet not all valuation methods are relevant to all the stages of a mine life. Although valuation methods are not specific to any of the stages of mineral development, few of them might just be applicable to specific stages.

14.2 PERIODIC CHANGE IN MINERAL PROPERTY VALUES

The unpredictability of mineral value implies that the price assessment of the mineral resource would differ on a daily, weekly, and yearly basis. Hence, even conventional techniques such as Discounted Cash Flow (DCF), Relative Multiples, and so forth cannot be of use without a few changes and outlines. Value forecast of a mining organization is an intricate matter. Different techniques exist, yet many are not of use. The sensitivity of a gold company value to changes in gold price has been assessed by Peter Tufano of Harvard University. He observed that a 1% change in the price of gold usually brought about a 2% change in the value of a gold mining company. Gold market values being unpredictable are associated with everyday changes of 1% in a typical occurrence. Hence, in mineral property valuations, particularly with reference to legal issue, it is essential to set a date for property assessment, and to utilize the business sector's (in the case of fundamental analysis) or purchaser's and merchant's (in the case of fair market value) figures as of that date in the valuation (Davis, 2002).

Roscoe (2007) and Baurens (2010) have described in greater detail about various types of mineral assets that need specific techniques of appraisal. Various classes of mineral properties include: (1) Production properties, (2) Development properties, (3) Marginal development properties, (4) Exploration properties, and (5) Speculative properties. The classification is based on technological information instead of the type of mineral deposit. Production properties are mineral properties that are actually undergoing production to deliver marketable material on a daily basis from the mine on a business scale, which consists of mining and treating prior to selling in market. Development properties are those which are economic and feasible and are known to occur, including mines under production. These mineral properties are at an adequately progressed stage in which case sufficiently dependable information exists for mineral property valuation by DCF technique, with a sensible degree of sureness. Broadly, these information includes reasonably assured mineable assets, effective mining plan and production rate, metallurgical test results and process recoveries, capital and operating cost estimates, environmental and reclamation cost estimates, and projected product price (Roscoe, 2007). The net present value (NPV) of a sequence of estimated cash flows discounted at a rate to adequately highlight the risk of the mining project determines the worth of the property under this category. Marginal development properties are those which fall in between the exploration and development properties. These are properties with all around characterized mineral assets which could change to economically exploitable reserves under better conditions, and consist of adequate dependable information in revealing of financial aspects being peripheral with winning circumstances when the valuation commences. Better conditions may incorporate material values, advancements in scientific know-how, nearby facilities development, and so on. These additionally incorporate mines that are incidentally shut down because of low commodity values.

Exploration properties include those deposits that have not been demonstrated as economically viable. Estimation of asset under this category depends on the likelihood of occurrence together with demonstrating reasonable asset as economic. Eventually, only few of such properties would likely to be mineable assets. Exploration properties can also be with and without measurable mineral assets. Speculative properties include those mineral properties which have next to no certification of mineral prospective.

14.3 EXPLORATION ASSETS AND THE EXPLORATION PROCEDURE

Exploration assets are the naturally occurring potential mineralized volume of ground that ultimately leads to the stage of a mine. State-of-the-art exploration is a procedure which works in phases. Broadly, every phase of the exploration is intended to lead to the succeeding choice point as to whether or not to proceed further with exploration of a property, in view of consequences of the past stage. Every progressive phase becomes, by and large, further costly, because of the continuously increasing point-by-point activities. At whatever point an exploration activity passes onto the following phase, the estimation of a property might be improved, lessened, or continue as before, contingent upon how aftereffects of the project influence the potentiality.

Key aim of the exploration procedure includes recognizing together with focusing effort at the assets that further guarantees the exploration potential. Clearly, it is those exploration assets exhibiting greater potential that are of higher importance in mineral industry. Business analysts would have every kind of property assessed by means of an essential examination, discounting an anticipated stream of future revenue. This becomes problematic, yet, for exploration and speculative assets, as data on production potential and the scheduling of that generation, if available, are so inadequate that providing of revenue forecasts to a great extent becomes unimportant. Income methods become unsuitable during assessing of those choices, and thus it is needed to utilize genuine choices or choice examination systems in assessing these sorts of assets, or look to real exchanges to construe their qualities (Kilburn, 1990; Agnerian, 1996; Lawrence, 1998).

An individual can self-assertively separate the mineral resource's aggregate worth between the two sorts of benefits, normally deducting the expense of introduced capital from the aggregate resource worth to determine the estimation of the exploitable mineral. In any case, such bookkeeping is discretionary since the exploitable mineral merits nothing without the introduced capital, and the introduced capital merits nothing without the exploitable mineral. In other words, there is nothing characteristically profitable around a drill hole and as a consequence valuation on expenditure is not possible (Carson and Landefeld, 1994; BEA, 2000).

14.4 VALUATION TECHNIQUES, APPROACHES, AND METHODOLOGY

Different subdivisions of mineral properties are assessed for various motives which include merging and acquiring, property purchase transactions with a relationship or business affiliation between the seller and the buyer of the property, stock valuing of new securities issue to the public, backup for assets contracts, lawsuits, take-over, and coverage rights. "Value" refers to "Fair Market Value," which provides amount payable with respect to mineral asset by a ready seller to an eager purchaser in the open business sector in terms of real money or proportional on a specific date. Compelling date of valuation is one of the essential ideas, since mineral property estimations change over time, contingent on occasions on adjacent properties, market premium, ware costs, and so on. The estimation of a mineral task can be resolved utilizing an assortment of valuation systems and related approaches, namely, (1) market, (2) cost, and (3) income. Choice to apply any valuation approach would be contingent essentially on the phase at which the task has been created. Valuation approach to deal with a greenfield undertaking will be significantly unique in comparison to the one with highly investigated mineral asset. Moreover, an activity of valuation might create diverse results for the same venture contingent upon which technique has been employed. Most basic part incudes the ability of the competent person performing the valuation in order to distinguish notable concerns with an assurance of integrating in the valuation procedure. The vulnerability of a straightforward application of automated keys for appraisal devoid of a clear comprehension of the consideration factors and the field of probability need to be stressed. Experience, knowledge of best practice, and the capacity to perceive of the valuator and the necessity of adherence to in the minerals business constitute the utmost imperative attributes in respect of a valuator (Lawrence, 1989, 1998; Davis, 1998, 2002; Cairns and Graham, 1998; CIM, 2000; Domingo and Lopez-Dee, 2007; Roscoe, 2007).

The three primary ways to deal with valuation of mineral properties are (1) Cost, (2) Market, and (3) Income. Distinctive methodologies apply to various sorts of mineral properties as do diverse methods, and are shown in Table 14.1. The appraised value method (AVM) can be connected to exploration properties and as a rule to marginal development properties. Income approach are not thought to be suitable for exploration properties.

TABLE 14.1 Valuation Approaches to Various Kinds of Mineral Assets

		Properties		
Approach	**Technique Employed**	**Development**	**Marginal Development**	**Exploration**
Cost	Appraised value	Not applicable	Certainly	Certainly
	Geoscience factor	Not applicable	Likely	Certainly
Market	Comparable transaction analysis	Certainly	Certainly	Certainly
	Option agreement expressions	Certainly	Certainly	Certainly
Income	Discounted cash flow	Certainly	Likely	Not applicable
	Option pricing	Certainly	Certainly	Not applicable

After Roscoe (2007).

Market approach techniques are ordinarily utilized as a cross-check on the appraised value technique (Roscoe, 2007).

14.4.1 Cost Approach: AVM

The AVM has been discussed in detail by Roscoe (1994, 1999), Agnerian (1996), Thompson (1991), Lawrence (1989, 1998), and Ellis (2011). The method is dependent upon the hypothesis of genuine estimation of an exploration property or a marginal development property reflecting its potential for the presence and revelation of a deposit that is considered to be economic. AVM expects the extent of spending on exploration necessary on a property is related to its worth. The cost methodology is given some legitimacy by the way that option agreements to mineral properties are frequently based on exploration spending requiring to earn interest. There is additionally frequently a reference to past exploration spending in option agreements that can be related to the residual interest value of the optionee (Roscoe, 1999).

The fundamental precept of the AVM is that an exploration property is justified regardless of the important past exploration spending in addition to justified future expenses. A critical component of this strategy frequently neglected in its application is just those past spending that are viewed as sensible and profitable are held as value. Beneficial implies that the aftereffects of the work give adequate consolation to warrant further work by distinguishing potential for the presence and revelation of an economic mineral deposit. Justified future expenses involve a sensible exploration spending plan to test the recognized potential that may be geophysical or geochemical anomalies, or favorable showings or known mineral bodies. As noted already, if exploration activity reduces potential, it is not gainful and its expense should not be held as worth or should be diminished. Clearly, if the property is considered to have insignificant exploration potential, it has practically no worth (Ellis, 2011).

The AVM is best suited to properties that are currently in the exploration stage. It is hard to apply this strategy to properties that have remained dormant for few years, particularly those which have had considerable spending before. Numerous such properties

TABLE 14.2 Guidelines for Retained Spending in Respect of Marginal and Idle Properties

Retained Fraction of Past Spending	Guidelines
0.75	Property with assets yet no work accomplished for a few years. Additional forthcoming work is justified. Generally a property with minimal assets and potential for all the more, however not exactly sufficiently energizing to pull in exploration spending effectively
0.50	Property with subeconomic assets, yet might have some potential in foreseeable future, restrictive on product costs, support facility, enhanced innovation, economic conditions, and so forth. Further wok not suggested at time of valuation
0.25	Idle property with subeconomic assets with almost no desire for advancement, however cannot discount them totally
0.0 to 0.1	Idle property with no exploration potential or without any assets
Nominal value of $5000 to $10,000	Dormant property having vague yet very little or insignificant investigation

Modified after Roscoe (2007).

have subeconomic or marginal resources demarcated by the past work, and few succeed as marginal development properties. The way to the valuation of dormant properties is a practical evaluation of the balance exploration potential that include green exploration targets, potential to raise the grade or tonnage of the current asset, or potential for development with changes in innovation or commercial conditions (Kilburn, 1990). In respect of marginal development properties and idle exploration properties, a set of guidelines relating to the fraction of the past spending to retain as value is given in Table 14.2.

Utilization of the AVM requires an intensive comprehension of the exploration process, business standards, and unit costs of exploration. It requires that the valuator understands possible outcomes and suitable methods of exploration with reference to geology and favorable domains for exploration. All the necessities are satisfied by a trained person having a good combination of knowledge, experience, and specialized ability to judge, and in addition knowledge about certifiable property transaction values. One point of preference of the AVM is that exploration cost data and specialized information are promptly accessible to properties under the categories of exploration and marginal development. It is a good method for looking at the relative estimations of mineral properties.

14.4.2 Market Approach

Strategies utilizing a market approach find relevance to a wide range of mineral deposits. Strategies explained under this approach include (1) comparable transaction method and (2) option agreement terms method (Roscoe, 2007). The second method mentioned here is in regular use to put a worth on mineral property transactions utilized for comparison, because of the fact that most mineral property transactions are not money deals. The operative date of the valuation is vital for these techniques so that comparable transactions ought to be within a stipulated time from that date.

14.4.2.1 Comparable Transaction Method

The technique utilizes the transaction value of equivalent properties to build up a worth for the mineral asset (Thompson, 1991; Roscoe, 1994, 1999). The constraint of this methodology in mineral business lies in the nonavailability of genuine equivalents, since every property is novel as for key variables, for example, geology, mineralization, costs, phase of investigation, and support structure. In general, there are moderately couple of businesses for mineral properties as contrasted with the number of landed property business, by and large. At whatever time mineral business transactions of such nature do happen, they seldom include stringently involvement of cash money, thereby giving the only option to the valuer to changing over pieces of stocks, dividends, bonds among others into existing cash proportionate.

Despite the aforesaid specifications, business prices of comparable assets may demonstrate a scope of qualities in respect of a specific property. Transactions of exploration property likewise provide a sign of how dynamic the market might behave at any point of instance. To illustrate this, it can be said that lately there have been generally few exploration property exchanges across North America, in view of the discouraged condition of the exploration and mining commercial enterprises, resulting into comparably low market values.

The worth of an exploration property relies upon its latent about the presence together with finding of a deposit that turns to be economic. Exploration potential of a mineral property is dependent to certain degree on its geographical location, and depends to a more degree on its geological characteristics, mineralization showing, exploration findings and potential locales, details of contiguous areas, among others. Mineral properties under exploration with the leasehold of existing areas of mining frequently have a superior worth due to greater anticipated prospective for locating orebodies, along with the presence of existing support structures. The fundamental point of preference of this technique is for reason of validating the worth of mineral properties determined with the application of different strategies, which gives a general measure of comparable property worth. The primary detriment is the nonavailability of genuine equivalents since every such property is different from others.

14.4.2.2 Option Agreement Terms Method

The method is applicable when a mineral premises is liable to a prevailing option agreement. Characteristically in an option agreement, a plan of conferred and voluntary money disbursements and work responsibilities applies over a span of few years. An estimation of the worth of the property is reflected in the disbursements made and work responsibilities satisfied, in addition to the subjective likelihood of the optionee making the remaining of the disbursements and attaining the rest of the exploration programs (http://www.cim.org).

This strategy is best suited to properties being effectively explored amid the early period of the option duration. The strategy is by and large not extended to properties on which the option has been practiced by satisfaction of the disbursement terms and work responsibilities. Around then, the property worth as a rule surpasses the disbursements made. One point of interest of this strategy is that it has some true legitimacy in the early years of the option period. One drawback is that the valuation is important just amid the early years of the option period. With time as more exploration results are gathered, the property worth is liable to deviate up or down from the option agreement terms. Either the outcomes would not legitimize further spending and the option is abandoned, or results will be adequate that further spending and disbursement terms will appear to be a deal in comparison to the property worth. The strategy can be utilized to decide the value of comparative dealings, because most exploration property transactions are option or joint endeavor procure in understandings (Roscoe, 2007).

14.4.3 The Income Approach

The approach is employed as an intermediary measure of market value or as an indirect method of utilizing market values, where these are not present. In all, there are five methods available as given below and attempted that consider future advantages/ income streams which can be obtained from the mineral resources as supplier of capital services (http://unstats.un.org).

14.4.3.1 Net Price Method

The technique can be a distinct option for NPV on the basis of the Hotelling rent model which accepts that under certain economic situations nonrenewable asset rent will increase at a rate equivalent to the rate of discount (rebate) as the asset becomes scare. The worth of the asset stock can be computed essentially as the present rent per unit of asset times the span of the stock. Since rent increases with time at a rate precisely adequate to counter-balance the rebate rate, there is no compelling reason to markdown future asset income. While this strategy is easy to apply, this has been observed to overestimate the market value of beneath the ground resources. While Canada applied the NPV technique as well as the net price strategy, Philippines applied the net price strategy in its beginning mineral records arrangement and is looking at the possibility of shifting to NPV technique (Domingo and Lopez-Dee, 2007).

The drawbacks on the applicability of the method include:

- The supposition of the Hotelling model that under impeccable rivalry, the rents would ascend in accordance with the rate of premium may not hold as a general rule because of business sector defect.
- The rent utilized might incorporate different types of rent, for example, rent because of variances in the production cost, cartel rents due to the implementation of business power in fixing price, notwithstanding the genuine asset rent.
- The global prices of minerals are not controlled by flawless contest.

14.4.3.2 User Cost Method or "El Serafy" Method

This technique is applicable to environmental assets that create marketed facilities. It differentiates between the actual income and the gross receipts created by an asset. It characterizes actual income as the measure of income that would be maintained forever irrespective of fixed lifetime of the asset by aptly contributing a bit of the gross receipts created which can be the consumption cost, generally alluded to as the client cost. The relationship between the actual income and the gross receipts is expressed as:

$$N - X = N/(1+i)^{n+1}$$

where X is the actual income that can be spent, N is the aggregate annual receipts, i is the rate of discount, and n is the further period in years for which current extraction rates could be maintained (Domingo and Lopez-Dee, 2007).

The drawbacks on the applicability of the method include:

- With a specific end goal to figure the client costs, a few presumptions are required that are prone to bias the estimates.
- With respect to aggregate annual receipts for a given year t, (N_t), the present level of receipts is held steady amid the lifetime of the asset.
- The rate of extraction is additionally held constant until the last fatigue of the asset, accordingly the life expectancy of resource in the present year, n, is not permitted to alter with time.
- It presumes a steady rebate rate.

14.4.3.3 Net Present Value

Numerous organizations prescribe the utilization of NPV for appraising mineral asset stocks. In light of the investigation directed by the underground resources, nearly all nations accepted and implemented the NPV technique with the exception of the Philippines. It is an accepted strategy which calculates the net revenue streams of a resource extending over the cost-effective lifetime of the asset. The method involves predicting flow of forthcoming total income an asset would create during optimal extraction, followed by decrementing the income flow by applying a suitable cost of capital. Under specific conditions, for example, taxes being assumed as nil, the aggregate of the discounted values of income for a given period of time would break-even with the asset worth in the market. This approach is reliable with maximizing the worth of a company and is utilized by financial specialists as a part of the assessment of an organization or in capital planning when looking at the value of various ventures.

Four advantages of using NPV for asset appraisal include:

- *Concept of Money Value With Time:* It perceives the idea that money received now is of more worth than money received after some time.
- *Revenue Streams*: It computes an asset's anticipated revenue streams together with incorporating risks associated. It aids in removing discrepancies in financial accounting, because the income streams includes all benefits.
- *Risks*: NPV fuses the risks connected with an asset by means of the anticipated revenue streams with or without the rate of decrement.
- *Adaptability*: It gives adaptability and profundity, owing to change in the mathematical statement of NPV on account of price hike.

In spite of its many advantages, NPV has few restrictions of which valuators should be conscious of. They include:

INCOME FLOWS AND DISCOUNT RATE

An analysis of NPV is associated with two primary requirements, viz., Revenue stream and Rate of Return. Revenue stream speaks about expected net benefits amid the lifetime of the asset. Bearing in mind the presumptions and expectations underlying real-income stream, deciding expected net benefit is difficult in most practical situations. Decision on a suitable rate of decrement is critical to the scheming of NPV. Approximation of rate of decrement can be made using various approaches for estimating the rate of decrement; however, the one widely recognized is to search for an equivalent risk associated investment of known rate of decrement. One simple method for choosing a suitable rate of decrement is to consider the return on investment of another project where the principal required for the exploration could provide return when invested. Clearly, NPV attained utilizing changing rates of decrement over the asset life span is comparatively sensible over the one computed utilizing a fixed rate of decrement for the lifetime of the asset. The rate of interest considered for decrementing expected revenue streams to respective present worth remains one vital information in the procedure. Maximum of the companies possess an all-around characterized strategy with respect to their capital structure. Likelihood of future changes in tax rates, prices, inflation, etc., are matters of perceptions, and one party's perception could be challenged by the other. On the other hand, higher discount rates can be utilized for ventures associated with high risk. A higher assumption of this rate would be in favor of the buyer and a lower one in favor of the seller.

CAPITAL AVAILABILITY

Assumption of a value of NPV exceeding zero is made irrespective to the requirement of principal. Regardless of what amount is required or what are the limitations, capital is often presumed to be promptly available. In practical situation, this is not so since access to capital markets is restricted by general performance of the organization.

OPTION INVESTMENTS

The capital prerequisites might undergo change with time, requiring choices that might alter the risk scenario. NPV utilizes data identified during the fruition of the investigation. It is ascertained in a nondynamic way so as to not to take into account any expected variations. The unbending nature of this supposition may prompt values to be underestimated.

RATE OF REINVESTMENT

Rate of reinvestment is presumed at the decrementing rate applied in the NPV scheme, which is alright when capital apportioning is not required. However, IRR scheme does not make any supposition regarding the rate of reinvestment that has a tendency of misrepresenting estimated worth.

14.4.3.4 Appropriation Method

In numerous nations, governments are the essential proprietors of the country's common assets. As landowners, governments collect the whole rent resulting from extraction of the assets they claim. Asset rent is ordinarily collected by governments through fees, taxes, and royalties exacted on organizations that are involved in extraction. One method for evaluating the monetary rent owing to an asset is to compare this with the fees, taxes, and royalties collected from the organizations that had already carried out asset extraction. Be that as it may, by and by, fees, taxes and royalties have a tendency to downplay asset rent as they might be set by governments in view of different needs; for example, implicit price subsidies to extractors, and encouraging employment in the industry. Additionally, the rate of installments to government may not move in accordance with market price for the extracted item; however, one would expect the actual economic rent to do as such. At the point when these information are not independently identifiable, or suitable, asset rent must be ascribed utilizing different roundabout strategies. Be that as it may, if the two arrangements of information are available, an examination of the values might be useful for analysis of economic policy (Domingo and Lopez-Dee, 2007).

14.4.4 DCF Method

DCF valuation method is the most normally utilized valuation instrument (Van Horne, 1977; Schwab and Lusztig, 1969). The strategy has particular merits over the other techniques applied in the market and cost approaches. These incorporate its capacity to consider the impacts of royalties, leases, taxation, and financial outfitting on the subsequent income. What is more, the useful effect of unredeemed capital balances assessed losses, depreciation, and amortization on free cash streams can likewise be demonstrated. Aggregating cash streams on assets classes as inferred, or those with further lower geological certainty, is forbidden by some global codes. It is just under special circumstances that numerous securities trades will acknowledge such cash streams and the impact of cash stream commitments from inferred resources on project functioning should be established independently from those obtained from other resource and reserve classes. The method can be utilized in creating various quantifiable outcomes. All alone and as an investment device, it depends on the rule that for an initial investment, the investor would look to the future cash streams of that unit to give a base return, which will be no less than a foreordained return over the investor's hurdle rate for that venture. The hurdle rate speaks to the base return of a venture, less than which the choice to invest or develop a fresh venture will be negative, or more than which the project will be developed. The hurdle rate ought to dependably be higher than the capital cost for the investor (http://diversity.org.za/saimm2/Journal/v105n01p029.pdf).

In a macroeconomic domain that is adequately promising and steady for the DCF strategy to be applied to a mining venture, the critical information for the most part will be included in a Life of Mine (LoM) scheme. The LoM scheme which is associated with a pre-feasibility, feasibility, or a bankable feasibility study would comprise (1) estimates of mineral resources and mineral reserves compliant with international reporting codes; (2) report of predicted grade, tonnage, and associated recovery on daily, monthly, or annual basis; (3) projected cost of mining, milling, and refining; (4) estimated profiles of capex over the mine life; and (5) social responsibilities and contributions and other factors that may affect the expenditure or income.

Variations in balances of working capital are by and large computed taking into account history of balance proportions, connected to predicted incomes and working expenses. They effect on short-term cash streams and accordingly should be demonstrated into the cash streams. Actually, any working capital locked up amid the life of the mine operation will be discharged toward the end of the operation life. Once the financial inputs have been accepted, the DCF is derived. The resultant cash stream is then used to determine the net present worth (NPV) of the operation at a foreordained discount rate or a variety of discount rates. The determined NPV, on which the rate of return on investment can be computed, is utilized as an intermediary for the operation's verifiable worth. This is frequently contrasted and the worth or returns the market ascribes to the mine operation, in the event that it is a recorded entity, or contrasted and other investment opportunities so as to optimize investment or development plans. In any cash stream calculation, the effect of inflation on the last outcome cannot be exaggerated. One just needs to consider the impact of tax assessment as connected to real taxable income rather than being levied against general taxable income (Lilford and Minnitt, 2005).

The essential valuation technique for development properties and operating mines is DCF. The basic valuation principle is that value reveals the existing net cost benefit of the net cash streams that are required to be produced over the project life. Evaluating this benefit includes scheming of expected after-tax cash streams and changing over these cash streams into a present worth or net asset value (NAV) by method of discounting, a procedure that introduces a discounting factor to result in a rate of return on investment considering the time value of money as well as risk criteria. DCF NAV estimations have numerous utilizations in the mining business, including legitimizing project scheme decisions or investment choices, and giving value estimations to accounting checks. Although the DCF strategy is broadly acknowledged, it is liable to constraints and occasionally is inconsistent. Valuators at the cutting edge of mine valuation are looking to enhance DCF scheme through three critical improvements over a static DCF approach, viz. (1) the dynamic modeling of uncertainty; (2) the recognition of the cash flow effects and contingent payoffs; and (3) the use of market-based risk discounting methods such as real options (http://www.canadianminingjournal.com/features/mining-valuation-three-steps-beyond-a-static-dcf-model/).

14.4.4.1 Dynamic Modeling of Uncertainty

Normal DCF schemes are static and hence include a specific net cash stream with expected values of mining venture and commercial variables, for example, grade of metal. Specialists perceive that these variables are uncertain and might evaluate how uncertainty affects NAV with a sensitivity analysis where variables, for example, price of metal are made to vary over a limit. Monte Carlo simulation technique may be adopted for analysis of uncertainty, where a probability distribution depicts probable changes in a particular variable over the life span of the mine operation. However, these methods do not contemplate how projections are revised and unreliability is settled over the life span of the operation. Incorporating stochastic procedures into Monte Carlo simulation can give an improved portrayal of the dynamics of uncertainty and its impact on cash stream. A log-normal procedure may be applied to model back propagation in base metal prices. Such schemes are applied to signal additional features, for example, uncertainty in long-term equilibrium value levels and the onward nature of curves (http://www.canadianmining-journal.com/features/mining-valuation-three-steps-beyond-a-static-dcf-model/).

14.4.4.2 *Flexibility and Contingent Payoffs*

A static DCF model depends on a cash stream situation based on one policy, jointly with foreordained financing and tax payouts. However, this situation might give a defective assessment of cash stream, since working approach might be modified and the structure of financing and tax payoffs may be changed in light of changes in the venture and business setup. This impediment of static DCF is especially tricky when endeavoring to assess the worth of subeconomic resources, the financial effect of extra duties, or the actual expense of a financing game plan with embedded commodity derivatives. A normal procedure to balance the DCF deficiency is to build a set of schemes for various designs and venture setups. Notwithstanding, this methodology leads to a dispersing of NAVs that frequently give little knowledge into how flexibility and contingent payoffs interrelate with uncertainty to influence venture NAV (http://www.canadianminingjournal.com/features/mining-valuation-three-steps-beyond-a-static-dcf-model/).

14.4.4.3 *Market-Based Valuation Methods*

The configuration of discounting risk in a static DCF NAV process suggests an uncertainty description in which net cash stream uncertainty develops with time in an all-around manner. Be that as it may, for most ventures, net cash stream uncertainty is not all around carried on as it changes with year in an inconsistent way amid the venture because of components, for example, changes in metal price, grades, and in the fatigue of tax shields. The irregularity between an undertaking's real-risk description and the description suggested by the DCF risk discounting function is especially difficult during valuation of long-life base metal ventures where vast forthright capex are needed prior to cash streams are created over duration greater than 30 years. Business economics has created market-based techniques, for example, option pricing to value business assets which have dynamic risk features. These techniques, when applied to the mining business, utilize the same data and have a comparable way to deal with building a cash stream scheme as the normal DCF strategy. Be that as it may, the DCF and market-based NAV techniques are distinguished by the manner they apply to risk tuning. While the DCF approach applies collective risk tuning to the net cash stream, market-based techniques apply a customized risk tuning to individual uncertainty sources in light of their systemic risk structure before establishing a risk tuned net cash stream (http://www.canadianminingjournal.com/features/mining-valuation-three-steps-beyond-a-static-dcf-model/).

14.4.4.4 *Tail Margin Method*

An augmentation of the DCF/NPV technique for valuation prompted the advancement of the tail margin valuation strategy (Lilford, 2002). Prior to the application of this approach, an assumption is made that the LoM plan or mining profile can be assessed with a certain degree of assurance since it depends on mineral resource. Be that as it may, the mining profile does not consider the extraction of the entire mineral resource. Then again, if mineral rights are to be valued, these rights ought to be situated nearby or extensional to a current mine working, which would permit a valuer to suggest vital notions with respect to its working factors.

With the cash stream description of a current operation accomplishing a stable state, the free cash stream is rated as unit value. Stable state alludes to the process working within sustainable, steady levels regarding mine output, operating costs, capex, working capital, and tax structure. For application purposes, stable state can be taken as 3- to 5-year period, in which the margin does not differ by more than 20% between the two end periods and by not more than 10% between two progressive years. The annual rates are ascertained over a foreordained stable state duration and subsequently averaged to indicate a deflated and discounted unit rate. Cash streams from these periods are deflated to actual monetary expressions terms and discounted at a suitably decided rebate rate. A unit rate is then extended to the metal units, being the expected tail quantity that fall beyond the LoM plan. Forecast of the recoverable metal units depend on the working's current mining and metallurgical components. Applicability of tail margin valuation can be made in most situations, though with slight limitation in the valuation of mineral properties. The technique has a tendency to be limited to one or the other situations, viz. (1) where the resources of a working are vast to the point that a DCF examination can actually value a bit of the life of the asset. As a result, the discount rate in the operation's subsequent period shrinks additive value with time; (2) where the LoM scheme does not include mining of total reserve of a currently working mine; (3) where the valuation of a sector of mineral rights in the bordering property is being mined. The bordering operation will have adequate information accessible for a valuer to arrange a DCF examination and subsequently decide a material unit value rating. This rating can then be extended to the mineral rights sector; (4) where a mineral property owes to its resemblance in terms of salient attributes to a current working or project but not necessarily situated in proximity to the property.

Besides, it is improbable that the entire mineral reserves that could be mined would be accurately assessed during the feasibility study. The economic mineral body on which the mine operation is initially started is liable to be represented by a part of what may be mined when the operation ends. For this situation, mineral production past the anticipated LoM at the feasibility stage is relied upon to be sourced from extensions of the same economic mineral body or new bodies. Extraction from extensions to known mineral bodies would need the services of infrastructure that are existing, and for new findings of economic mineral body new infrastructure would be required. Accordingly, a LoM is calculated. The idea offers consent to the intrinsic uncertainty associated with any resource/reserve estimation and the belief that data gathered during mining might add resources/reserves to the LoM scheme. This would especially be noteworthy when depletion of mineral reserves in a working mine is imminent and the later findings of mineralized extensions (the tail margin) adds value and life to the existing mining venture. It is this specific occasion that the tail margin technique for valuation refers to. If a mineral body is geologically heterogeneous, the lease area might need to be dealt with as more than one separate valuation pieces. Every piece will be certified with a sustainable unit value for every volume of item. Further, valuation will depend on the knowledge and experience of the competent valuator (Lilford and Minnitt, 2005).

14.5 MINERAL VALUATION CODES

Mineral resource/reserve valuation codes have endeavored to institutionalize how the valuation of mineral assets is carried out. The codes have likewise expanded people's reporting responsibility through professional member bodies. Globally, the International Accounting Standards Board issues paper for comment on the extractive industries and the International Valuation Standards Committee rules should be followed. The motivation behind valuation codes is to guarantee reliable reporting of data and appraisals. Most codes offer fundamental valuation standards yet differ as far as techniques acceptable for valuation, meanings of value, and specific substance necessities. The implementation of codes vary with security exchange jurisdictions and is dependent upon selection level by stock exchanges, watchdogs, and certified bodies. These codes layout best practices, which legitimate prerequisites for competent persons carrying out the job of valuing and preparing documents for public reporting.

Internationally, the major national valuation codes are as follows: The Code for the Technical Assessment and Valuation of Mineral and Petroleum Assets and Securities for Independent Expert Reports (The VALMIN Code, 2005) developed by a joint committee of the Australasian Institute of Mining and Metallurgy with a number of other bodies, for Australasia; The Standards and Guidelines for Valuation of Mineral Properties (The CIMVAL Code, 2003) developed by a Special Committee of the Canadian Institute of Mining, Metallurgy and Petroleum (CIM) on Valuation of Mineral Properties for Canada; and The South African Code for the Reporting of Mineral Asset Valuation (The South African Mineral Asset Valuation (SAMVAL) Working Group, 2009) developed by the South Africa Mineral Asset Valuation (SAMVAL) Working Group under the joint auspices of the Southern African Institute of Mining and Metallurgy (SAIMM) and the Geological Society of South Africa (GSSA), for South Africa (Njowa et al., 2014). In April 2012, CRIRSCO-equivalent committee for valuations, named the International Mineral Valuation Committee (IMVAL), was created in Brisbane soon after the VALMIN Seminar Series. IMVAL intends to provide a platform for the harmonization among CIMVAL, SAMVAL, and VALMIN codes to promote best practices in the international reporting of mineral valuation results. Abergel (2014) has provided a summary of key features of CIMVAL, VALMIN, and SAMVAL valuation codes.

14.6 CONCLUDING REMARKS

The most basic restriction controlling all valuation techniques rests with the valuer. The capacity to effectively decipher the majority of the accessible data so as to choose a favored valuation strategy is imperative. Once the requisite data are obtained, valuer must know about the setbacks of each of the techniques accessible for use. Obviously, new techniques can be produced, yet they are not effortlessly inspired to acceptance by the valuation association or community. The market approach depends on past mineral property transactions to give a best gauge to the present estimation of a property. The most generally acknowledged valuation techniques in this approach incorporate the value per unit strategies and Lilford TEM strategy, Kilburn, US$ per ounce, market capitalization per ounce, and comparable asset valuation techniques. The cost approach is a standout among the most straightforward valuation approaches available. It depends on the reason that the estimation of a property must be worth in any event that sum consumed on the property to accomplish a specific level of geological comprehension. Attributable to its effortlessness, the methodology disregards a large portion of the basic quality characteristic in any mineral property. The two critical strategies here are the multiples of exploration spending and the analysis techniques. The income approach exhibits the most broadly utilized and comprehended valuation systems accessible to valuers. The premise for a considerable lot of the particular techniques under this heading is DCF strategy. With the appropriate rate of discount, the DCF strategy helps the developer in the contemplations for initiating project development, augmenting the life of a project, or/and growing the size a venture. Obviously, lenders and different partners can likewise consider DCF valuations for their particular purposes. With this as a base, sensitivity analyses and simulation analyses can be applied to survey the vigor of a venture under various situations. Extending the power of DCF strategies, the tail margin and option pricing techniques of valuation are recognized. Advances in technology in the course of recent decades have helped with making the use of formerly unsound valuation techniques more esteemed. The probability technique for analysis of mineral property and the Monte Carlo simulation strategy have been developed to be computer dependent for use. It must be recognized that valuation approaches are not characteristically secret elements that anybody can use to produce a response. Experienced valuers are equipped with the knowledge of best practice and accordingly will choose the most pertinent strategies as required. They might likewise modify strategies to suit particular circumstances, yet will must have the capacity to legitimize their deviations from the standard. It is fundamental again to emphasize the significance of the valuer's experience involved in guaranteeing that the most relevant valuation procedures are considered for valuation endeavors. Wherever conceivable, one can utilize diverse valuation techniques to expand the validity of the outcomes (http://diversity.org.za/saimm2/Journal/v105n01p029.pdf).

No valuation strategy can be said to be correct, at the same time, no technique is incorrect either. The techniques portrayed ought not to be seen as being exclusive of one another. The basic thought is that they ought to supplement the findings of one another. All valuations of organizations today contrast from one another on the grounds that the organizations are diverse as well as in light of the fact that distinctive individuals with varying information and foundation carry out valuation exercise. Particularly in the valuation of organizations in mining business, an extremely tough job is assessing the quantity of future annual ore produce from mine because of the association of element of uncertainty in the assessment. The price predictions of mineral commodities are also extremely difficult to foresee and will vary even among professional experts. Additionally, different inputs, figures, as factor of discount, costs and technique utilized, change from expert to expert. The organization valuation gives one conceivable conjecture for an organization and ideally gives a decent sign without bounds, however with time, there could be a probability of demonstrating it incorrect. In any case, basic philosophy on valuing together with application of the fundamental schemes can be made owing to the fact estimations can be different. Fig. 14.1 shows the applicability of different valuation methods with respect to the stages of development of a mineral property.

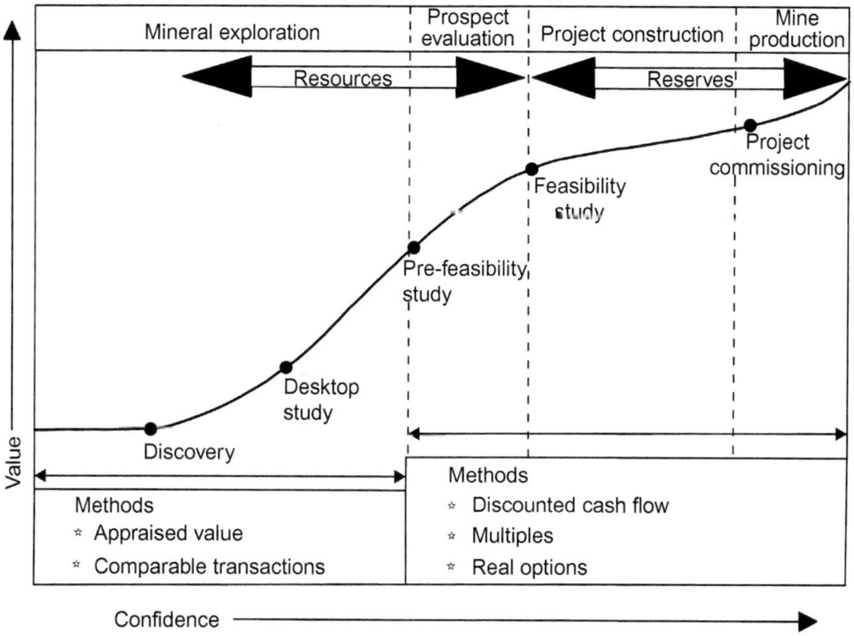

FIGURE 14.1 Different valuation methods depending on the stages of development for a mineral property.

References

Abergel, K., 2014. Mineral Asset Valuation Codes: Towards an International Standard (MASc thesis). Dept. of Mining Engineering, Queen's Univ., Kingston, Ontario, Canada, pp. 9–46. <https://qspace.library.queensu.ca/bitstream/1974/12364/1/Abergel_Keith_201408_MASc.pdf>.

Agnerian, H., 1996. Valuation of exploration properties. CIM Bull. 89 (1004), 69–72.

Baurens, S., 2010. Valuation of metals and mining companies. Collaborative Study of Univ. of Zurich, Swiss Bankng Inst., and Dr. T. Hens., November 2010, 81 p.

BEA, 2000. Bureau of Economic Analysis. Accounting for Subsoil Mineral Resources. Survey of Current Business, February 2000. Found at: <http://www.bea.gov./scb/pdf>.

Bhappu, R.R., Guzman, J., 1995. Mineral investment decision making: a study of mining company practices. Eng. Min.J., pp. 36–38., July 1995.

Cairns, R., Graham, A.D., 1998. On using current information to value hard-rock mineral properties. Rev. Econ. Stat. LXXX (4), 658–663, 1998.

Carson, C.S., Landefeld, J.S., 1994. Accounting for Mineral Resources: Issues and BEA'S Initial Estimates. Bureau of Economic Analysis. Survey of Current Business 74 (4), 50–72.

CIM, 2000. CIM Standards on Mineral Resources and Reserves: Definition and Guidelines, August 2000.

CIMVAL, 2003. Standards and Guidelines for Valuation of Mineral Properties, Special Committee of the Canadian Institute of Mining, Metallurgy and Petroleum on Valuation of Mineral Properties (CIMVAL, 2003). <http://web.cim.org/committees/CIMVal_Final_Standards.pdf>.

Davis, G.A., 1998. A review of option valuation. In: Torries, Thomas F. (Ed.), Evaluating Mineral Projects: Applications and Misconceptions. Society for Mining, Metallurgy, and Exploration, Inc., Littleton, CO, pp. 117–123.

Davis, G.A., 2002. Economic methods of valuing mineral assets. In: ASA/CICBV 5th Joint Business Valuation Conference, Orlando, Florida, October, 27 p.

Domingo, E.V., Lopez-Dee, E.E.P., 2007. Valuation methods in Mineral Resources. In: 11th Meeting of the London Group on Environmental Accounting Johannesburg, March 26–30 ,2007, 13 p.

Ellis, T.R., 2011. A review of mining cost approach methods for mineral evaluation. In: SME-AIMA Annual Meeting, Denver, March 2011.

Kilburn, L.C., 1990. Valuation of mineral properties which do not contain exploitable reserves. CIM Bull. 83, 90–93.

Lawrence, R.D., 1989. Valuation of mineral assets: accountancy or alchemy? In: Paper Presented at the CIM Annual General Meeting, Quebec, May 2, 1989.

Lawrence, R.D., 1998. Valuation of mineral assets: an overview. In: Paper Presented as Part of a Course Offered by the Geological Association of Canada and the Prospectors and Developers Association of Canada, May 17, 1998.

Lilford, E.V. 2002. Methodologies in the valuation of mineral rights. Project Report submitted to the Faculty of Engineering, University of the Witwatersrand, Johannesburg.

Lilford, E.V., Minnitt, R.C.A., 2005. A comparative study of valuation methodologies for mineral developments. J. S. Afr. Inst. Min. Metall., pp. 29–41, January.

Njowa, G., Clay, A.N., Musingwini, C., 2014. A perspective on global harmonization of major national mineral asset valuation codes. Resour. Policy 39, 1–14.

Roscoe, W.E., 1994. The value of mineral properties as raw material for new mines. In: Presentation to the CIM Mineral Economics Society Discussion Group, October 1994.

Roscoe, W.E., 1999. The valuation of mineral properties for compensation. In: Presentation to the British Columbia Expropriation Society, Fall Seminar, Vancouver, October 1999.

Roscoe, W.E., 2007. Valuation of mineral exploration properties using the cost approach. <http://www.cim.org/mes/pdf/VALDAYBill_Roscoe.pdf>.

Schwab, B., Lusztig, P., 1969. A comparative analysis of the net present value and the benefit-cost ratio as measures of the economic desirability of investments. J. Finance, 507–511, June 24, 1969.

The South African Mineral Asset Valuation (SAMVAL) Working Group, 2009. The South African Code for the Reporting of Mineral Asset Valuation (The SAMVAL Code), 2008 Edition as amended July 2009. <http://www.samcode.co.za/downloads/SAMVAL2009.pdf>.

Thompson, I.S., 1991. Valuing mineral properties without quantifiable reserves. In: Paper presented at CIM Mineral Economics Symp., Toronto, January 17, 1991.

The Valmin Code, 2015. Code for the Technical Assessment and Valuation of Mineral and Petroleum Assets and Securities for Independent Expert Reports. <http://www.valmin.org/valmin_2005.pdf>.

Van Horne, J., 1977. Financial Management and Policy, fourth ed. Prentice/Hall International editions, London, pp. 84–95.

Appendix: Case Study of Rampura-Agucha Zinc-Lead Deposit, India*

Rampura-Agucha mine is the world's largest zinc mine, operated by Hindustan Zinc Limited (HZL), of Sterlite Industries, which is a part of UK based, Vedanta Group. It is the largest and richest, sediment-hosted, strata-bound zinc-lead deposit in India. As is the case with many rich mineral deposits elsewhere in the world, this is a "chance discovery." A routine inspection of a garnet pit near Rampura village, in August 1977, ended in the discovery of world-class zinc-lead deposit. The discovery of this deposit aggrandized India's domestic zinc-lead resource and the large reserves with high metal content, lying at shallow depth, could be exploited by relatively low-cost open-pit operation. The availability of all infrastructural facilities in the vicinity of the deposit was an added advantage.

This area had a long, linear, shallow depression, debris dumps, old workings, and slag spreads in the vicinity, indicating the mining and smelting activities by the metal workers in this part of ancient India. Hence, "*sensu stricto*," Rampura-Agucha is a "rediscovered deposit." Archaeometallurgical investigations in and around this deposit area indicated the exploitation and extraction of "silver and lead" by the ancient metal workers, c.4th century BC. The details on these investigations are described elsewhere in this section.

This super giant deposit was explored by HZL, systematically, in a highly professional manner, which was authenticated and appreciated by many multinational exploration companies. One could have never imagined that an unimpressive linear surface depression of this area, noticed in 1977, when it was pristine, would end up as such an important and the largest zinc (lead) mine in the world.

Good fortune and hard work have put HZL in an enviable position at Rampura-Agucha. This world-class super giant ore body has at least 40 years of reserves, at the current production rate (open pit + underground), with a potential to add additional resource. Operations at Rampura-Agucha are now world-competitive, being one of the low-cost producers of zinc concentrates. Besides catering to the needs of the company's smelters, the concentrates have been exported for quite some years. From the beginning, HZL believed in adopting in its operations, a sustainable development that included proper balancing of economic, environmental, and social goals, through a commitment to efficient and effective operations.

The first author of this book had the privilege of working as the Project Geologist, from the inception of exploration, through the completion of bankable feasibility study stage,

*World's largest zinc mine.

till the start of pilot-pit operation, at Rampura-Agucha. The route from discovery to commercial operation was a long and winding road, and the history of progressive activities that were seminal to make the deposit to a stage, where it is now, are detailed in the following pages.

A.1 RAMPURA-AGUCHA ZINC-LEAD DEPOSIT—*FROM DISCOVERY TO DEVELOPMENT SAGA*

A.1.1 Location and Access

The Rampura-Agucha deposit (N25°50′00″; E74°44′15″) is located 15 km southeast of Gulabpura in Rajasthan state and is connected by an all-weather road via Hurda (Fig. A.1). Gulabpura is located some 65 km north of the district headquarters, Bhilwara, and 75 km from Ajmer on Bhilwara-Ajmer State Highway No. 4. Bijainagar, another industrial town is about 3 km away from Gulabpura. Both the towns are connected by rail-road linking Ajmer and Chittorgarh. The nearest seaport is at Kandla in Gujarat state, some 700 road-kilometers away from the deposit. Rampura-Agucha deposit is about 220 and 230 road-kilometers away from Jaipur and Udaipur, respectively, and both these cities are connected by air with the international airports of New Delhi and Mumbai.

The village Rampura, which was located originally adjoining the western flank of the deposit, has been rehabilitated with all basic infrastructural facilities in 1990, prior to the commencement of open-pit operation. Agucha village is about 1.5 km southwest of the deposit. The proximity of these two villages to the zinc-lead deposit was seminal in naming it as "Rampura-Agucha Deposit."

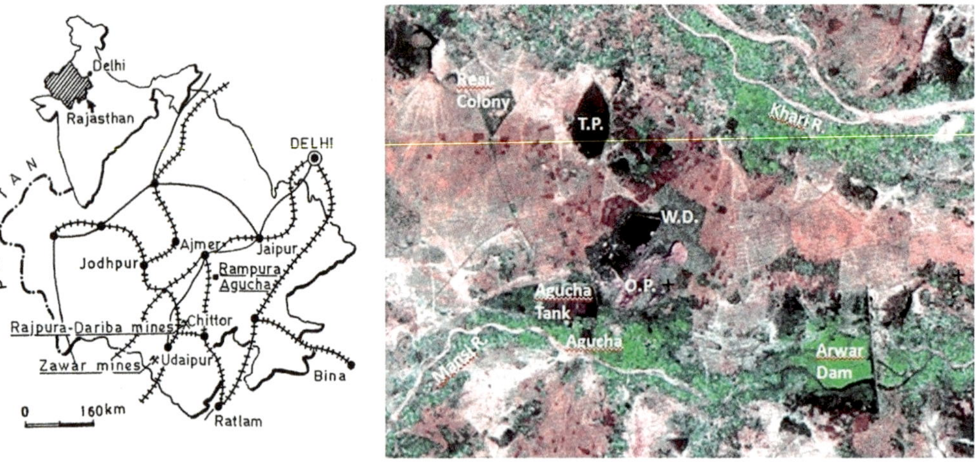

FIGURE A.1 LEFT: Location map. Right: Satellite imagery of Rampura-Agucha and its environment. *O.P.*, Open Pit; *T.P.*, Tailing Pond; *W.D.*, Waste Dump.

A.1.2 Topography, Drainage, and Climate

Rampura-Agucha area represents a gentle peneplained flat terrain most of which is covered with a thin mantle of soil cover. A shallow depression of about 1.6 km trending approximately northeast-southwest with small debris heaps on either side was a noticeable topographic feature in the area. The ground elevation at Rampura-Agucha area is about 386 ± 2 m. The topography gently slopes toward the rivers Mansi and Khari, located at about 2 km in south and 11 km toward northeast, respectively. The irrigation tanks around this area essentially cater to the agricultural needs. In winter months, innumerable migratory birds take refuge in Agucha tank, making it an ornithologist's paradise.

The Rampura-Agucha deposit area falls under semiarid, subtropical climate. The area enjoys a short period of rainfall (annual rainfall, about 360 mm), much of which is in July–September during southwest monsoon season. The winter season is pleasantly cool with temperature touching down to 3°C. The summer is hot and dry with day temperature reaching up to 46°C in mid-summer, but the nights are quite pleasant. During the summer months, this area experiences heavy dust storms with gale-force wind, especially in the afternoons and nights.

A.1.3 History of Discovery of the Deposit

Garnet is found to be ubiquitous in almost all rock types around Rampura-Agucha deposit area and has been worked in the past for semiprecious stones. Many pits had been dug in the country rocks all around the deposit area. Before commercially exploiting, it was customary that such sites had to be inspected and cleared by Provincial Department of Mines & Geology (DMG), Government of Rajasthan, India, for exploitation. Some parties had already licenses to work in this area. Since a new party from Ajmer, applied for prospecting license for garnet in an area west of Rampura village, DMG got the potential of the area assessed and wanted to inspect to know the approximate ratio of gem and abrasive varieties in and around Rampura-Agucha area. During the course of inspection in August 1977, DMG geologist, on the fortune-favored day, noticed and collected gossan and slag samples. Chemical analyses of gossan samples indicated the presence of lead (0.1–4.0%) and zinc (0.5–4.0%); and slag samples around 5.0% lead and 3.0% zinc. The presence of such significant metal values aroused DMG's interest for further sampling in this area, culminating in the discovery of Rampura-Agucha zinc-lead deposit.

After further reconnaissance, preliminary geological, geophysical, and geochemical surveys were taken up in the field seasons of 1978 and 1979, which resulted in indications of good potential prompting DMG to initiate surface drilling campaign in May 1979. The intersection of 67 m of mineralization with plus 10% Pb + Zn in the first drillhole, strengthened the prospects of establishing a very promising deposit.

A.1.4 Exploration by Hindustan Zinc Limited

Such indications attracted the attention of Hindustan Zinc Limited (HZL) and in order to explore expeditiously, HZL, proposed a scheme of detailed exploration envisaging 15,500 m of surface drilling and 500 m of underground development and submitted to its Board which was approved in February 1980, for an estimated expenditure of Rs. 12.5 million. The programme was based on a potential of 26 million tonnes grading about 10% (Pb + Zn), estimated by HZL from the data of 4 boreholes and surface geological features. HZL also applied for Mining Lease for 12 sq. km in December 1979 and the same was granted in March 1980. HZL moved into this area in February 1980 to initiate surface drilling campaign and DMG was requested to continue drilling on contractual basis. By August 1980, the number of drills at site was increased to 10 (7 of HZL and 3 of DMG) in order to complete the exploration work within 24 months, in line with the direction of HZL Board and Government of India.

The senior writer had the privilege of associating himself right from the inception of exploration campaign till the preparation of full feasibility study in Rampura-Agucha.

A.1.5 Exploration Scheme of HZL

The scheme envisaged (Table A.1) was a four-stage exploration with an in-built advantage of sequential evaluation to modify the quantum of work at every stage, enabling collection of desired data in minimum time at a minimum cost. The broad objectives of work proposed under different stages were:

- 200 m-spaced drilling to establish the potential of ore reserves
- 100 m-spaced drilling to estimate ore reserves grade and metallurgical characteristics
- 50 m-spaced drilling for detailed mine design and pilot plant beneficiation tests
- advance exploration to gain knowledge about the depth extension of ore body.

The very first borehole drilled by HZL intersected an "ore zone of 75 m" establishing the strike extension of 1.4 km and confirming the projected width in this part, which gave good indications that the potential could sustain a large industrial venture. In order to ascertain fuller potential and the configuration of the ore body, deep drilling envisaged in the last stage was advanced before taking up the work in the second and third stages. The surface exploration was limited to a strike length of 1.6 km as the drill holes on the northern and southern extensions proved to be negative.

In order to ensure QA/QC standard in data collection, norms for data recording were adopted right from the beginning of exploration. To generate confidence of the investors, HZL introduced various checks and counter-checks like sampling bias, analytical bias, reliability of analytical technique, laboratory bias, and reliability of grade and tonnage estimates by conventional statistical tests from the inception. The good potential of lead-zinc resources equally drew the attention of the Department of Mines, Government of India, which resulted in the appointment of a committee to review the progress, supervise, and guide the exploration work in Rampura-Agucha.

TABLE A.1 Exploration Scheme Conducted at Rampura-Agucha

Stage	Drilling Interval	Meters/ No. of Boreholes	Objectives	Activities	Months/Cost (Rs. in million)
1A	200 m × 50 m	2400/12	Establish broad potential over the Strike length and metallurgical characteristics	Surface mapping and contouring, diamond drilling, sampling, check studies, petro-mineralogy data base creation, reserves estimation	10/2.2
1B	200 m × 50 m	750/7	Knowledge of gossan Old workings		
DECISION POINT					
2	100 m × 50 m	3560/36	Firmly establish grade, metallurgical characteristics, enable preparation of DPR for conceptual mine planning, and investment decision	Reserve estimation, bench-scale beneficiation, geotechnical, groundwater potential, environmental interim report	9/1.9
DECISION POINT					
3	50 m × 50 m UG exploratory mine development	5660/31	Precise ore reserves for open-pit design	-do-	8/5.9
DECISION POINT					
4	Close-spaced, near surface, wide-spaced at lower levels	3130/31	Precise delineation of gossan, extension of ore body up to 380 m vertical depth	Final report submission	9/1.6

A.1.6 Drill Core Recovery

- Drill holes were mostly drilled in NX and BX sizes to maximize core recovery and minimize borehole deviation. Deviation survey done by *Multi-shot Borehole Camera or Tropari*; to minimize errors in estimation.
- This provided adequate quantity of samples for physico-mechanical, petro-mineralogical, and beneficiation test studies.
- Ensured maximum core recovery (>95%) in mineralized zone.
- Concore (continuous core) drilling through reverse circulation. Used impregnated with synthetic diamond bits with reduced loading for economy; wire-line drilling, etc.

A.1.7 Drill Core Length

- Choice of sample length is important, as, it is the basic unit for reserve calculation.
- In order to simplify the computation, for statistical analysis, evaluation, etc., equal core length samples of about 0.5 m or 1.0 m were made, and codified.
- Samples, drawn mainly from ore zones; wherever disseminations of minor amount of mineralization noticed, complete host rock was sampled with large sample interval.

A.1.8 Core Logging

While core logging, standard proforma was devised for documentation and development of computer file for better menstruation. The details include: run length, % core recovery; rock quality designation (geotechnical); rock type, structural and textural aspects; distribution of Rock Forming Minerals (RFM) and Ore Forming Minerals (OFM); specific gravity; visual estimates of metal content; any other special feature observed in the core sample. Borehole deviation measurements by depth are also recorded (see chapter: Drilling in this book).

A.1.9 Bulk Density

- Ore bodies vary in mineral content and physical character, resulting in a range of *in situ* bulk density.
- Bulk density determined for all samples, all grade ranges, ore types, and internal waste, footwall and hanging wall rocks.
- A regression relationship between *in situ* bulk density and metal content is computed, for using in block-tonnage estimation.

A.1.10 Sampling, Analytical, Laboratory Bias, and Reliability of Estimates

Each step in sampling and assaying procedure introduces an amount of error in estimation. These errors get compounded and produce erroneous values. Routine check tests were carried out during exploration and evaluation stages. Any significant bias, was corrected before ore reserves calculation was attempted. HZL introduced various checks and counter-checks like:

- **a.** *Sampling bias* (First half and corresponding second half of core, sampled and analyzed)
- **b.** *Analytical bias* (Every 10th sample reassayed to check the bias)
- **c.** *Reliability of analytical technique* (by Conventional, AAS and XRF methods)
- **d.** *Laboratory bias* (three sets of same sample assayed in three different laboratories) and
- **e.** *Reliability of grade and tonnage estimates by conventional and statistical methods.*

A.1.11 Sampling Bias

No. of sample pair	123	123
Mean grade of first half	1.43% Pb	10.99% Zn
Mean grade of second half	1.30% Pb	11.11% Zn
Mean difference	+0.13%	−0.12%
"F" value	1.10	1.02
Paired "t"	2.70	0.66
Correlation coefficient "r"	0.95	0.97

The relative variation in assays of both the halves was significantly of low order and error on account of half-core sampling on global basis was insignificant.

A.1.12 Analytical Bias

All the borehole samples in the reserve estimates were analyzed at HZL's Central R&D Laboratory by AAS and XRF methods. Every 10th sample from each borehole was analyzed in the same laboratory and 990 pairs of samples were generated for statistical tests.

No. of sample pair	990	990
Mean grade of original assay	1.88% Pb	11.04% Zn
Mean grade of duplicate assay	1.83% Pb	11.04% Zn
Mean difference	0.05%	0.0
"F" value	1.04	1.00
Paired "t"	1.89	0.09
Correlation coefficient "r"	0.93	0.97

The mean difference was negligible. The variation between two assays was statistically insignificant and within acceptable limit. It could be safely concluded there was no bias in original and duplicate assay.

A.1.13 Estimation Parameters

Cut-off: Cut-off grade of 2% (Pb + Zn), was taken for the estimation. However, in exceptional cases, barren/lean grade zones having no trend, were included within the overall zone, as there was impracticability of those being left unmined.

Specific gravity: Specific gravity determinations were carried out in many samples representing different levels.

Level (mRL)	No. of Borehole Intersection	Total No. Samples	Average Specific Gravity	Maximum	Minimum
360–260	72	4267	2.984	3.67	2.27
260–0	12	506	3.046	3.55	2.17
360–0	84	4773	2.991	3.67	2.17

A.1.14 Cross-Sections and Level Plans

Geological cross-sections were prepared at 50 m interval along the series of boreholes on 1:1000 scale. Similarly, level projection plans were drawn at 360, 310, 260, 110 mRL on 1:1000 scale.

A.1.15 Ore Reserves Classification

HZL's, in-house "4P" classification of ore reserves, as adopted in the earlier explorations of the company, was followed, to categorize different degrees of confidence, as per details below:

Proved ore (P2): Proved ore extends 25 m up or down dip from borehole intersection toward the point where the ore body is known to exist. An area of influence of 25 m was extended in the strike of both the directions for the purpose of estimation of proved reserves. An area of influence of 12.5 m was restricted on the extremities of the lode.
Probable ore (P3): Probable ore was estimated as adjacent to the proved ore for a further distance of 25 m in all directions, except the extremities of the lode, where the area of influence was restricted to 12.5 m from the proved limit.
Demonstrated ore (P2 + P3): The sum of both proved and probable (P2 + P3) was termed as "Demonstrated category" of ore reserves.
Possible reserves (P4): Possible reserves included the area adjacent to possible (P3) ore up to the assumed ore body limits.

The principal method employed in ore reserve estimation was conventional "Long-section method," while the same was checked by "Cross-sectional" and "Statistical methods." The ore reserves were computed by category, level, and section.

A.1.16 Grade Estimation

Conventional method: Assay zones in each borehole at 2% (Pb + Zn) combined cut-off were calculated by weighting sample length with specific gravity. The assay zone average grade represented the individual block grade in respective levels. Global grade was estimated by weighting individual block average grade by their respective tonnage. The global grade for 52.02 million tonnes of demonstrated category of ore reserves (out of total reserve of 61.103 million tonnes), was estimated at 1.93% Pb and 13.48% Zn. The grade of possible reserves was not calculated separately, as it formed outermost limit of the ore zone intersections, hence the average grade of the demonstrated ore reserves was assigned for the total ore reserves.

A.1.17 Confidence Limit

Confidence limits of the mean grade are calculated taking into consideration of (1) variability in mineralogy, (2) metal distribution pattern, and (3) number of samples considered for estimation. Taking into account of the Pb and Zn frequency distributions, mean grade and confidence limits were estimated employing normal and lognormal statistics. The frequency plot of Zn and its coefficient of variation indicated that the samples conformed to a normal distribution, whereas the positively skewed Pb conformed to a lognormal distribution.

A.1.18 Estimate of Average Grade Between 360–0 mRL

No. of boreholes	84	84
No. of samples	5087	5087
Average grade	1.52% Pb[a]	13.32% Zn
Variance (S^2)	3.16	63.00
Standard deviation (S)	1.78	7.94
Coefficient of variation	0.93	0.59
Confidence limit at 95% level	±0.03[a]	±0.22
Percentage variation of mean grade	±1.97[a]	±1.65

[a] *Average grade by lognormal statistics.*

A.1.19 Specific Gravity Weighted Average Grade

Due to variation in the specific gravity, weighted mean grade was worked out by regression equation as shown below:

%Zn (Specific gravity weighted)	0.279 ± 0.99% Zn (length weighted)
%Pb (Specific gravity weighted)	0.05 ± 1.002% Pb (length weighted)

Applying these equations, the mean grade of "Zn" 13.32% was likely to be enhanced to 13.57% and that of "Pb" (lognormal) from 1.52% to 1.57%. Since "Pb" distribution was lognormal, lognormal statistical mean grade has been accepted.

A.1.20 Comparison of Grades by Various Methods

	Grade	
Method	**%Pb**	**%Zn**
Conventional		
Long-section	1.93	13.48
Cross-section	1.95	13.42
Statistical Estimates at 95% Confidence Level Using Long-Section Method:		
Normal statistics	–	13.32 ± 0.22
Lognormal statistics	1.52 ± 0.03	–
Corrected with specific		
Gravity weightage	1.57 ± 0.03	13.57 ± 0.22

A.1.21 Grade Acceptance

Lead grade: Since the "Pb" distribution is skewed, the average grade of 1.57% ±0.03 using lognormal statistics at 95% confidence level could be accepted as the grade of the deposit.

Zinc grade: Since there is insignificant difference between conventional method and the one obtained using normal statistics (corrected with specific gravity), the grade of 13.48% by Long-section method was accepted as the grade of the deposit.

A.1.22 Trace Element Studies

Twenty-five borehole composite samples were analyzed for elements "S, Cd, Sg, Co, Ni, Mn," besides Pb, Zn, and Fe. The range and mean of minor element contents are given below:

Element	Minimum	Maximum	Mean	Confidence Limits at 95% Level
S	11.60%	19.00%	14.51%	±0.90
Mn	0.07%	0.20%	0.12%	±0.01
Cd	250 ppm	780 ppm	452 ppm	±50
Ag	27 ppm	102 ppm	54 ppm	±8
Co	25 ppm	41 ppm	34 ppm	±2
Ni	50 ppm	124 ppm	87 ppm	±6

A.1.23 Ore Reserves

Continuous integration and evaluation of the geological exploration data and in-built flexibility in the exploration approach, resulted in outlining a total *in situ* ore reserves of 53.13 million tonnes with 1.5% Pb and 13.9% Zn, spread over a strike length of 1375 m and to a depth of 360 m in June 1981. Out of these reserves, the demonstrated category constituted 36.24 million tonnes representing 68% of the total reserves. With the contribution from Rampura-Agucha deposit, the total lead-zinc resources of the country showed a big leap, both in terms of grade and quantity.

The Rampura-Agucha deposit with its large reserves and high metal content (plus 15% Pb + Zn) was amenable to open-pit mining. One of the major problems in the planning of an open-pit mine was the choice of suitable slope parameters. In order to configure slope with respect to geological structure, the other technical aspects like geotechnical (rock quality designation, compressive and elastic properties of rock types, rate of penetration, and physico-mechanical tests on different rock samples), detailed petro-mineralogical, groundwater potential, surface contouring, and environmental studies have been carried out over a period of time to provide reliable and accurate basic information for feasibility studies and preparation of a Detailed Project Report. To complete these studies, HZL had availed the expert services of different organizations like Geological Survey of India, Survey of India, Central Groundwater Board, Central Mining Research Station, Indian Bureau of Mines, and University of Rajasthan.

Since the expertise for the working of open pit, base metal hard rock mines in the country was limited, and due to anticipated complex metallurgy of ore for beneficiation, SNC, (Canada), was commissioned to carry out Definitive Feasibility Study for this deposit in 1982.

A.1.24 Further Investigations

Further drilling envisaged in the exploration scheme and additional drilling work proposed to probe the strike extensions were continued. The ore reserves were recalculated and updated incorporating the results of the additional work in February 1982. The total ore reserves of the deposit, based on 18,000 m of drilling in 133 boreholes covering a strike extent of about 1400 m and vertical depth of 370 m, at 2% cut-off, were estimated at 61.10 million tonnes averaging 1.57% Pb, 13.48% Zn, and 9.58 Fe. The demonstrated reserves constituted 52.02 million tonnes representing 85% of the total reserves, which included 3.12 million tonnes of mixed sulfide and nonsulfide zone (transition zone) between fresh sulfides and gossan. An additional 2.6 million tonnes of gossan have also been estimated above the transition zone. The ore reserves by level and category are given in Table A.2 (HZL, 1982).

TABLE A.2 Ore Reserves by Level and Category as on 1982

	Ore Reserves (in Million Tonnes)					
Levels (mRL)	**Demonstrated**	**Possible**	**Total**	**%Pb**	**%Zn**	**%Fe**
380–370	Gossan		2.604			
370–360	3.067	0.005	3.122	1.96	14.06	8.32
360–310	13.988	0.129	14.117	2.06	13.69	8.91
310–260	12.268	0.092	12.360	2.04	13.77	9.64
260–210	10.217	0.185	10.402	1.74	12.24	9.82
210–160	7.515	0.589	8.104	1.48	11.91	10.39
160–110	3.651	2.374	6.025	2.14	16.77	10.58
110–60	1.134	2.978	4.112	2.14	16.79	9.83
60–00	0.185	2.676	2.861	1.93	13.48	9.78
Total (370–00)	52.023	9.060	61.103	1.57	13.48	9.58

A.1.25 Ore Reserve Acceptance

Demonstrated category of ore reserves using statistical method was computed between 360 and 260 mRL and the same was compared with other methods:

Long-section method	26.256 million tonnes
Cross-section method	26.902 million tonnes
Statistical method	25.22 ± 2.29 or 9.07% variation

Comparison of the estimates indicated that the variation between statistical and Long-section method was about 4%, which is within acceptable limits. Therefore, the error of estimation of ±9.07% as obtained using Statistical method for the first 100 m block could be accepted. The global estimate indicated a confidence limit of ±11.70% which is close to previous estimate. Since the estimates for block between 360 and 260 mRL by different methods had close agreement, the total ore reserves of 61.10 million tonnes by Long-section method (370–0 mRL) was accepted as realistic global reserves.

HZL commissioned SNC Engineering and Project Management Services Inc. (SNC), Canada to carry out Definitive Feasibility Study for this deposit in 1982. As a part of data validation process, SNC independently calculated the total geological ore reserves of the deposit using geostatistical methods and estimated at 60.36 million tonnes averaging 1.93% Pb, 13.48% Zn, and 9.47% Fe. SNC's ore reserve estimate not only confirmed but compared well with the manual estimate by HZL.

TABLE A.3 Ore Reserves by Blocks as on November 1988

	Ore Reserves (in Million Tonnes)				
Block	**Demonstrated**	**%Pb**	**%Zn**	**Possible**	**Total**
North Block (00 to N800)	12.402	2.71	15.16	3.235	15.637
South Block (S712.5 to 00)	40.552	1.73	13.03	7.442	47.994
Global reserves	*52.954*	*1.96*	*13.53*	*10.677*	*63.631*

Subsequently, HZL drilled four deep boreholes in the northern section of the deposit, in order to establish the plunge and continuity of the ore body. Deep drilling in the north, established the plunge of the ore body and opened a new vista for additional resource potential of the deposit along northeastern direction. After incorporating the results of the drilling campaign, HZL estimated the ore reserves in November 1988 and are given in Table A.3 (HZL, 1988).

A Longitudinal Vertical Projection of the Rampura-Agucha ore body showing the intersections of diamond drill holes along with ore reserves, by category, as on April 2001 is shown in Fig. A.2.

A.1.26 Exploratory Underground Work and Bulk Sampling

The prime purpose of any bulk sampling is to confirm grade indicated by prior sample data from drill holes; establishment of metallurgical criteria, to collect more data on geological aspects and investigation of rock mechanics properties. In order to collect an unoxidized, representative fresh bulk ore sample for pilot plant metallurgical tests, an exploratory underground development of about 567 m was carried out. The work involved sinking an incline from hanging wall side of the ore body, at 30 degree to a depth of about 40 m below the surface (350 mRL), the development from it of a main crosscut up to the westernmost limit (footwall contact) of the ore body and 256 m of underground drilling. An exploratory drive was developed toward the north with crosscuts at every 50 m interval covering a strike length of 150 m. The underground development confirmed the width and grade of the ore body, predicted by surface boreholes and provided bulk sample for beneficiation tests and fine-tune the process parameters for improving the grade and recovery.

A.1.27 Beneficiation Tests

Preliminary bench-scale beneficiation studies conducted in HZL's Central R&D Laboratory showed that the ore was amenable to differential flotation. Due to the presence of deleterious contaminants like graphite/mica, the lead content in the concentrate and the recovery posed problems. Portions of composite drill core samples representing 100 m

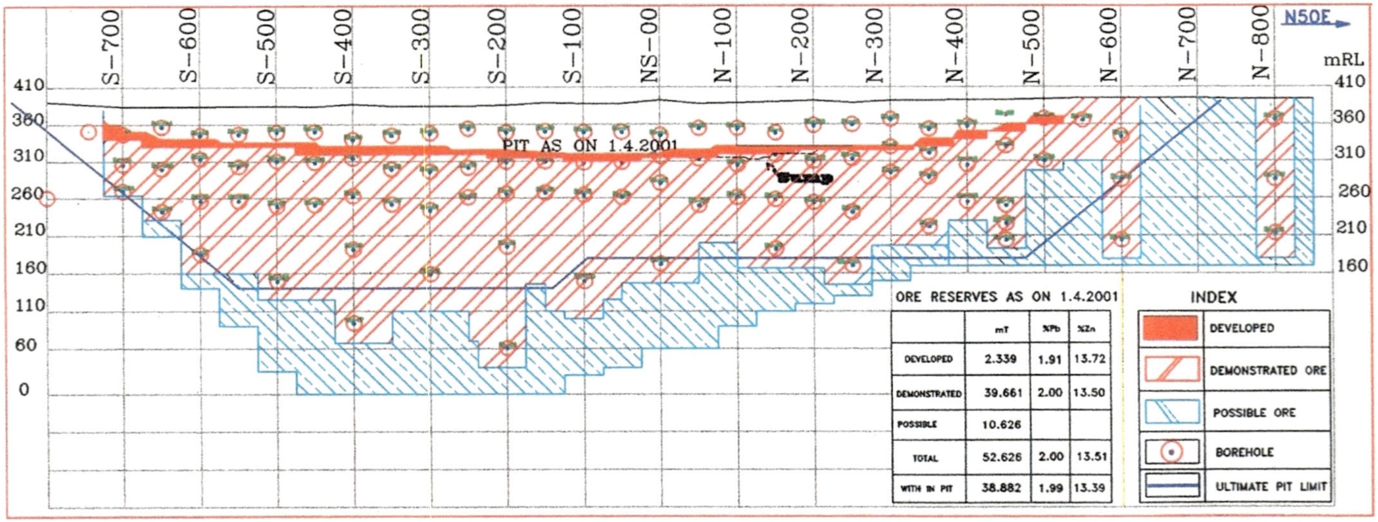

FIGURE A.2 Longitudinal vertical projection of Rampura-Agucha ore body.

TABLE A.4 Chemical Analysis of Sulfide and Nonsulfide ore From Rampura-Agucha

	Sulfide Ore			Nonsulfide Ore	
Element	**1**	**2**	**3**	**1**	**2**
Zn%	12.20	12.01	12.23	2.40	2.55
Pb%	1.74	1.62	1.75	1.40	1.50
Fe%	10.30	9.50	9.87	9.70	9.40
Cu%	0.01	0.04	0.03	0.01	0.02
Cd%	0.04	0.036	0.03	0.02	0.021
S%	12.10	12.10	12.05	2.70	1.30
SiO_2%	38.60	39.80	38.20	49.20	50.10
Al_2O_3%	10.70	11.00	12.46	13.50	13.10
MgO%	2.10	2.08	2.73	0.84	0.95
CaO%	1.30	1.39	1.82	0.73	0.78
BaO%	0.10	0.05	Nd	0.10	0.05
Ni%	<0.01	Nd	0.007	<0.01	Nd
As%	0.04	0.07	<0.01	0.04	0.04
Sb%	<0.01	0.01	<0.01	<0.01	0.01
Mn%	0.10	Nd	Nd	0.01	Nd
Sn%	<0.01	Nd	Nd	<0.01	Nd
C%	2.76	2.55	7.20	3.22	3.18
Hg (ppm)	10	Nd	Nd	8	Nd
Ag (ppm)	40	43	20	40	37

Analysis by: (1) LURGI, Germany; (2) SALA, Sweden; (3) Indian Bureau of Mines, Nagpur

depth of the deposit and nonsulfide ore sample were sent to M/s Lurgi, Germany, M/s Sala International, Sweden, and Indian Bureau of Mines, Nagpur, for conducting beneficiation test works. The chemical analysis of the composite sulfide and nonsulfide ore samples is given in Table A.4. All the above three laboratories reported very similar metallurgical results. Pilot plant tests on 3550 tonnes of fresh bulk ore sample recovered from underground were done, in SALA pilot plant located at Rajpura-Dariba mine, under the guidance of SNC in 1982. These tests helped to evolve various design parameters for ore beneficiation plant and fine-tune the process operations. Replicate tests in the pilot plant scale demonstrated that it should be possible to obtain an 85% recovery of zinc at 51% grade and about 60% recovery of lead at 55% grade.

Beneficiation tests on ore samples from Rampura-Agucha, combined with mineralogical studies, had given sufficient indications of the quality of concentrates likely to have high silica and other impurities which led to the conclusion that the Imperial Smelting Process (ISP) with its flexibility to tolerate higher levels of impurities and lower energy requirements would be the techno-economically advantageous process route for the proposed new smelter. Chanderiya near Chittorgarh, in Rajasthan, India,was considered to be the techno-economically advantageous site for the smelter because of the availability of rail-road link, a reliable water source, and its central location with respect to other mines around Udaipur area. MECON (India) was commissioned to prepare a project report for establishment of 70,000 tonnes per year zinc and 35,000 tonnes per year primary lead smelter at Chanderiya.

A.1.28 Government Approval

The potential of Rampura-Agucha provided a firm basis for establishing a new smelter. An integrated proposal for an open-pit mine at Rampura-Agucha and a Zinc-Lead smelter at Chanderiya (near Chittorgarh) was submitted to Government of India in January 1983. This proposal, after due scrutiny, by the Planning Commission, Bureau of Public Enterprises, Plan Finance and clearance of the Department of Environment, Civil Aviation (All Federal Government agencies), and the Provincial Government, was approved by the Public Investment Board in April 1983.

In 1985, the Overseas Development Administration of United Kingdom indicated its willingness to provide some finance and the following year the Government of India decided that the project should go ahead in two phases. The first phase called for a Detailed Project Report with firm cost estimates, plus some development of infrastructure, while the second was the construction phase proper. The Government of India approved the implementation of the Integrated Project for the development of Rampura-Agucha zinc-lead mine and Chanderiya smelter on the basis of firm cost estimates of Rs. 6172 million, submitted by HZL. The project was to be completed within 30 months from the date of sanction that is, November 1988. Approval was received to finance the above investment through equity funds of Rs. 1860 million (comprising aid-funds from United Kingdom of Rs. 1610 million to be routed through Government of India and Owner's Equity of Rs. 250 million). Internal resources were to be generated by HZL to the tune of Rs. 1270 million and the balance was the loan component, to be raised through public issue of Bonds.

The development of Rampura-Agucha mine started in November 1988 and the first ore production was in March 1991.

A.1.28.1 Summary of Investigations

The extent of exploration and summary of investigations carried out in this deposit till the late 1980s, are given here under.

1.	Regional mapping	25 sq. km
2.	Detailed mapping on 1:10,000 scale	5 sq. km
3.	Surface diamond drilling	24,900 m in 229 boreholes
4.	Underground diamond drilling	567 m
5.	Total samples collected	13,835 Nos.
6.	Samples collected from mine	1744 Nos.
7.	Borehole composite samples	
	Sulfide zone	6 Nos. (1613 kg)
	Gossan zone	1 Nos. (450 kg)
8.	Geotechnical samples	148 Nos.
9.	Check samples	123 Nos.
10.	Analytical check samples	990 Nos.
11.	Laboratory check samples	46 Nos.
12.	Beneficiation test	
	Bench-scale and Pilot plant	10 Nos.
13.	Pilot plant test	
	At Rajpura-Dariba mine	3550 tonnes
14.	Other studies	
	a. Slope stability tests for open-cast mining Central Mining Research Institute, Dhanbad b. Environmental studies in and around the area Geological Survey of India, Western Region, Jaipur c. Groundwater potential around the belt, other sources of water Central Ground Water Board, Western Region d. Topographic survey in detailed scale of the deposit area Survey of India, Dehradun e. Petro-mineralogical studies of drill core/ore samples University of Rajasthan	

A.1.29 The Technology Tie-ups

The contracts for supply of technology were entered into during the first cost estimates' preparation phase of the project. The technology adopted for the mine and the smelter is different to what is being followed in the existing units of the company. The technologies were not available with any one agency or even with number of agencies within any one country which necessitated dovetailing of multiengineering process technologies. Close coordination was, therefore, required for procurement of basic engineering and know-how from different know-how suppliers.

Accordingly, RTZ Technical Services (UK) was selected for the beneficiation plant at Rampura-Agucha and M/s Davy McKee (Stockton) Ltd. (UK); as the main contractor for Chanderiya Smelter Plant with many consortium members ISPL (UK); Lurgi GmbH (FRG); MECHIM, Belgium; MIM Technology Marketing Ltd. (UK); Norzink (Norway). It was decided to divide the construction into turn-key areas. As for as Rampura-Agucha was concerned, the work involved (1) open-pit development, (2) concentrator plant, (3) 132 kV substation and power distribution, and, (4) water supply. Similarly, for Chanderiya Smelter many areas were identified. Based on prequalification bids, turn-key contractors were selected on the basis of decision matrix which took into account their past experience, technical and financial capabilities, and track record. Offers were invited from short-listed parties for each area, based on detailed specification provided by the respective know-how suppliers through Davy Mckee.

The UK-India Aid Agreement envisaged that for environmental aspects, HZL will adopt British or Indian Standards, whichever are more stringent. M/s W.S. Atkins and Cramer Warner (both from UK) were commissioned for the preparation of environment management plan for Mine and Smelter complexes, respectively.

A.1.30 Preliminary Activities

Subsequent to the approval of the project by the Government of India, in November, 1988, many actions were initiated. Land acquisition was completed and the site was made available for the developmental activities in the shortest possible time. Some 850 hectares of land out of 929 hectares at Rampura-Agucha mine area was made available or acquired within 1 month of project approval. Similarly, HZL ensured the availability of water and power required for the construction at the site, on priority, to support the site activities. For Rampura-Agucha mine, 56 km pipeline for water was laid from the identified Banas River. HZL also got the clearances from the Environmental Committees of both Government of India and Government of Rajasthan.

A.1.31 Project Beginning

The first blast at Rampura-Agucha mine was in November 1988, while the foundation stone laying for Chanderiya Smelter was in February 1989. All the turn-key contracts had been finalized and the lowest bidders had been called for final negotiations. All major contracts were concluded by mid-January 1989. The payments during the contracts were linked with accomplishment of major tangible milestones of construction and procurement activities and not on partial completion of jobs (85% of the payment for procurement was linked to receipt of materials at the site and no to other intermediate stages). This has certainly motivated the parties to ensure quick availability of equipment at the site.

The balance 15% of the payments was linked with mechanical completion to provide adequate incentive to the contractors for speedy completion of all jobs up to the commissioning stage.

Davy Mckee, as a Managing Contractor, assisted and coordinated HZL and other consortium members to import equipments timely, from various sources in the world. HZL, on its part, got the necessary approvals from concerned government agencies for the import of equipments and subsequent custom clearances. The requirements of steel and cement for construction were procured and supplied to contractors, by HZL. The contractors were allowed to procure nonstandard, special steel items, needed for their construction.

A.1.32 Fine-Tuning the Recoveries

In the first phase of development of Rampura-Agucha pit, it was decided to gain quick access to the ore so that a larger quantity of ore could be treated in a 100 tonnes per day pilot plant for refining the parameters of beneficiation. These tests indicated that it would not be possible to produce the clean zinc concentrate required for existing smelters without sacrificing recovery.

The requirement of clean concentrate to augment the supply to existing electrolytic smelters is, however, limited. Therefore it was decided to try to produce part of the zinc concentrate as a clean one and the remaining part with higher levels of impurities acceptable to Imperial Smelting Furnace (ISF). Detailed tests established that without sacrificing recovery, 30% of the zinc concentrate could be produced with 52% zinc and less than 2.5% silica, the remainder having 48% zinc and 4% silica. This hypothesis was accepted and incorporated in the beneficiation flowsheet.

The incidental production of ore from development of the Rampura-Agucha pit, was treated in the company's existing beneficiation plants over 2 years. This has permitted further refinement of the metallurgy besides catering a nucleus of trained manpower. This opportunity was utilized to experiment with new reagents. With the use of locally available reagent "Amalgum P-10D"—the lead grade had improved to 48% from 40% with reduction of graphite from 9% to 4% and a marginal improvement in recovery. Treatment of over 400,000 tonnes of Rampura-Agucha ore in the existing beneficiation plants of HZL, had given confidence that the commissioning of the Rampura-Agucha concentrator should be possible with minimum teething troubles.

A.1.33 Training

In terms of knowledge and technical know-how, the mining at Rampura-Agucha is to have features different from existing units of the company units. The mines currently under production are underground mines whereas Rampura-Agucha is an open-pit mine. In view of the above, the company had drawn up a comprehensive training scheme for its employees to ensure requisite know-how transfer and smooth commissioning.

A.1.34 Environmental Aspects

The environment and development are two sides of the same coin. Associated with any development, some amount of environmental impact is inevitable. It should be "environment" and "sustained development" based on scientific principles to ensure sustainability bereft of greed and profit. The extent of damage caused by mining is scarcely perceptible to highly obtrusive. It is not possible to stop mining activities for sake of keeping environment intact. The crux of the task is, how one can minimize impact on the ecosystem and still achieve and advance in the industrial development of the nation.

Environmental compliance and employee safety are the goals for today's progressive, worldwide mining operations. Following this tradition, a well-designed Environmental Impact Assessment (EIA) and Environmental Management Plan (EMP) were prepared by M/s W.S. Atkins, United Kingdom, in 1989-90, prior to mining operation at Rampura-Agucha. The studies involved, land use, water use, socioeconomics, soils, hydrology, water quality, meteorology, air quality, ecology, waste management, and noise. The report was based on base-line data generated over an year, identification of impacts, developments of EMP, and recommended mitigation measures and formulation of EMP. During Project Construction and commissioning stages, HZL installed well-designed pollution control and effluent treatment facilities and other environmental measures to maintain ecology and present environmental quality.

Owing to the proximity of the mining complex to the villages, Rampura, Bheru Khera Agucha, and productive agricultural fields, HZL has adopted a "good neighbor" policy and considerable effort has been expended in reducing the impact of operations on the community. Conscious of their rehabilitation, the company, with the help of provincial government, has not only provided alternative land holdings to its population at its cost, but also built modern village complexes for them.

A.1.35 Land Use

Rampura-Agucha has a leasehold area of 1200 hectares (Ha) out of which 850 Ha have been acquired for industrial purposes which includes mining activity, tailings dam, waste dump, beneficiation plant, offices, and workshops. Another 74 Ha has been acquired for township, which is outside the mining lease.

A.1.36 Mine Area

- Garland drains to prevent flooding of mine and contamination of natural water courses (Fig. A.4)
- Pit dewatering facility with treatment facilities for pumped-out water
- Regular water sprinkling on haul roads, with application of chemical stabilizers to minimize airborne dust
- A 16-km long, 1.5 m high perimeter wall around the entire mining complex to avoid movement of cattle in the safety zone.

A.1.37 Plant Area

- Dedusting system associated with wet scrubbers for dust removal
- Waterjet sprays for dust suppression at locations where enclosure and fugitive dust collection is impractical
- Cover and side panels for all conveyors
- A 3.6 m high boundary wall around the plant area.

A.1.38 Solid Waste Handling

The two types of solid wastes generated in Rampura-Agucha operations are: (1) overburden waste rock and (2) tailings. These are disposed off in specific ways.

A.1.38.1 Waste Rock Dump

The 40.5 million m^3 capacity waste dump to keep waste rocks from the mine has been designed to have a 10 m lift with a maximum height of 40 m. The overall final slope of 27 degree is being maintained to limit the surface air velocity and dust entrainment and to aid slope stability. It can be approached by short and straight haul roads from the pit. The waste is stacked systematically in layers and leveled by dozer and soil is spread for vegetation on inactive waste dump benches. Plantation is done on the benches as and when they are completed (Fig. A.5). Another two sites having a capacity of 18.52 million m^3 are also available for the purpose. The area likely to be available from this dump for stabilization and biological reclamation at four different periods is about 47 Ha over a period of 10 years. During monsoon season, grass seeds are spread and saplings are planted on the inactive permanent slope of the dump. It is intended to plant about 16,000 saplings in the coming 2 years in and around this area.

A.1.38.2 Tailings Dam

Tailings of the beneficiation process are stored in tailing dams for long-term disposal. The tailings dam occupies on area of 0.86 sq. km within the lease area and is lined with impervious soil of 130 mm thick at the bottom of inside wall to avoid seepage to underground. The sides of the dam are also sealed by impervious soil. Thickened tailing is discharged into the dam through a pipeline. The water in the dam is allowed to settle down and the clear water is recycled back to the plant for use. No water is allowed to be discharged outside the dam. The project was designed for "Zero discharge" and now operates on this strategy.

The quality of water in piezometer wells is within the permissible limits.

A.1.39 Air Quality

The mining activity contributes Suspended Particulate Matter (SPM) and dust apart from oxides of nitrogen (NO_x) and sulfur dioxide (SO_2) from DG set and automobile exhausts. Dust at the mine site is generated during surface clearing, removing overburden mining, crushing waste rock disposal, and earthworks for rehabilitation. Dust can contaminate the atmosphere and invade the surrounding area and cause hazardous working conditions for vehicle drivers and plant operators. Dust can also contaminate machinery such as bearings, air filters, and various engine parts. In Rampura-Agucha mine, the SPM has decreased during the last 10 years.

The air quality within a zone of 10 km radius around the mine has been monitored for pollutants like RPM, SPM, SO_2, NO_2, and CO. The values of all the air pollutants are well within the prescribed limits as specified by Central/State Pollution Control Board.

The measures adopted for this accomplishment are:

- Wet drilling provision on all drilling machines
- Regular maintenance of vehicles and machineries
- Wet venturi scrubbers on crushers
- Regular water sprinkling on the haul roads by automatic sprinkling system
- Enclosed conveyor belts
- Maintenance of 8–10% moisture content in the concentrates
- Regular ambient air quality, dust fall, and stack monitoring.

A.1.40 Water Quality

The surface water run-off during monsoon causes water pollution due to wash-offs from dumps, concentrate storage area, etc. It also causes soil erosion. The flooring and upstream sides were sealed with impervious soil to reduce permeability. Drainage channels and buffer ponds were provided to collect run-off from all potential sources of contamination from the oxidized ore dump. Sealed installation of piezometers around the waste dump, the tailing pond, and gossan dump helped to monitor the infiltration of potentially toxic materials. Extensive green-belt development with a system for surface water collection and its pumping to the tailings dam prevents this pollution. The waste water generated in the process is also routed to the tailings dam, where the solids are allowed to settle. The clarified water is recycled. Make-up fresh water is added as per requirement. Waste water generated from sanitary uses is treated in septic tanks and soak pits and subsequently used for irrigation and gardening.

Analyses of selected water quality parameters for surface and groundwater resources within 10 km of the mine area have established that the water quality is well within the limits prescribed by IS: 10500 and IS: 2296. Garland drain has been laid all around the mine for retaining suspended solids and collection of run-off water.

A.1.41 Noise and Vibrations

Blasting, drilling, removing overburden and waste rock dumping are the primary sources of noise in the mining site. Blasting generates ground vibrations and impulsive noise. Ground vibrations are controlled by limiting charge per delay and are regularly monitored in the nearby areas using seismographs. The maximum noise level of 130 dB(A) has been recorded at a distance of 550 m from blasting location. It has been observed that workers in general are exposed to a noise level of 55–85 dB(A) during the

8-hour shift. Measures taken to keep the effect of occupational exposure within permissible limits are:

- Judicious selection of machinery and equipment, their mountings and ventilation systems
- Provision of noise enclosures or padding wherever necessary
- Regular maintenance of vehicles and machineries
- Boundary walls (Baffle walls), waste banks, and dense belt of trees as acoustic barriers
- Provision of protective gears to those exposed to high noise levels
- Exposure within the limits prescribed undermine rules/bye laws
- Suitable blasting parameters to minimize ground vibrations; within safety limits

A.1.42 Green-Belt Development

HZL has given due importance to green-belt development program since inception of the mine by designing proper layout of plants, roads, drainage, etc. Further, green belt around the periphery of boundary wall, green lush lawns, flower beds, and roadside plantations have been developed and are well maintained regularly. These things helped to improve the overall status of the flora in the area. The plantation is done at the peripheral areas of tailings dam, waste and gossan dump, plant area, along the roads and township. HZL has planted more than 2.5 lakh trees in this belt with a survival rate of about 80%.

The green-belt development program was started in 1988 by planting 5000 saplings covering an area of 4.5 Ha. Presently, 222.22 Ha land has been brought under this program and by the end of 1999 nearly 2,14,000 saplings have been planted in industrial and residential areas. In addition, 31,850 plants have been planted in the township covering an area of 29.76 Ha. A variety of plant species have been selected on the basis of climatic conditions and its suitability to survive. The species include *Citrus lemon, Tamarindus indicea, Euginea jambolana, Dalbergia sisso, Poinciana regia, etc.*

Screen plantations have also been developed to reduce the visual impacts due to waste dump and other industrial structures.

A.1.43 Monitoring

Regular monitoring of surface and groundwater quality and air quality in and around the mine complex. Monthly monitoring for air quality at three selected sites around the complex is being carried out. The water quality of 20 wells located in a radius of 8 km around the mine and 11 piezometer wells are being monitored regularly on monthly basis. The analytical results indicated that the values are well within prescribed limits. Soil samples for fly-off dust and dust fall rate, are also monitored regularly.

HZL is committed to uphold its rich tradition of a pollution-free environment and has got certified the Environmental Management System as per ISO.14001.

Continuous improvement is the essence of HZL's commitment to environmental management. A waste management plan for industrial and domestic wastes, chemical and hazardous wastes, and used oil and tyres, is practiced based on the principles of sustainability that is, reduce, reuse, recycle, and recover.

A.1.44 Initial Mining Operations

A.1.44.1 The Ore Deposit

The Rampura-Agucha deposit lying as it does at comparatively shallow depth is ideally suited for open-pit mining using conventional equipment. The ore body at Rampura-Agucha is the largest single zinc-lead ore body and has a strike extent of about 1550 m with a northerly plunge of about 65 degree at the northern end. The dip of the ore body varies from near vertical at the surface to about 60 degree southeast. The width of the ore body varies from a few meters to 110 m. The persistence of ore has been established up to a depth of 350 m and is likely to extend further down. The lead:zinc ratio in the deposit is 1:7. The ore body is comparatively narrower and richer in grade in the northern end of the pit and wider in the southern side of the pit. Hence, the open-pit operations can go deeper only in the southern end of the pit.

A.1.45 Design of the Pit

The open pit, at Rampura-Agucha, was originally designed by SNC-Lavalin (Canada) in 1982, the economic limits of which were estimated to a depth of 220 m below the surface (up to 160 mRL). The design parameters of the open pit included an incremental stripping ratio of 10:1 with final hanging wall and footwall slopes of 48 and 38 degrees, respectively, and a minimum mining width of 50 m. Based on these criteria, the total pit reserves up to the ultimate pit were calculated as 45.54 million tonnes assaying 13.27% Zn and 1.94% Pb. To permit quicker return in the early phase, the pit was to be located in the thickest area of the deposit allowing most of the ore to be exposed with least amount of stripping. For the first 20 years, stripping ratio worked out to 4.4 tonnes of waste per tonne of ore. The bench height planned in ore was 5 m to curb dilution and 10 m in overburden. The mine was planned to produce and treat 900,000 tonnes per annum (tpa) ore with a mill-head grade of 12.1% Zn and 1.66% Pb. The mine layout plan is shown in Fig. A.3.

A.1.46 Pilot Pit

In the early 1987, a pilot pit was opened (Fig. A.8) to confirm the design parameters, the ore/gossan contacts and to provide ore for continuing beneficiation tests to refine design parameters for the commercial beneficiation plant. The surplus ore from the pilot pit was treated in the existing concentrator plants at HZL's operating mines. The experience gained helped in early stabilization of the concentrator plant at Rampura-Agucha. During prefeasibility stage, some 468,000 m^3 wastes were stripped by contractors, who later removed a further 2,390,000 m^3 of waste during preproduction stage. At the same time some 247,000 m^3 of ore were also stockpiled. HZL commissioned a 3000 tpd open-pit lead-zinc mine and matching beneficiation plant associated with infrastructural facilities at Rampura-Agucha in March 1991 (HZL Staff, 1992).

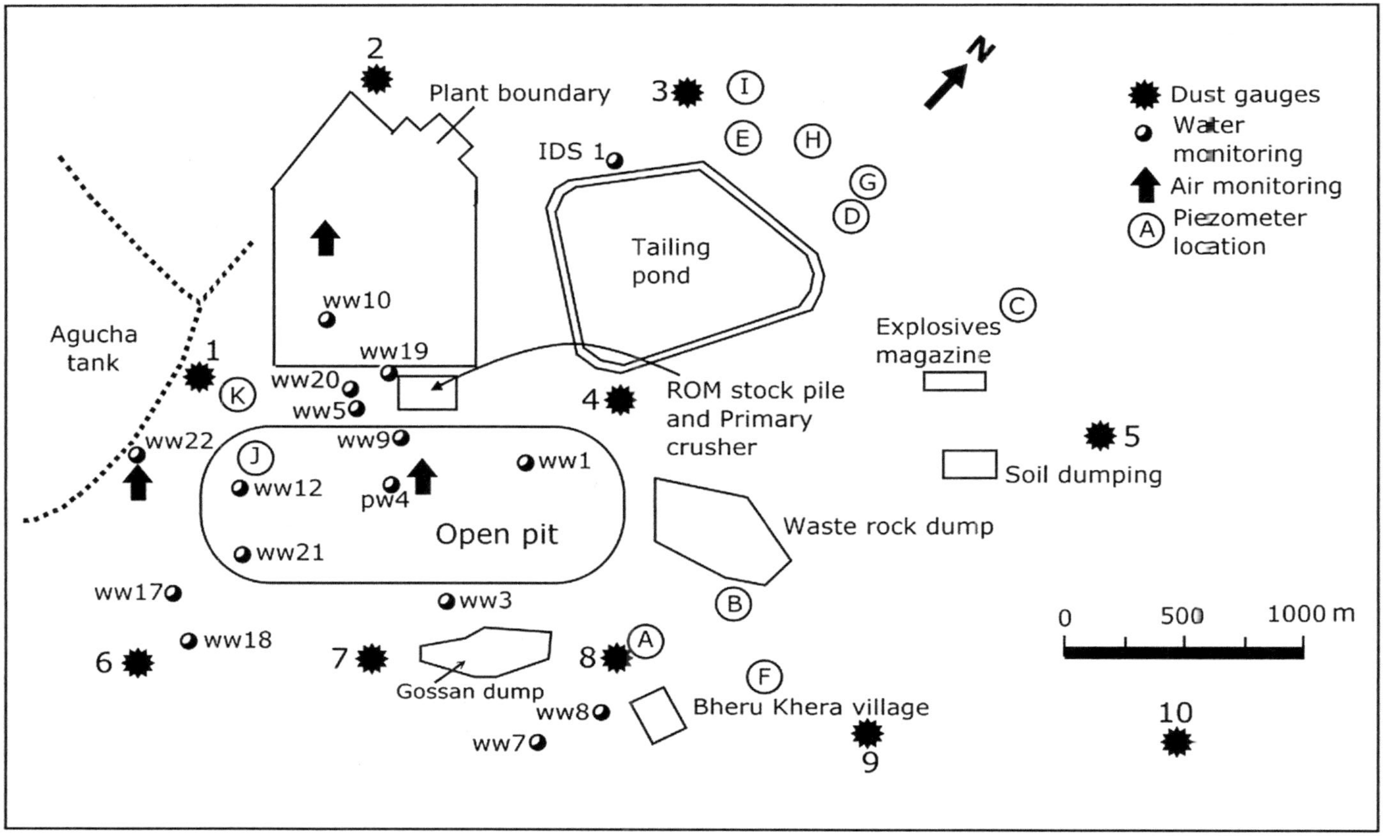

FIGURE A.3 The mine layout plan of Rampura-Agucha, with locations of monitoring stations for ground pressure, air, airborne dust, and groundwater qualities.

A.1.47 Pit Design by British Mining Consultants Limited

The open pit designed earlier was modified subsequently by British Mining Consultants Ltd. (BMCL) in 1992. The pit designed by BMCL had an ultimate pit bottom at 140 mRL in the south and 180 mRL in the north. The estimated total reserves contained in the open pit at the time of design were 48.55 million tonnes of ore with 12.81% Zn and 1.88% Pb. The corresponding waste excavation would be 234.21 million tonnes resulting in an overall stripping ratio of 4.82. The pit designed by BMCL was based on the principle of first designing a starter pit and incrementing the pit on plan and along depth in steps of 10 m till the stripping ratio for the incremental pit reached the limiting incremental stripping ratio of 14, based on the economics of operating an underground mine, prevailing at that time. After accounting for the excavations already made till then, BMCL estimated the balance reserves in the pit as 37.103 million tonnes of ore with 176.54 million tonnes of waste.

A.1.48 Mine Expansion, Phase-I

In its endeavor to reduce the demand–supply gap, the company had expanded the mining and beneficiation capacities at Rampura-Agucha mine from 3000 to 4500 tpd in November 1997 and since then both the mine and the beneficiation plant are operating at full capacity since January 1998.

The pit was worked in 10 m benches, using a berm width of 6.5 m in the hanging wall and side wall, and 8 m in footwall. The face inclination on individual benches is 70 degree while overall pit slopes are 36 degree on the footwall side 40 degree in hanging wall and 42 degree in side slopes. The mining operation, started from 390 mRL in 1991, is at 310 mRL in 2001. Thus the average rate of descent has been about 8 m per year. The present extent of the open pit is 1400 m long by 600 m wide by 60 m deep. The planned ultimate pit will be 1600 m long, 700 m wide, and 280 m deep from surface (390 mRL). Owing to the dip of the ore body, the footwall side of future expansion follows the ultimate pit limit and the hanging wall was progressively pushed back until in the last expansion, it attains its ultimate position. The stripping ratio achieved so far was 5.289. The ultimate stripping ratio was likely to be 1:5.

A.1.49 Mine Expansion, Phase-II

The expansion of smelting capacities of Debari Zinc Smelter from the existing 49,000 to 59,000 tpy and Vizag Lead–Zinc Smelter from 30,000 to 40,000 tpy and setting-up of a Greenfield Zinc Smelter of 100,000 tpy capacity, were under progress. Thus in order to meet the concentrate requirement, HZL was considering a second phase expansion in capacity of Rampura-Agucha mine and beneficiation plant from the existing 4500 to 6000 tpd ore through process upgradation, wherever required, optimization of existing equipment and provision of balancing facilities.

Engineers India Ltd. (EIL) was engaged as consultants for the second phase of mine expansion. EIL redesigned the ultimate pit taking the pit slope angles as that of BMCL pit, using an optimization software (Whittle-3D) by generating 3D block model with possible open pit. Accordingly, EIL estimated the ore reserves in the pit at 43.25 million tonnes with 12.53% Zn and 1.84% Pb. Upon expansion, the mine would have an annual production rate of 1.8 million tonnes. With the expanded capacity of operation, the estimated life of the mine will be about 23 years. The mine was operated on three shifts of 8 hours each, for 6 days a week, and the total manpower then, was 730 employees.

A.1.50 HZL, Under the New Management

With the liberalization policy of Government of India, since the 1990s, different proportions of equity capital of many of its enterprises, were disinvested to strategic partners/private parties, with an appropriate role in management. Consequently, 26% of the equity capital of HZL was disinvested to M/s Sterlite Industries in early 2002. Subsequently, this company acquired additional equity capital from market/public up to 64.9% of HZL. Since M/s Sterlite Industries, which is part of UK based, Vedanta Group, held a major equity portion, it took over the management, by retaining the name of the company. Presently, Government of India retains the equity capital of 29.4% and the rest with the private investors.

Rampura-Agucha owes its present stature, due to quick action. With its future assured, the company quickly adopted an aggressive philosophy of exercising downstream control over as much of its own products as possible. A vigorous policy of vertical integration from mine to consumer was instituted by building smelters and refinery. HZL had vertically integrated operations as near as possible to the end user. The new management embarked on massive developmental work and expansion activities to a place where it is now. The company galvanized for new levels of production by aggressive increase of production rate in mining and matching concentrator operations. HZL quickly dominated in size and earnings, in zinc/lead/silver sector.

The company carried out drilling campaign periodically for deeper intersection of ore body and incorporating all these data, they have recalculated the ore reserves and resources.

The Reserves and Resources as on "March 2014" were:

Reserves	Million Tonnes	Grade (%)	
Proved and probable		Zn	Pb
Open-pit mine	*12.7*	13.1	1.9
Underground	*36.8*	14.4	1.8
RESOURCE			
Measured and indicated	*18.3*	15.0	2.0
Inferred	*35.2*	9.9	2.1

The operating performance details of Rampura-Agucha mine for three financial years as per HZL's Annual Report to Shareholders, are given in Table A.5.

TABLE A.5 The Operating Performance Details of Rampura-Agucha Mine (kt = kilo tonne)

	FY 2012	FY 2013	FY 2014
Ore mined (kt)	5947	6149	5451
Ore milled (kt)	5982	6168	5394
ZINC			
Feed grade (%)	*12.0*	*12.3*	*13.0*
Mined metal (kt)	649.6	677.3	640.8
LEAD			
Feed grade (%)	*1.7*	*1.8*	*1.7*
Mined metal (kt)	59.5	65.6	57.4

A.1.51 Rampura-Agucha Mine

A.1.51.1 Overview

Rampura-Agucha is an open-pit mine (RAM), commissioned in 1991, and is the world's largest zinc mine. This ore is a stratiform, sediment-hosted, high-grade zinc-lead deposit and the ore body is massive and lens shaped. The ore grade is fairly consistent and is not deteriorating as one moves toward depth. The method of working is: drill, blast, load, and haul besides underground operation. The utilization of heavy earth mining machineries is at par with the global standards. The mine is equipped with the world-class infrastructural facilities including the latest generation slope monitoring radar system; truck dispatch system; simulators to enhance the operator's skill. The management system of Rampura-Agucha comprises of the Quality System ISO 9001:2008, the Environmental System ISO 14001:2004, and the Occupational Health, Safety Management System OHSAS 18001:2007, SA 8000:2008, and 5S Certifications. RAM has an ore production capacity of 6.15 million tpa, with best-in-class zinc-lead reserves grade of 15.4%. Reserves and resources of Rampura-Agucha mine as on March 31, 2014 are 103 million tonnes.

The open-pit mining operation at the RAM is likely to cease in 2018. The establishment of an underground mine is continued to exploit the deposit below the open pit at a depth of 372 m below surface beyond that it will go underground. HZL is contemplating the economic viability to further increase in depth of the ultimate pit. The underground operation is likely to continue for about 30 years. The initial ore extraction will be from declines. HZL is currently using a spiral ramp and is planning a vertical shaft (involving sinking of 950 m hoisting shaft), at RAM. The isometric view of the final open pit and the proposed underground mine at Rampura-Agucha, as per HZL and A.C. Choksi Institutional Research and HZL, is shown in Fig. A.4.

The RAM open-pit mining operation is mined at a stripping ratio of 1:13 (1:14), much higher than the present mine strip ratio of 1:7. It is at its peak. This is one of the reasons for going underground, which shall not be cost accretive. The equipments planned for mining

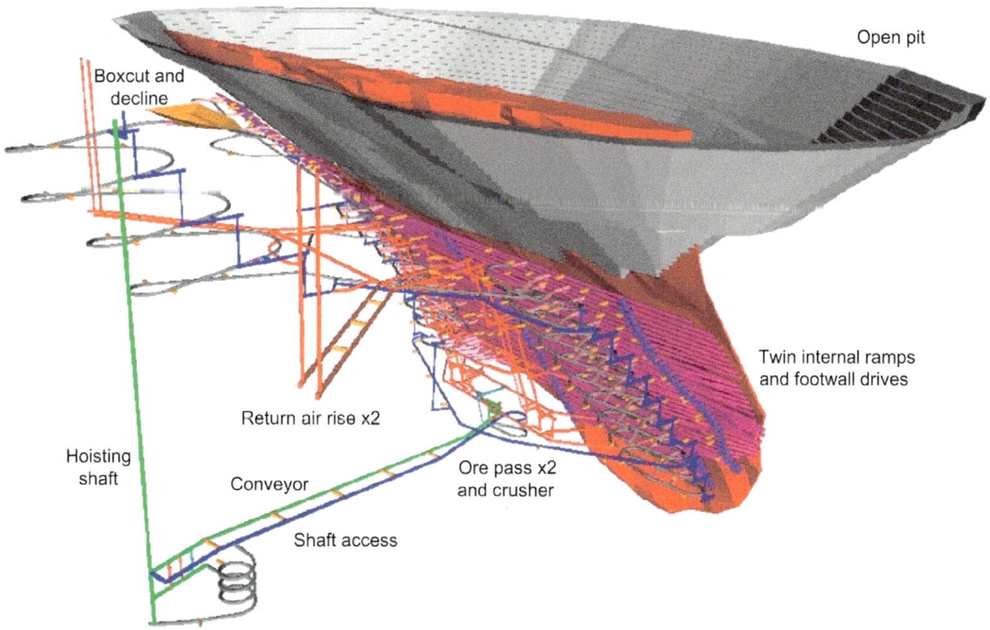

FIGURE A.4 The isometric view of the open pit and the proposed underground mine at Rampura-Agucha.

will use a combination of 34 m^3 excavator, 220 tonnes and 95 tonnes class mining trucks. The footwall slope of the open pit has been excavated to its final position and due to the dip of the ore body future waste stripping is focused on the hanging wall side of the deposit. Pit overall slope angle of the footwall and hanging wall slopes are 35 degree and 42 degree, respectively. Slopes are monitored using a Ground Probe Radar and no major material failures are reported till date.

The production forecast for the underground mine is planned to build up to a level of 2 Mtpa during the period of concurrent open pit and underground mining and expanded to 3.5 Mtpa once the open pit is complete. The principal access for underground mine is planned via a 12-km long surface decline excavated to a dimension of 5.5 m × 5.0 m, which splits into two ramps to provide a north and south access to the ore body. The ramps will be supported by a 7.5 m diameter surface shaft for ore hoisting together with two 7.5 m diameter return air ventilation shafts. An additional underground ramp is also planned together with establishing access to surface via the base of the open pit, which will provide additional ore haulage capacity until the surface shaft completion. The proposed underground operation is a fully mechanized operation principally using Long Hole Open Stoping (LHOS) with cemented paste fill and cemented rock fill to maximize extraction in an overhand manner. There are some mining difficulties associated with the nature of the ore and geotechnical conditions that have impacted on the ability of the mine to ramp up production which principally comprise of the presence of ore body shears and poor ground conditions.

The LHOS mining method will be undertaken from levels spaced at 25 m vertical intervals either on a longitudinal or transverse basis for ore body widths below or above 15 m, respectively. All development is planned in the footwall with drives and crosscuts providing access from the surface ramps. Long-hole drilling will be employed and large 17 tonnes class LHDs in conjunction with 60 tonnes class LPDT trucks will be used for stoping with smaller 30 tonnes units used for development. The trucks will either be loaded directly or via ore passes for transport to surface or the surface shaft once commissioned.

In the event of an emergency, there are two ramps each with access to surface and connections between footwall drives will be equipped with ladders and refuge chambers are planned at strategic locations. A 2.5 Mtpa cemented paste fill plant is built. The current focus is with the trial mining block which is located between the 60 m high crown pillar and the base of the final open pit and 23 m in height. The first main mining block is 100 m in height and located immediately below the crown pillar.

The main ventilation return air from underground mine will be located at ends of workings in south and north. The main intake roots will be through ramp and shaft. The air will be distributed to each stope from lower footwall drive and cross through stopes to upper level and return to main circuit. Necessary regulators will be provided to control the flow as per requirement. Two ventilation return shafts of 7.5 m diameter, 500 m depth will be equipped with exhaust fans. Planned total quantity of air is based on combined requirement of person and equipment deployed in that area. For dilution of diesel fumes in work area, quantity of air based on 0.05 m^3/s per KW is considered. As a part of HZL's expansion programme, underground mine will be ramped up to an ore production capacity of 3.75 million Mtpa, which will involve sinking of a 950 m hoisting shaft.

Processing operations initially started at RAM in 1992 through a single processing stream at a throughput of 0.9 Mtpa. The Run Of Mine ore, of about 1.2 m, is fed to primary gyratory crushers. The primary crusher product, of 150 mm size, is conveyed to coarse ore stockpile and from there it is conveyed to subsequent beneficiation process. Zinc and lead concentrates are separated by using conventional froth flotation process and the remaining material, slurry, is disposed of as tailing. The treatment capacity was later increased as further equipment has been added to the original circuit and additional processing streams have been installed. Stream-1 was expanded to achieve a current throughput of 2.5 Mtpa while Stream-2 was installed in 2005 and partially integrated with Stream-1 and has a throughput of 1.5 Mtpa. The relatively independent Streams-3 and Streams-4 were brought on line in 2008 and 2010, respectively with treatment capacities of 1.0 Mtpa and 1.5 Mtpa, resulting in an overall treatment capability of 6.5 Mtpa. All the four streams are relatively highly instrumented with multistream metal analysis and a high degree of monitoring, automation, and control including froth monitoring in the flotation circuits.

As the production from RAM open pit declines, further material will be sourced from underground operations, and also from HZL's other captive zinc-lead mines nearby, in the years to come.

Pictures depicting different stages of development of Rampura-Agucha mine are shown in Figs. A.5–A.17.

FIGURE A.5 Left: Rampura-Agucha prospect: shallow depression filled with water, in its pristine stage, in December 1979. The twin trees on the debris mound was a landmark. Right: Where it all began!—Rampura-Agucha exploration camp in February 1980. *Source: Photo courtesy by Gandhi*

FIGURE A.6 The shallow depression, as viewed from south end, in Rampura-Agucha, tilled for cultivation. In the background, drilling operation was on. The drilling camp on the left flank, as it was in April 1980. *Source: Photo courtesy by Gandhi.*

FIGURE A.7 Rampura-Agucha prospect, in March 1981. The well is in the contact of ore body and hanging wall. The head frame of exploratory incline is in the background. *Source: Photo courtesy by Gandhi.*

FIGURE A.8 Left: Pilot-pit operation in Rampura-Agucha in December 1988. Right: Aerial view of Rampura-Agucha open-pit mine, as in October 1999. *Source: Photo courtesy by Gandhi.*

FIGURE A.9 Rampura-Agucha open-pit operation, in December 2003. *Source: Photo courtesy by Gandhi.*

FIGURE A.10 The Rampura-Agucha open-pit mine in 2013. *Source: Photo courtesy by Gandhi.*

FIGURE A.11 A view of Rampura-Agucha open-pit mine in 2014. *Source: Photo courtesy by Kavdia.*

FIGURE A.12 Panoramic view of Rampura-Agucha open pit with waste dump and tailing pond in 2015. *Source: Photo courtesy by Kavdia.*

FIGURE A.13 Sprinkling of water (by sprinklers and water tank) on the haulage way, to suppress the dust in Rampura-Agucha mine complex. *Source: Photo courtesy by Gandhi.*

FIGURE A.14 (Left) Garland drains to prevent flooding of mine and contamination of natural water courses. (Right) Plantation near waste rock dump area. *Source: Photo courtesy by Gandhi.*

FIGURE A.15 Panoramic views of Concentrator Plant Complex, Rampura-Agucha. *Source: Photo courtesy by Gandhi.*

FIGURE A.16 Ball mills and flotation cells in the concentrator plant, Rampura-Agucha. *Source: Photo courtesy by Gandhi.*

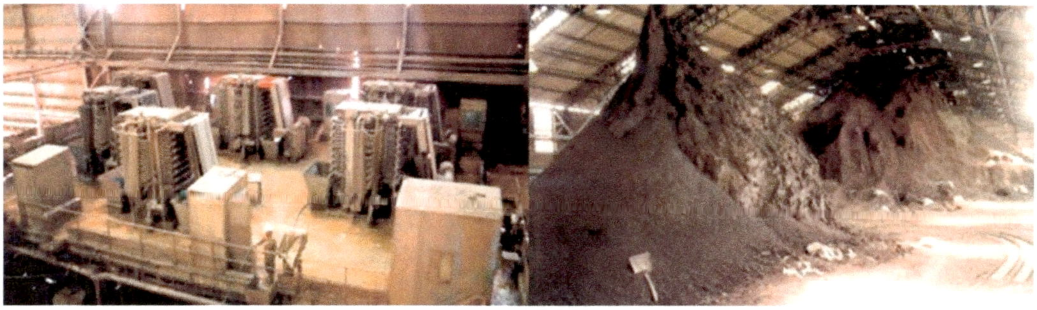

FIGURE A.17 Concentrate filters and the concentrate storage yard, Rampura-Agucha. *Source: Photo courtesy by Gandhi.*

A.2 RAMPURA-AGUCHA ORE DEPOSIT CHARACTERISTICS AND ITS GENESIS

Many exploration teams and university research groups, both from India and abroad, have carried out investigations on various aspects of Rampura-Agucha deposit, which is one of the best-studied zinc-lead deposits. Detailed descriptive accounts on various aspects of this deposit are given in the memoir on this deposit by Gandhi (2003). However, for the benefit of readers to understand the deposit, a brief account on the ore body characteristics and its genesis are given below.

Rampura-Agucha deposit occurs in the oldest part of Bhilwara Belt at the contact of with the Archean basement—Banded Gneissic Complex (BGC). The Bhilwara Belt consists of a pile of metasedimentary rocks intruded by igneous rocks. It developed as a result of crustal extension of the Archean basement about 2.0 Ga ago (Fig. A.18). Simplified map of Rampura-Agucha deposit is given in Fig. A.19.

The deposit occurs in a plunging isoclinal synformal structure of elliptical shape, comprising of sillimanite–graphite–mica schist, which hosts the mineralization, enclosed in garnet–biotite–sillimanite gneiss with minor bands of amphibolites and calc–silicate rocks, leucocratic rocks (pegmatite, aplite, and quartzites) and mylonites. The ore body is lens shaped with NE-SW strike length of 1600 m and a width varying from a few meters in the northeast to as much as 100 m in the central and SW sections and exhibits sharp contacts with the footwall and hanging wall rocks. The ore body dips between 50 degree and 80 degree SE and has been delineated by drilling up to a depth of 370 m from surface. The deposit had an oxidized gossan and a small zone of partially oxidized section between gossan and protore. In the gossan zone, there were sporadic high silver values (up to 200–250 ppm), which probably developed due to the dissolution of finely disseminated sulfosalt minerals in the deposit, migration of the silver within this zone and accumulation of the bulk of it in the lower part of the leached subzone. The calc–silicate bands of the hanging wall contact of the deposit acted, in places, as loci for coarse, recrystallized, argentiferous galena.

The mineralization occurs predominantly in the graphite–sillimanite–mica schist, which consists of quartz, alkali feldspar, plagioclase, sillimanite, graphite, and various micaceous minerals like muscovite, biotite, and chlorite. Sphalerite, the predominant ore

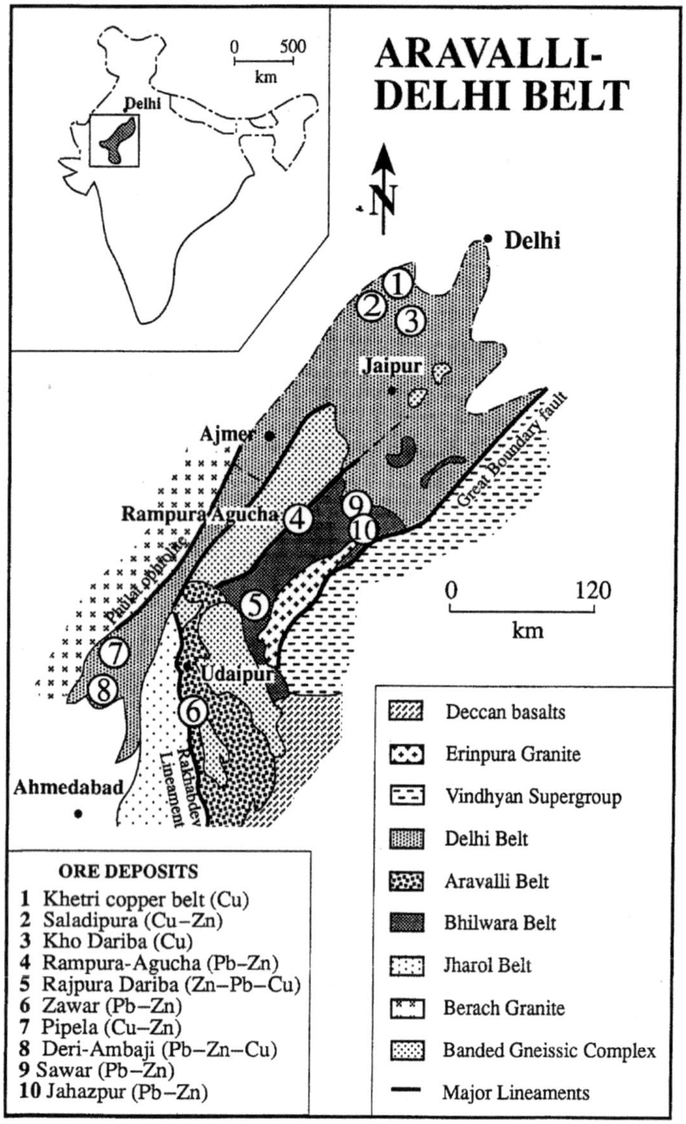

FIGURE A.18 Generalized geological map of Aravalli-Delhi fold belt including the main ore deposits. *Source: Gandhi (2003).*

mineral occurs with galena, pyrite, and pyrrhotite in varying proportions. A large variety of minor sulfide phases have been identified within the ore, especially chalcopyrite, arseopyrite, and Ag–(Pb)–Sb sulfosalts. Ore microscopy of samples from many drill core intersections from the ore body did not reveal any metal zonation within the ore body, although the Zn/Pb/Fe ratios varied in places within meters. Although sphalerite is the most important base metal sulfide at Rampura-Agucha, small sections of the ore body can be dominated by galena and pyrrhotite.

In the Lower Proterozoic, the crustal rocks of the BGC were thrust over the Western margin of the Bhilwara Belt, resulting in high-grade metamorphism in the Rampur-Agucha area.

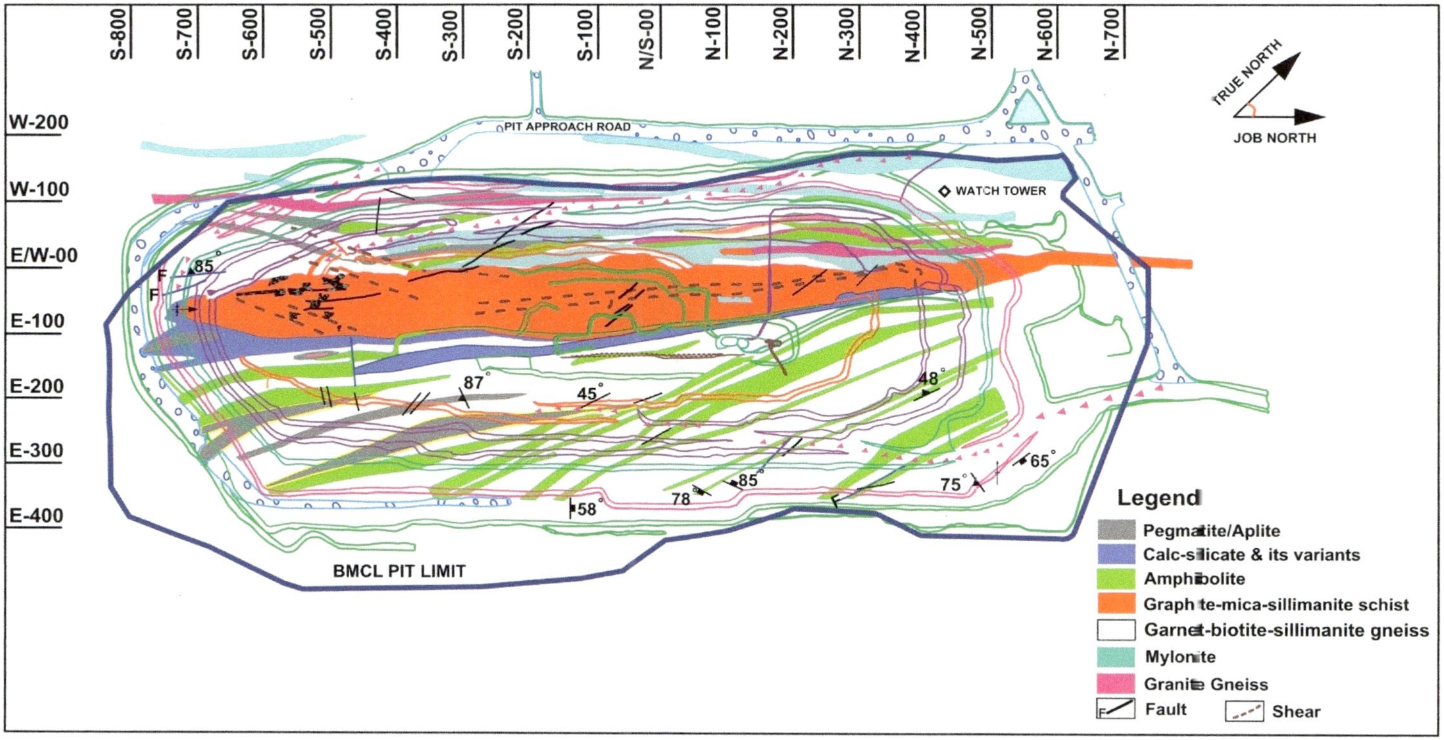

FIGURE A.19 Simplified geological map of Rampura-Agucha deposit. *Source: Gandhi (2003).*

Peak metamorphic conditions of upper amphibolites to granulite facies were estimated by garnet–pyroxene geothermometry (~700°C), sphalerite geobarometry (~7 kb), and fluid inclusion studies. Continuously zoned garnets with decreasing X_{Mg} toward the rim indicated a singular-phase metamorphic event. The mineral assemblages of the country rocks still represent high-grade metamorphic conditions: sericitization and chloritization are only indicators of retrogressive metamorphism. The high-grade metamorphic event resulted in a high degree of recrystallization of the ore (coarse-grained sulfide aggregates with grain sizes exceeding 3 mm) and obliteration of most of the primary sedimentary textures. Remobilization of galena and sphalerite and to a lesser degree of pyrrhotite into cracks of quartz and feldspars is widespread. In the course of this project, the approximate age of metamorphism of ~950 Ma has been determined by $^{39}Ar/^{40}Ar$ dating of metamorphic minerals: an amphibole age of 909 Ma, two muscovite ages of 888 and 874 Ma, two biotite ages of 797 and 788 Ma, vary due to different closure temperature of the three minerals.

Rampura-Agucha deposit contains assemblages rich in Ag–(Pb)–Sb sulfosalts. Freibergite, pyrargyrite, stephanite, argentite, dyscrasite, and various Pb–Ag–Sb sulfosalts occur either within or close to large aggregates of galena. Electron microprobe analyses reveal an average Ag content of 31% in freibergite, whereas galena was devoid of Ag. The sulfosalts are Sb-rich end members of the respective solid solution series with only limited As. The assemblages were affected by recrystallization and reequilibrium during high-grade metamorphism and subsequent cooling. Pyrargyrite presumably formed by replacement of freibergite and of Pb–Ag–Sb sulfosalts, stephanite, by decomposition of pyrargyrite and argentite and dyscrasite by exsolution from galena.

Rampura-Agucha deposit contains a number of rare oxide minerals. These have been formed as a result of high-grade metamorphism. Gahnite is a common minor phase in the ore body and has formed mainly by desulfurization of sphalerite. Pyrophanite-ilmenite is very rare and occurs intergrown with rutile, the major oxide mineral. Two new compositional varieties of rare oxide minerals, a Cr–V spinel and a Cr–V oxide ($(Cr-V)_2O_3$) have been determined. The compositions of these two minerals differ from known end members because of considerable substitution of Cr for V. Similar oxide minerals have been reported from several high-grade metamorphic deposits and area. The rare V oxide "schreyerite" ($V_2Ti_3O_9$) also occurs as exsolution lamellae in rutile at Rampura-Agucha. These unusual compositions are attributed to the locally high V contents in the precursor sediments; they further underline the isochemical nature of regional metamorphism of the ore body.

Dravite-rich tourmaline with Fe/(Fe + Mg) ratios, around 0.02 occurs at the hanging wall contact of the ore body with the paragneiss and is clearly associated with the mineralization. Tourmaline from the strata-bound ores is distinguished from schorl-rich tourmaline of two pegmatite samples which show Fe/(Fe + Mg) ratios of 0.43 and 0.62, respectively. At Rampura-Agucha, dravite-rich, premetamorphic tourmaline or its precursor mineral is very probably of exhalative origin, formed by the same hydrothermal fluid as the associated sulfide minerals and was later affected by recrystallization during high-grade metamorphism. The ore minerals of Rampura-Agucha deposit are given in Table A.6

Fluid inclusions in quartz of the country rock and dravite-rich tourmaline from Rampura-Agucha deposit have been investigated by microthermometry and Raman microspectrometry. Four different main types of fluid inclusions in quartz could be distinguished: (1) Gaseous (CO_2, partially mixed with CH_4-N_2), (2) low-salinity aqueous

TABLE A.6 The Ore Minerals of Rampura-Agucha Deposit

NATIVE METALS	
Native silver	Ag
Native antimony	Sb
SULFIDES/SULFOSALTS	
Sphalerite	Zns
Pyrite	FeS_2
Marcasite	FeS_2
Pyrrhotite	$Fe_{1-x}S$
Galena	PbS
Chalcopyrite	Cu,FeS_2
Arsenopyrite	FeAsS
Wurtzite	ZnS
Greenockite	CdS
Molybdenite	MoS_2
Lollingite	$FeAs_2$
Breithauptite	Ni,Sb
Gudmundite	FeSbS
Ullmanite	NiSbS
Tetrahedrite	$(Cu,Fe)_{12}Sb_4S_{13}$
Tennantite	$(Cu,Fe)_{12}As_4S_{13}$
Freibergite	$(Ag,Cu,Zn,Fe)_{12}(Sb,As)_4S_{13}$
Pyrargyrite	Ag_3SbS_3
Proustite	Ag_3SbS_3
Stephanite	Ag_5SbS_4
Acanthite	Ag_2S
Miargyrite	$AgSbS_2$
Argentite	Ag_4S
Dyscrasite	Ag_3Sb
Polybasite	$(Ag,Cu)_{16}Sb_2S_{11}$
Boulangerite	$Pb_5Sb_4S_{11}$

(Continued)

TABLE A.6 (Continued)

OXIDES	
Zincite	ZnO
Rutile	TiO_2
Karelianite-eskolaite	$(V,Cr)_2O_3$
Gahnite	$(Zn, Fe,Mg)\ Al_2O_4$
Magnetite	Fe_2O_3
Schreyerite	$V_2Ti_3O_9$
Dravite	$ZnOAl_2O_3$
Pyrophanite	$MnTiO_3$
CARBONATES	
Calcite	$CaCo_3$
Dolomite	$Ca, Mg\ (CO_3)_2$
Smithsonite	$ZnCO_3$
Cerussite	$PbCO_3$
SULFATES	
Gypsum	$CaSO_4 \cdot 2H_2O$
Anhydrite	$CaSO_4$
Goslarite	$ZnSO_4 \cdot 7H_2O$
Anglesite	$PbSO_4$
SILICATES	
Hemimorphite	$Zn_4(OH)_2O_7 \cdot H_2O$
Willemite	Zn_2SiO_4
PHOSPHATE	
Chlorapatite	$(CaCl)Ca_4(PO_4)_3$
GRAPHITE	C

inclusions (0–8 eq.wt% NaCl), (3) CO_2-H_2O inclusions in dravite, and (4) high-salinity aqueous inclusions. Low density CO_2-rich and low-salinity H_2O inclusions are contemporaneous and occur, as well as CH_4-N_2 inclusions, in close association with sulfide mineral inclusions. This indicates immiscibility between gaseous and aqueous phase and participation of these fluids during the remobilization of the ore. $H_2O-CO_2\ \pm CH_4N_2$ inclusions in dravite-rich tourmaline represent the metamorphic fluid, trapped during metamorphic recrystallization. Raman spectra of graphite indicated upper greenschist-facies metamorphic conditions which suggests that graphite reequilibrated with CO_2-rich phase during retrograde metamorphism.

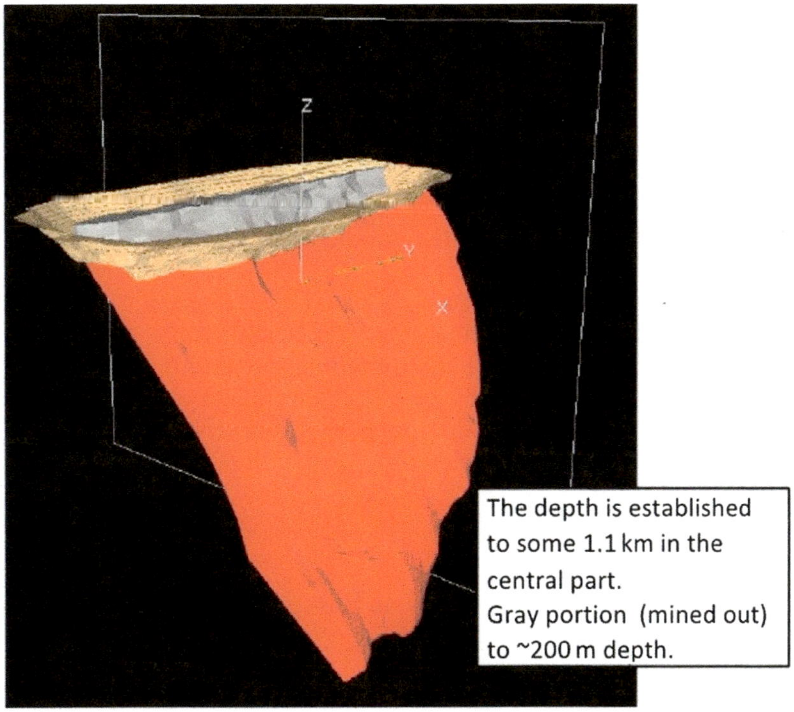

FIGURE A.20 Isometric view of the ore body at Rampura-Agucha as viewed from the south along with the present open-pit outline.

An isometric view of the ore body at Rampura-Agucha as viewed from the south along with the present open-pit outline is shown in Fig. A.20.

A.2.1 Genesis

The ore formation involves complex sets of interacting processes, and Rampura-Agucha is no exception. Taking into consideration of all the investigations carried out in this deposit, a comprehensive genetic model for the mineralization can be attempted. The Rampura-Agucha deposit is a stratiform sediment-hosted exhalative deposit. Although all relic of the hydrothermal fluid, responsible for the mineralization, disappeared during high-grade metamorphism and most synsedimentary structures were obliterated it can be assumed that the deposit is the product of hydrothermal convection during crustal extension (indicated by the occurrence of former subalkaline, tholeiitic ocean floor basalts at the deposit which may have contributed the heat for the convection cells). Indications of the composition of this hydrothermal fluid include tourmaline-rich samples, intergrown with sulfides and "Cl" contents in apatite and graphite. One may speculate that the low "copper" contents of the ore are due to precipitation of Cu in the feeder systems which, so far, have not been exposed by exploration or mining activities. The high amount of biogenically derived carbon (and low $\delta^{13}C$ values) suggests a strongly reducing environment in the basin. The predominance of former shales and minor carbonate-rich sediments at

Rampura-Agucha indicates a deeper, tranquil depositional environment (eg, a third order basin). The presence of a single giant ore body with rather high metal content and the occurrence of considerable amounts of magmatic rocks (now amphibolites) as possible heat source in the vicinity of this ore body, points toward intensive hydrothermal convection and a rapid precipitation of the sulfides. Rampura-Agucha deposit is of Mid-Proterozoic age (1.8 Ga), similar to the Australian deposits (eg, Mt Isa, MacArthur River, Broken Hill), Sullivan (Canada), and several smaller deposits in Rajasthan, India, such as Rajpura-Dariba, Zawar deposit, contemporaneously formed in the same, or adjacent metamorphic belt.

Formation of Rampura-Agucha deposit is envisaged to have taken place in oceanic basement within an extensional environment. Mid-oceanic-ridge basalt-type basalts have acted as a heat engine for submarine hydrothermal circulation; reducing conditions are indicated by the presence of carbon of biogenic origin. These observations are commensurate with a Red Sea-type setting. High-grade metamorphism was largely isochemical; during retrograde stages the complex and economically important "Ag" mineral association developed.

A.2.2 Special Features Observed Only at Rampura-Agucha

The deposit is associated with igneous rocks than other Sedex deposits. Mafic rocks (amphibolites) and felsic magmatites (aplites, pegmatites, and orthogneisses) comprise around 20% of the country rocks. The deposit comprises of a single giant ore body of 114 Mt with a considerable metal content (+15% Pb + Zn).

High-grade metamorphism (~700°C) obliterated most of the premetamorphic sedimentary features of the deposit. Due to the fact that the sedimentary environment of Rampura-Agucha was strongly reducing, oxide facies did not develop. This may be the explanation for the lack of barite, Fe- and Mn-oxides, which could not precipitate. However, not all characteristics, typical for Sedex deposits, are found at Rampura-Agucha and the deposit differs slightly from all other Sedex-type deposits. It is well known from the literature that there could be many subgroups in Sedex deposits depending upon various geological and geochemical characteristics.

Out of the important Sedex deposits worldwide, no one can be considered as identical with the Rampura-Agucha deposit. However, as a whole, host rock lithology, high-grade metamorphism together with the presence of pathfinder minerals such as gahnite, dravite, diopside, and the sulfidic paragenesis make Rampura-Agucha deposit very close to world-class deposits such as Broken Hill (Australia) or Gamsberg (South Africa).

It has features that distinguishes it from the other Sedex-type deposits such as a giant single ore body and the occurrence of considerable amounts of mafic and felsic magmatites and the lack of zonation, iron formation, and barite. Nevertheless Rampura-Agucha Zn–Pb–(Ag) deposit clearly belongs to the group of the Sedex-type deposits and therefore clearly of an synsedimentary origin. Since this deposit does not fit into any established Sedex deposit type, one is tempted to designate this unique deposit as "Rampura-Agucha Type," as a subgroup, in Sedex deposits.

A.3 ANCIENT MINING AND SMELTING FOR SILVER/LEAD IN RAMPURA-AGUCHA AREA

A.3.1 Introduction

The dumps of debris on either site of the shallow depression, slag dump, and some old workings in the wells around Rampura-Agucha area indicated the ancient mining activity but the extent of it could not be deciphered. When this deposit was discovered in the late 1977, it was felt that the ancient mining activity was not much in this area. But as the time passed by, and the exploration advanced, it became evident that this area had also been worked out by the ancients, to a reasonable degree, though not extensively as found elsewhere in Zawar or Rajpura-Dariba areas in Rajasthan, India.

HZL moved into this area to explore the deposit in detail, in February 1980, and the first author (SMG) had the privilege of associating himself right from the inception of exploration campaign in Rampura-Agucha. On one of the field visits in April 1980, SMG was in the vicinity of the drill site (BH 83 in S-200) where drilling was done to test the presence of ore at 100-m depth. A wooden piece of about 15 cm was recovered from the core barrel from a depth of about 80 m from surface level. The occurrence of wood in the boreholes, broken earthen potteries, slag and debris dumps stirred plethora of questions as to what, why, how, and where the ancients were working in this part of the country.

Over a period of time from March 1980, all the details relating to old working below ground were garnered as and when the open-pit operation advanced, to decipher the network of old working system. The writer had discussions with senior citizens of various villages located 10 km radius from the deposit area and also scanned accessible records and scriptures about the deposit area and its environs. SMG had the first-hand experience of entering into many of the old workings to decipher the layout of the old working system. The detailed account of old workings and the ancient history of the area have been given by Gandhi (1983), Tiwari and Kavdia (1984), Gandhi (1988, 2000, 2003, 2014), and Craddock et al. (1989).

The evidence of ancient mining activity in this area can broadly be grouped into two categories:

- Surface indications
  - Topographic depression
  - Debris dump
  - Slag dump
  - Surficial slumps
  - Crucible/retort/potteries
- Subsurface (below ground) indications
  - Old wells and their interconnections
  - Drill holes—gaps/wood/fluvial sand
  - Underground excavation

A.3.2 Surface indications

Shallow topographic depression extending to about 1.6 km trending northeast-southwest, indicated that some mining activity took place in the past. In central portion of

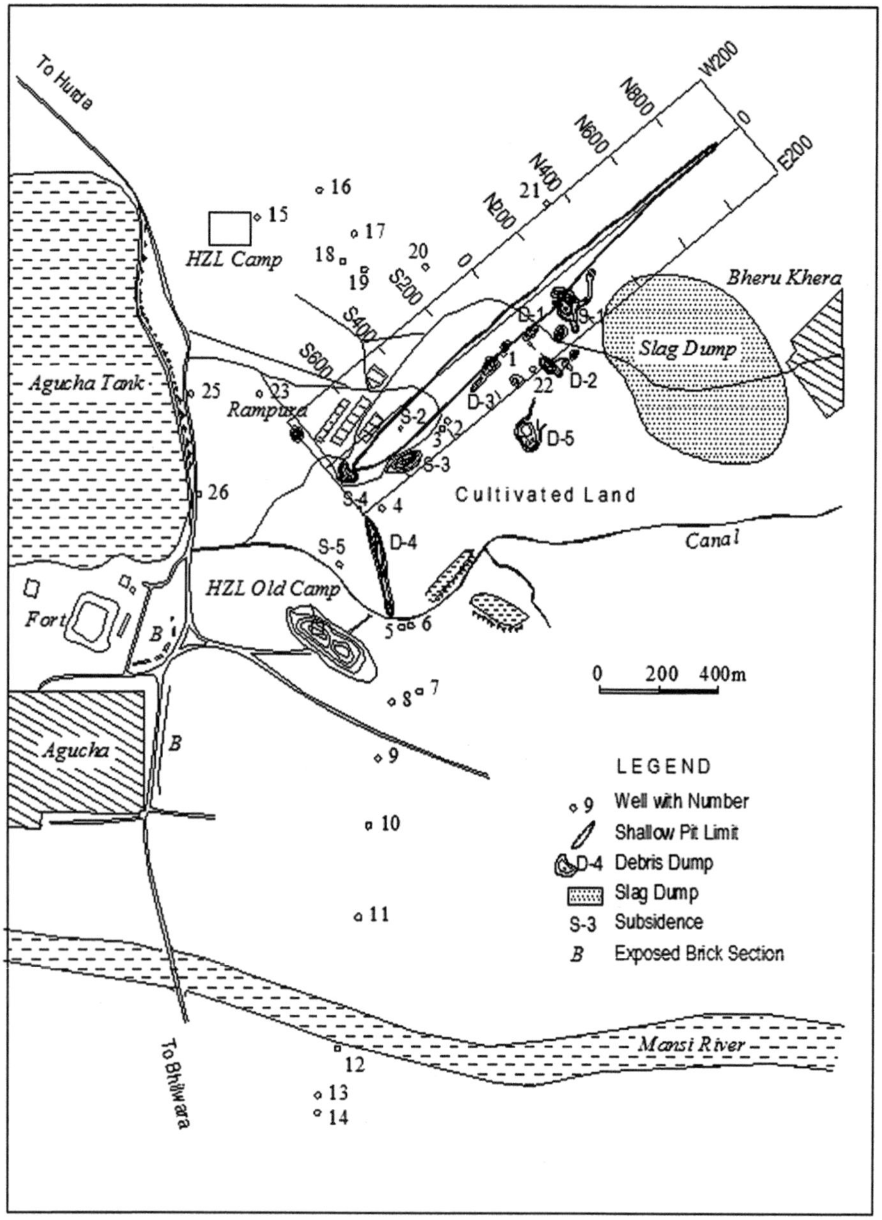

FIGURE A.21 Map of Rampura-Agucha deposit and its environs.

the depression, hump-like exposures were found which have been left out unmined by the ancients, probably, owing to its poor metal content.

The location of debris, slag dumps, and surficial slumps are shown in Fig. A.21. Debris dumps of different sizes and volume predominated the eastern portion (hanging wall) of the open pit. The examination of dump materials indicated that they constituted the mineralogy

TABLE A.7 The Composition of Slag Pieces from Rampura-Agucha

		Pb	Zn	Fe	Ag	Sb
			In wt%		In ppm	
1. Slag piece from South end of slag dump						
Dirty white portion		21.0	0.13	8.2	225	600
– Middle green layer		12.7	0.13	9.1	85	140
2. Slag pieces from eastern and near the Foot path to village Bheru Khera	(a)	7.9	0.17	3.3	106	–
	(b)	7.0	0.08	6.7	130	–
3. Composite samples of slags from western side of slag dump	(a)	4.4	2.80	8.3	30	–
	(b)	1.6	0.16	6.6	25	–
	(c)	2.0	0.03	5.2	20	–

Source: Gandhi (1988).

of garnet–biotite–sillimanite gneiss, which is the predominant rock type in the hanging wall of the deposit. Many of these dump mounds contain innumerable pieces of potteries of varying sizes and shapes, reflecting the existence of human inhabitation in these areas in the past.

Slag dumps and furnaces normally give evidences of extractive techniques adopted by the ancients. Slag dump, extending from the eastern side of north portion of the pit up to the boundary of the village Bheru Kheda, was noticed in the deposit area. Excavations done at random spots indicated that the depth of slag heap is not more than a meter deep with an aerial extent of about 1600 sq. m. Much of the slag was indurated, vitreous, and vesicular type. Some of the slag pieces analyzed had high lead, zinc values, indicating that the original metal values of the ore used must have been quite high. The composition of slag pieces of Rampura-Agucha area is given in Table A.7. Small specs of relict sulfides were noticed in many slag pieces. Some of the slag pieces were found to have charred wooden pieces entombed within them with vitreous and scoriaceous slag. Near the slag dump area, in the eastern part of shallow depression, a couple of broken retort-like pieces, tuyeres, the parts of crucible, and lead metal pieces had been recovered (Fig. A.22).

A.3.3 Subsurface Indications

There are number of wells on the eastern side of the shallow pit which have drive-like openings at different levels. According to local residents, these openings are interconnected and a tunnel-like opening extends as far as the Mansi River and is believed to extend further to the other bank of the river (Fig. A.23). It is rather enigmatic what this facility was used by ancients. Some of them are suspected to be ancient mining shafts and drives. As the mining operations at Rampura-Agucha advanced, some of these drive- and shaft-like openings were encountered at different depth levels (Fig. A.23).

In the boreholes drilled from hanging wall side wooden pieces ranging from a few cm to 20 cm had been encountered at various depths, the deepest recovered so far being from

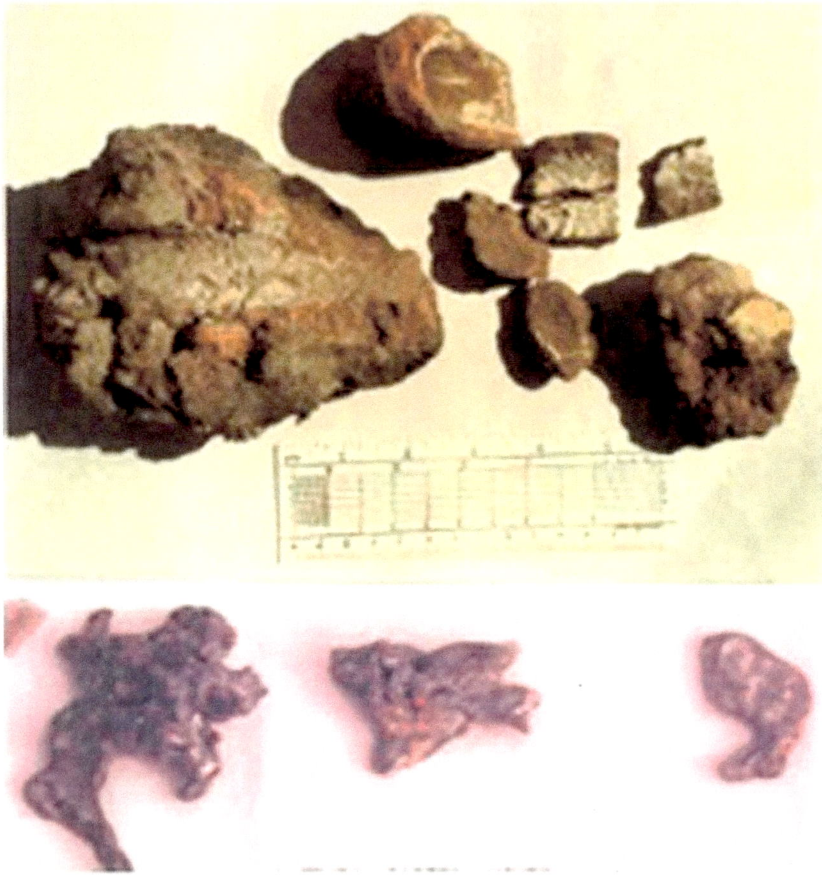

FIGURE A.22 (Top) Broken retort-like pieces, tuyeres; (Bottom) Lead metal pieces recovered from the hanging wall side of the deposit area.

160 m below surface. In the southern portion, the gaps, wooden pieces, and bamboo baskets (Fig. A.24) were encountered at the contact of ore and footwall ground. Radiocarbon dating of some of the wooden pieces recovered from borehole, mine opening, and incline indicated the following ages (from Craddock et al. 1989).

Location		Material	Age in Years Before Present
1.	Mine Gallery, 30 m below surface	Timber	2240 ± 60
2.	Basket, footwall area	Twigs	2380 ± 130
3.	Cupel debris area	Charcoal	2350 ± 40
4.	South wall, open pit	Wood	2320 ± 50
5.	Cupel debris area	Charcoal	2140 ± 50
6.	Old mine shaft	Wood	2860 ± 100

1–5, British museum, London; 6, Physical Research Lab., Ahmedabad, India.

FIGURE A.23 Shaft- and drive-like openings in different parts of the deposit.

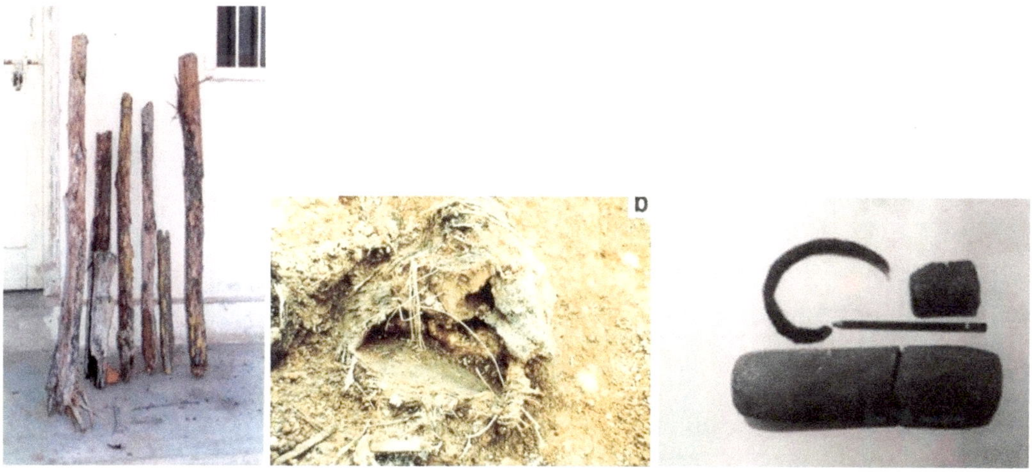

FIGURE A.24 Wooden pieces, bamboo basket remains, and grinding stone, recovered from underground workings of the deposit area.

During drilling, loose, fine fluvial sand was encountered in the northern sector of the deposit at different depths, indicating either flow of water from some source or some buried interconnected channel through which water might have flown. While sinking an exploratory incline in 1980, drive- and shaft-like openings were encountered at two places. The first one was at 20-m depth below surface and the second one was at 25-m depth,

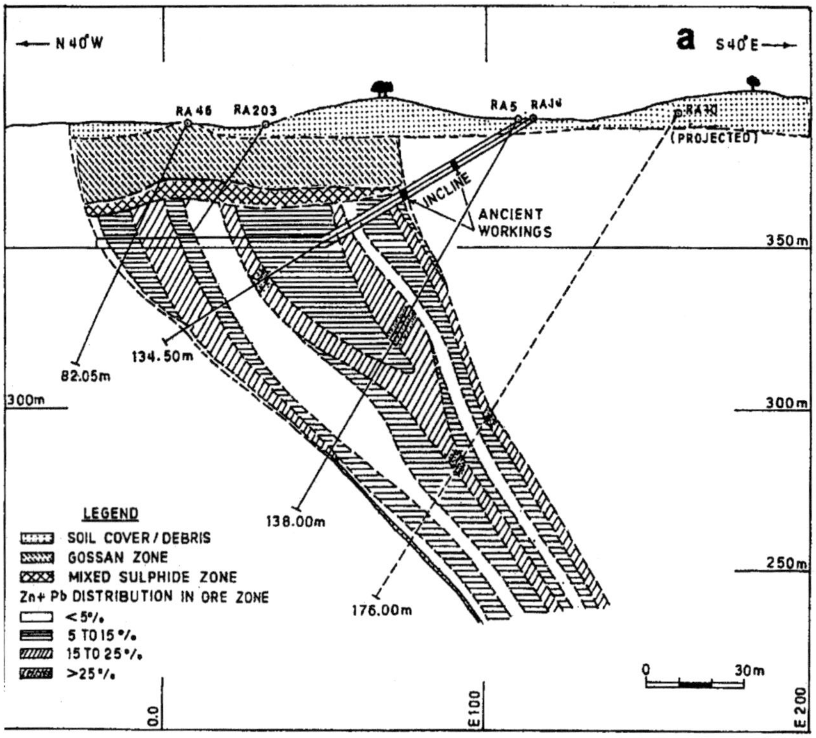

FIGURE A.25 Section along "0-0" showing ancient workings in the incline.

the latter at the contact of ore mineralization with the hanging wall (Fig. A.25). Timber pieces of varying lengths, pottery pieces, and grinding stones were recovered from underground workings (see Fig. A.24).

A.3.4 Ancient Smelting

The available metallurgical evidence indicates that in Rampura-Agucha, there was lead/silver production on a reasonable scale, principally during later part of first millennium BC. Lead ores are readily reduced and a roasting stage may not have been considered necessary. Fragments and tuyeres abound in the slagheaps that are found near the deposit area.

Enormous quantities of cupellation debris, particularly, fragments of refractories, were found at the ancient shallow pit area in the deposit. These were distinctive, both by being covered in lead glaze and by the fact that the body of the ceramic was bright red, showing that they had been in oxidizing atmosphere, unlike the black reduced ceramic form, from the usual reduction processes associated with smelting.

Excavation conducted at Agucha jointly by British museum and HZL in 1985–86, produced a large collection of small cupels, some with the tuyere still attached, pieces of

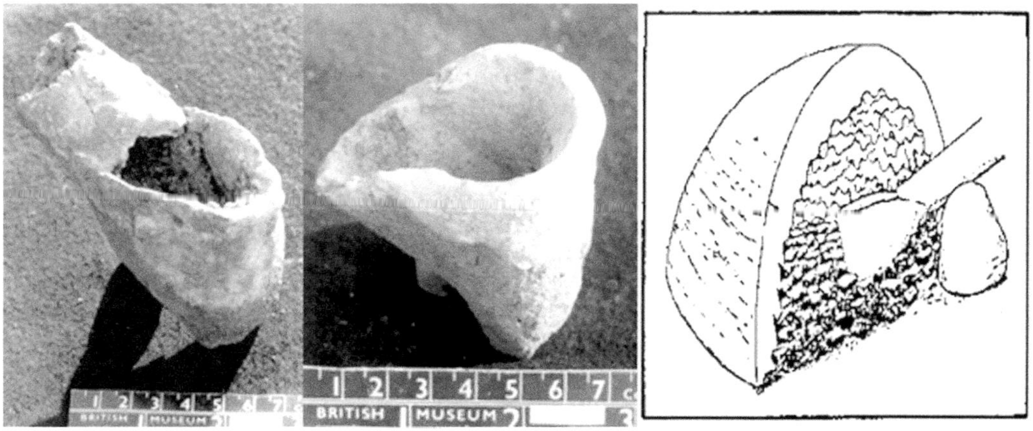

FIGURE A.26 (Left and Middle) Cupels, one attached with tuyere, used for silver extraction. (Right) Artist's impression of cupellation furnace setup for extraction of silver.

lead metal, tuyere fragments, small clay pillars with flat bases, and concave tops. The latter are believed to be have been tuyere supports, giving the arrangement seen in Fig. A.26 (Left and Middle). The tuyeres have thin walls (2–3 mm) with an internal diameter of about 25–30 mm and are set right down into the cupels, which are designed to accommodate them. The cupels have walls about 5 mm thick and are heavily tempered with straw. Their capacity is small, about 15 mL to the brim but only 8–9 mL to the end of the tuyere, as indicated by the rim of glazing inside the cupel. These cupels are too small and fine to be the standard vessels in which silver was refined on an industrial scale. The small vessels could probably represent the equipment for trial assays. The compositions of lead metal pieces recovered from the excavations, are given in Table A.8.

One large piece of litharge excavated in Agucha could probably represent an industrial-scale production. It is very possible that cupellation was generally carried out at Rampura-Agucha in large vessels or just in hollows scooped into the ground as found elsewhere in the world (Sardis in Turkey; Ramage, 1970). There is literary evidence for this in India: *Arthashastra* (c.320 BC), *Rasratnasammuchchaya* (13th century AD), and *Ain-i-Akbari*, etc. (Ray, 1956; Kangle, 1972; Kulkarni, 1982). In the case of *Ain-i-Akbari*, it is mentioned that a small hollow was scrapped in the ground and lined with a mixture of ashes of cow dung and of "babul wood," for extraction of silver from lead (Percy, 1870).

The product of the smelt was lead, which would have been rich in silver. The separation of silver from lead was done, as elsewhere, by cupellation. The argentiferous lead was melted in an open crucible or hearth and air blown across the exposed surface at about 900°C to 1000°C. This oxidized lead to lead oxide, some of which blew away as litharge to coat and glaze any exposed hot ceramic surfaces and some remained molten in the cupel itself. The silver remained unoxidized in the cupel and floated on the surface of the dense litharge, like oil on water. The scale of operation in this part seems to be very less because of the occurrence of very few crucibles in this area. Based on furnace fragments, cupels, tuyeres, and clay stands found at Agucha, a reconstructed "cupellation set-up" is shown in Fig. A.26 (Right).

TABLE A.8 Composition of Lead Metal Pieces Recovered From Rampura-Agucha

	Pb	Sb	Fe	Cu	Zn	Mn	Ag	Cd
Sample	**In wt%**				**In ppm**			
RA/M-1	96.94	0.41	0.04	0.03	30	60	1080	8
RA/M-2	92.98	0.45	0.03	0.17	60	35	1345	5
RA/M-3	96.29	0.31	0.35	0.04	60	–	1677	–
RA/M-4	96.76	0.31	0.41	0.03	890	55	1125	6
RA/M-5	96.73	0.42	0.04	0.03	300	60	1265	8
RA/M-6	99.60	–	0.05	–	110	55	1050	6
RA/M-7	95.29	0.38	0.01	0.11	190	–	1005	–

A.3.5 Archeological Investigations

Ruins of a partly buried village, temple, and fort with moat, near the village Agucha, indicate the existence of civilization in the past. Archeologists, basing their deduction on the clay artifacts from the ruins (earthen potteries, big bricks, beads, ancient coins, terracotta animal, and human figurines, stone plaques, etc.) believed in the existence of a flourishing civilization in this area during Maurya period (around 4th or 3rd century BC) and again in the Kushan period (1st and 2nd century AD). They also found "iron slags" and other materials, which belonged to pre-Maurya and Maurya period, in the elevated mound structure near Agucha.

A.3.6 Concluding Remarks

The early mines were worked principally for lead and silver in Rampura-Agucha. The indications are that silver extracted from the mines was not of great quantity, because this area had no appreciable primary silver ores, and whatever silver was extracted was only as a by-product. It is unclear, however, whether there was any industrial-scale production of silver. Much of the silver could have probably been used for coinage during the Mauryan period, when coinage was introduced and plenty of coins were minted c.4th millennium BC. Historical records (*Arthashastra* of Kautilya) indicate that academic and intellectual knowledge was highly advanced during Mauryan period. At a later date silver could have found use for both jewelry and coinage. Under the Mauryan Empire, much of Northern India became unified and thrived prosperously for more than a couple of centuries. The majority of the radiocarbon dates coincide with the Mauryan period, which only authenticates the flourishing mining and smelting activity that then existed in various places, viz., the Agucha and Dariba lead–silver mines, the Zawar zinc mine, etc.in Northwestern India. The scale and technical sophistication of mining and smelting activities were obviously at their zenith in the Mauryan Empire.

The Archaeometallurgical investigations in and around this deposit, unraveled the mysteries of exploitation and extraction practices for silver and lead metals by the ancients, c.4th century BC. The mining and smelting practices practiced by the ancients in this part of India, were very sophisticated and showed that there are no parallels or analogies anywhere in the world.

India scored "world's first" in the production of "pure metallic zinc," "high zinc, brass alloys," and the technology of producing "high quality steel." The consummate skill, expertise, and handiwork of the ancient metal workers in this part of the world, reflect yet another aspect of technical superiority of unknown miners and metallurgists of yesteryears.

References

Craddock, P.T., Freestone, I.C., Gurjar, L.K., Middleton, A.P., Willies, L., 1989. The production of lead, silver and zinc in early India. In: Hauptmann, A., Pernick, E., Wagner, G.A. (Eds.), Old World Archaeometallurgy. Bochum Bergbau Museum, pp. 51–70.

Gandhi, S.M., 1983. Rampura-Agucha zinc/lead deposit. Mining Mag. 149 (5), 315–321.

Gandhi, S.M., 1988. A Note on Old Workings and Ancient History in and Around Rampura-Agucha Zinc-Lead Deposit Area. Unpublished Investigation Report. RPD wing, HZL, 48 p.

Gandhi, S.M., 2000. Ancient mining and metallurgy in Rajasthan. Crustal Evolution and Metallogeny in the Northwestern Indian Shield. M.Deb, Narosa Publishing House, New Delhi, pp. 29–48.

Gandhi, S.M., 2003. Rampura-Agucha Zinc-Lead Deposit, vol. 55. Geological Society of India Memoir, 154 p.

Gandhi, S.M., 2014. Copper, Zinc, Lead Ores—their exploitation and metal extraction by the ancients in northwestern India. J. Geol. Soc. India 84, 253–266.

HZL, 1982. Addendum-I to Geology and Ore Reserve Estimates, Rampura-Agucha Lead-Zinc Prospect, Bhilwara District, Rajasthan, Unpublished HZL Report, 65 p. (June 1981).

HZL, 1988. Addendum-II to Geology and Ore Reserve Estimates, Rampura-Agucha Lead-Zinc Prospect, Bhilwara District, Rajasthan, Unpublished HZL Report, 65 p. (June 1981).

HZL Staff, 1992. Rampura-Agucha Mine. Min. Mag. v. 167, 372–375.

Kangle, R.P., 1972. The *Kautilīya Arthasāstra,* Text, English translation and Study, 3 vols. Motilal Banarasidass Pvt. Ltd., Delhi.

Kulkarni, D., 1982. Rasratnasamuchchaya, Vol 1, Ch. 1-11, Hindi translation and commentary on original Sanskrit text. Meharchand Lachmandas, New Delhi.

Percy, P., 1870. Lead. John Murray, London.

Ramage, A., 1970. Sardis: Pactolus North. Bull. Am. Schools Orient. Res. 199, 18–26.

Ray, P., 1956. History of Chemistry in Ancient and Medieval India. Indian Chemical Society, Calcutta.

Tiwari, R.K., Kavdia, N.K., 1984. Ancient mining activity around Agucha village, Bhilwara Dist., Rajasthan. Man Environ. 8, 81–87.

Index

Note: Page numbers followed by "*f*" and "*t*" refer to figures and tables, respectively.

P

Q

R

T